17

CRM SERIES

Centro di Ricerca Matematica Ennio De Giorgi

Vladimir I. Bogachev
Department of Mechanics and Mathematics,
Moscow State University, 119991 Moscow, Russia
and
St.-Tikhon's Orthodox Humanitarian University
Moscow, Russia

Roberto Monti
Dipartimento di Matematica
Università di Padova
Via Trieste, 63
35121 Padova, Italia

Emanuele Spadaro
Max-Planck-Institut
für Mathematik in den Naturwissenschaften
Inselstrasse 22
D-04103 Leipzig

Davide Vittone
Dipartimento di Matematica
Università di Padova
Via Trieste, 63
35121 Padova, Italia

Geometric Measure Theory and Real Analysis

edited by
Luigi Ambrosio

EDIZIONI
DELLA
NORMALE

ISBN 978-88-7642-522-6
ISBN 978-88-7642-523-3 (eBook)

Contents

Preface

In 2013, a school on Geometric Measure Theory and Real Analysis, organized by G. Alberti, C. De Lellis and myself, took place at the Centro De Giorgi in Pisa, with lectures by V. Bogachev, R. Monti, E. Spadaro and D. Vittone.

The lectures were so well-organized and up-to-date that we suggested publishing them as Lecture Notes. All lecturers kindly agreed to this project.

The book presents in a friendly and unitary way many recent developments which have not previously appeared in book form. Topics include: infinite-dimensional analysis, minimal surfaces and isoperimetric problems in the Heisenberg group, regularity of sub-Riemannian geodesics and the regularity theory of area-minimizing currents in any dimension and codimension.

Sobolev classes on infinite-dimensional spaces

Vladimir I. Bogachev

Contents

Introduction

Sobolev classes of functions of generalized differentiability belong to the major analytic achievements in the XX century and have found impressive applications in the most diverse areas of mathematics. So it does not come as a surprise that their infinite-dimensional analogs attract considerable attention. It was already at the end of the 60s and the beginning of the 70s of the last century that in the works of N. N. Frolov, Yu. L. Daletskiĭ, L. Gross, M. Krée, and P. Malliavin Sobolev classes with respect to Gaussian measures on infinite-dimensional spaces were introduced and studied. Their first triumph came with the development of the Malliavin calculus since the mid of the 70s. At present, such classes and their generalizations have become a standard tool of infinite-dimensional anal-

This work was supported by the RSF project 14-11-00196.

ysis. They find applications in stochastic analysis, optimal transportation, mathematical physics, and mathematical finance.

The aim of this survey is to give a concise account of the theory of Sobolev classes on infinite-dimensional spaces with measures. We present a number of already classical cornerstone achievements, some more recent results, and open problems with relatively short formulations. There are already some books presenting elements of this rapidly developing theory (mostly in the Gaussian case), see Bogachev [13, 16] (see also [12]), Bouleau, Hirsch [25], Da Prato [34], Fang [41], Janson [58], Malliavin [66], Malliavin, Thalmaier [67], Nourdin, Peccati [73], Nualart [74], Shigekawa [83], and Üstünel, Zakai [87]. There is also another direction developing Sobolev classes on the so-called measure metric spaces, see Ambrosio, Di Marino [7], Ambrosio, Tilli [10], Cheeger [30], Hajłasz, Koskela [53], Heinonen [54], Keith [59], Reshetnyak [77–79], Vodop'janov [88], which is quite different from the topics discussed here.

The survey is based on several courses I lectured at the Scuola Normale Superiore di Pisa in the years 1995–2013.

Over the years I have had a splendid opportunity to discuss problems related to Sobolev classes in infinite dimensions with many experts in this field, including H. Airault, L. Ambrosio, G. Da Prato, D. Elworthy, S. Fang, D. Feyel, M. Fukushima, M. Hino, A. Lunardi, P. Malliavin, P.-A. Meyer, D. Nualart, M. Röckner, I. Shigekawa, S. Watanabe, N. Yoshida, and M. Zakai.

1 Measures on infinite-dimensional spaces

Given a topological space X we denote by $\mathcal{B}(X)$ its Borel σ-field. Bounded measures on $\mathcal{B}(X)$ (possibly, signed) will be called Borel measures. Such a measure μ can be uniquely written as $\mu = \mu^+ - \mu^-$, where μ^+ and μ^- are mutually singular nonnegative measures called the positive and negative parts of μ, respectively. Set

$$|\mu| = \mu^+ + \mu^-, \quad \|\mu\| = |\mu|(X).$$

The class of all μ-integrable functions is denoted by $\mathcal{L}^1(\mu)$ and the corresponding Banach space of equivalence classes (where functions equal almost everywhere are identified) is denoted by $L^1(\mu)$. Similar notation $\mathcal{L}^p(\mu)$ and $L^p(\mu)$ is used for the classes of μ-measurable functions integrable to power $p \in (1, \infty)$ and the respective spaces of equivalence classes. For a Hilbert space H, the symbol $L^p(\mu, H)$ is used to denote the L^p-space of H-valued mappings.

If a measure ν on $\mathcal{B}(X)$ has the form $\nu = \varrho \cdot \mu$, where ϱ is a μ-integrable function, which means that

$$\nu(A) = \int_A \varrho(x)\,\mu(dx), \quad A \in \mathcal{B}(X),$$

then ϱ is called absolutely continuous with respect to μ, which is denoted by $\nu \ll \mu$, and ϱ is called its Radon–Nikodym density with respect to μ. A necessary and sufficient condition for that, expressed by the Radon–Nikodym theorem, is that ν vanishes on all sets of μ-measure zero. If also $\mu \ll \nu$, which is equivalent to $\varrho \neq 0$ μ-a.e., then the measures are called equivalent, which is denoted by $\nu \sim \mu$.

A nonnegative Borel measure μ on a topological space X is called Radon if, for every set $B \in \mathcal{B}(X)$ and every $\varepsilon > 0$, there is a compact set $K_\varepsilon \subset B$ such that $\mu(B \backslash K_\varepsilon) < \varepsilon$.

Theorem 1.1. *Each Borel measure on any complete separable metric space X is Radon. Moreover, this is true for any Souslin space X,* i.e., *the image of a complete separable metric space under a continuous mapping.*

In particular, this is true for the spaces $C[0, 1]$, $\mathbb{R}^\infty$, and all separable Hilbert spaces.

For the purposes of this survey it is sufficient to have in mind the space $\mathbb{R}^\infty$, the countable power of the real line with its standard product topology (making it a complete metrizable space). The Borel σ-field in this space coincides with the smallest σ-field containing all cylinders, *i.e.*, sets of the form

$$C_B = \{x\colon\ (x_1, \ldots, x_n) \in B\}, \quad B \in \mathcal{B}(\mathbb{R}^n).$$

The measure μ_n defined on $\mathbb{R}^n$ by the formula

$$\mu_n(B) = \mu(C_B)$$

is called the projection of μ on $\mathbb{R}^n$. These projections are consistent in the sense that the projection of μ_{n+1} on $\mathbb{R}^n$ equals μ_n.

By Kolmogorov's theorem, the converse is true: given a consistent sequence of probability measures μ_n on the spaces $\mathbb{R}^n$, there is a unique probability measure μ on $\mathbb{R}^\infty$ with these projections (and there is a natural extension of this result to the case of signed measures, where in the inverse implication the uniform boundedness of μ_n is required).

There is a dual concept to that of projections: conditional measures. Let us consider the one-dimensional subspace $\mathbb{R}e_1$ generated by the first coordinate vector e_1 and its natural complementing hyperplane Y_1 consisting of vectors x with $x_1 = 0$. Let ν^1 be the projection of $|\mu|$ to Y_1, *i.e.*, its image under the natural projecting to Y_1. It is known (see [15, Chapter 10]) that there are Borel measures $\mu^{1,y}$, $y \in Y_1$, on the real line (these measures are probability measures if so is μ) such that for every bounded Borel function f, writing x as $x = (x_1, y)$ with $y = (x_2, x_3, \dots)$ and identifying y with $(0, x_2, x_3, \dots)$, one has

$$\int f(x_1, x_2, \dots)\, \mu(dx) = \int\int f(x_1, y)\, \mu^{1,y}(dx_1)\, \nu^1(dy),$$

where the function defined by the integral in x_1 is Borel measurable in y. Similarly, there exist conditional measures $\mu^{n,y}$, $y \in Y_n$, corresponding to the nth coordinate vector e_n and its natural complementing hyperplane Y_n consisting of vectors with zero nth coordinate. Unlike finite-dimensional distributions, conditional measures (even regarded for all n) do not uniquely determine the measure; the problem of reconstructing a measure from its conditional measures is the subject of the theory of Gibbs measures. Of course, it is not essential that we have considered basis vectors. For a general Radon measure μ (possibly, signed) on a locally vector space X that is a direct topological sum of two closed linear subspaces Z and Y, letting ν be the image of $|\mu|$ under the projection on Y, one can find Radon measures μ^y, $y \in Y$, on Z such that for each bounded Borel function f on X one has

$$\int_X f\, d\mu = \int_Y \int_Z f(z, y)\, \mu^y(dz)\, \nu(dy),$$

where we write elements of X as $x = (z, y)$, $z \in Z$, $y \in Y$, the function $y \mapsto \|\mu^y\|$ is ν-integrable, and the inner integral is also ν-integrable.

Finally, note that sometimes it is more convenient geometrically to define the conditional measures μ^y on the straight lines $\mathbb{R}h + y$ rather than on the real line. In that case the previous equality reads simply as

$$\int_X f\, d\mu = \int_Y \int_Z f(x)\, \mu^y(dx)\, \nu(dy).$$

It will be useful below to represent different measures μ and σ via conditional measures using a common measure ν on Y that may be different from their projections on Y. This is possible if take ν on Y such that both projections are absolutely continuous with respect to ν. Indeed, if $\mu = \mu^y\, \mu_Y(dy)$ and $\sigma = \sigma^y\, \sigma_Y(dy)$, where $\mu_Y = g_1 \cdot \nu$, $\sigma_Y = g_2 \cdot \nu$, then we obtain the representations

$$\mu = g_1(y)\mu^y\, \nu(dy), \quad \sigma = g_2(y)\sigma^y\, \nu(dy)$$

or

$$\mu = \mu^{y,\nu}\, \nu(dy), \qquad \sigma = \sigma^{y,\nu}\, \nu(dy),$$

where ν in the symbol $\mu^{y,\nu}$ indicates that the disintegration is taken with respect to the measure ν on Y in place of μ_Y.

We recall the definition of variation and semivariation of vector measures (see Diestel, Uhl [38] or Dunford, Schwartz [39]). Let H be a separable Hilbert space. A vector measure with values in H is an H-valued countably additive function η defined on a σ-algebra $\mathcal{A}$ of subsets of a space Ω. Such a measure automatically has bounded semivariation defined by the formula

$$V(\eta) := \sup \Bigl| \sum_{i=1}^{n} \alpha_i \eta(\Omega_i) \Bigr|_H,$$

where sup is taken over all finite partitions of Ω into disjoint parts $\Omega_i \in \mathcal{A}$ and all finite sets of real numbers α_i with $|\alpha_i| \le 1$. In other words, this is the supremum of variations of real measures $(\eta, h)_H$ over $h \in H$ with $|h|_H \le 1$. However, this does not yet mean that the vector measure η has finite variation which is defined as

$$\mathrm{Var}(\eta) := \sup \sum_{i=1}^{n} |\eta(\Omega_i)|_H,$$

where sup is taken over all finite partitions of Ω into disjoint parts $\Omega_i \in \mathcal{A}$. The variation of the measure η will be denoted by $\|\eta\|$ (but in [39] this notation is used for semivariation).

By the Pettis theorem (see Dunford, Schwartz [39, Chapter IV, §10]), an H-valued mapping Λ is a vector measure of bounded semivariation provided that $(\Lambda, h)_H$ is a bounded scalar measure for each $h \in H$.

The sets of measures of bounded variation and bounded semivariation are Banach spaces with the norms $\eta \mapsto \|\eta\|$ and $\eta \to V(\eta)$, respectively. It is easy to give an example of a measure with values in an infinite-dimensional Hilbert space having bounded semivariation, but infinite variation: consider the standard basis $\{e_n\}$ in l^2 and take Dirac's measures $\delta(e_n)$ in the points e_n and the vector measure $\eta = \sum_{n=1}^{\infty} n^{-1} \delta(e_n) e_n$. Its semivariation equals the sum of the numbers n^{-2}, but it is of infinite variation.

The space of all continuous linear functions on a locally convex space X is denoted by X^* and is called the dual (or topological dual) space.

Let $\mathcal{F}\mathcal{C}^\infty$ denote the class of all functions f on X of the form

$$f(x) = f_0(l_1(x), \dots, l_n(x)), \quad f_0 \in C_b^\infty(\mathbb{R}^n),\ l_i \in X^*,$$

where $C_b^\infty(\mathbb{R}^n)$ is the class of all infinitely differentiable functions on $\mathbb{R}^n$ with bounded derivatives. In case of $\mathbb{R}^\infty$ we obtain just the union of all $C_b^\infty(\mathbb{R}^n)$.

2 Gaussian measures

A *Gaussian measure* on the real line is a Borel probability measure which is either concentrated at some point a (*i.e.*, is Dirac's measure δ_a at a) or has density $(2\pi\sigma)^{-1/2}\exp\big(-(2\sigma)^{-1}(x-a)^2\big)$ with respect to Lebesgue measure, where $a \in \mathbb{R}^1$ is its *mean* and $\sigma > 0$ is its *dispersion*. The measure for which $a = 0$ and $\sigma = 1$ is called *standard Gaussian*.

Similarly the standard Gaussian measure on $\mathbb{R}^d$ is defined by its density

$$(2\pi)^{-d/2}\exp(-|x|^2/2)$$

with respect to Lebesgue measure.

Although below a general concept of a Gaussian measure on a locally convex space is introduced, we define explicitly general Gaussian measures on $\mathbb{R}^d$. These are measures that are concentrated on affine subspaces in $\mathbb{R}^d$ and are standard in suitable (affine) coordinate systems. In other words, these are images of the standard Gaussian measure under affine mappings of the form $x \mapsto Ax + a$, where A is a linear operator and a is a vector. A bit more explicit representation is provided by the Fourier transform of a bounded Borel measure μ on $\mathbb{R}^d$ defined by the formula

$$\widetilde{\mu}(y) = \int \exp\big(i(y,x)\big)\,\mu(dx), \quad y \in \mathbb{R}^d.$$

In these terms, a measure μ is Gaussian if and only if its Fourier transform has the form

$$\widetilde{\mu}(y) = \exp\Big(i(y,a) - \frac{1}{2}Q(y,y)\Big),$$

where Q is nonnegative quadratic form on $\mathbb{R}^d$.

The Fourier transform of the standard Gaussian measure is given by

$$\widetilde{\gamma}(y) = \exp(-|y|^2/2).$$

The change of variables formula yields the following relation between A and Q if μ is the image of γ under the affine mapping $Ax + a$:

$$\begin{aligned}\widetilde{\mu}(y) &= \int \exp\big(i(y, Ax + a)\big)\,\gamma(dx)\\ &= \exp\big(i(y,a)\big)\int \exp\big(i(A^*y, x)\big)\,\gamma(dx)\\ &= \exp\big(i(y,a) - |A^*y|^2/2\big),\end{aligned}$$

that is, $Q(y) = (AA^*y, y)$. It is readily verified that μ has a density on the whole space precisely when A is invertible.

The vector a is called the mean of μ and is expressed by the equality

$$(y, a) = \int (y, x)\, \mu(dx).$$

For the quadratic form Q we have the equality

$$Q(y, y) = \int (y, x - a)^2\, \mu(dx).$$

These equalities are verified directly (it suffices to check them in the one-dimensional case).

Let us define Gaussian measures on general locally convex spaces.

Definition 2.1. Let X be a locally convex space with the topological dual X^*. A Borel probability measure γ on X is called Gaussian if the induced measure $\gamma \circ f^{-1}$ is Gaussian for every $f \in X^*$. If all these measures are centered, then γ is called centered.

In the case of the space $\mathbb{R}^\infty$ the space $\mathbb{R}^\infty_0$ of finite sequences coincides with the dual space. Hence Gaussian measures on $\mathbb{R}^\infty$ are measures with Gaussian finite-dimensional projections.

Example 2.2. An important example of a Gaussian measure is the countable product γ of the standard Gaussian measures on the real line. This measure is defined on the space $X = \mathbb{R}^\infty$. This special example plays a very important role in the whole theory. In some sense (see Bogachev [13] for details) this is a unique up to isomorphism infinite-dimensional Gaussian measure.

Another important example of a Gaussian measure is the Wiener measure on the space $C[0, 1]$ of continuous functions or on the space $L^2[0, 1]$. This measure can be defined as the image of the standard Gaussian measure γ on $X = \mathbb{R}^\infty$ under the mapping

$$x = (x_n) \mapsto w(\cdot), \quad w(t) = \sum_{n=1}^{\infty} x_n \int_0^t e_n(s)\, ds,$$

where $\{e_n\}$ is an arbitrary orthonormal basis in $L^2[0, 1]$. One can show that this series converges in $L^2[0, 1]$ for γ-almost every x; moreover, for γ-almost every x convergence is uniform on $[0, 1]$.

We recall that a countable product $\mu = \bigotimes_{n=1}^{\infty} \mu_n$ of probability measures μ_n on spaces $(X_n, \mathcal{B}_n)$ is defined on $X = \prod_{n=1}^{\infty} X_n$ as follows: first it is defined on sets of the form $A = A_1 \times \cdots \times A_n \times X_{n+1} \cdots$ by

$$\mu(A) = \mu_1(A_1) \times \cdots \times \mu_n(A_n),$$

then it is verified that μ is countably additive on the algebra of finite unions of such sets (called cylindrical sets), which results in a countably additive extension to the smallest σ-algebra $\mathcal{B} := \bigotimes_{n=1}^{\infty} \mathcal{B}_n$ containing such cylindrical sets.

The standard Gaussian measure γ on $\mathbb{R}^\infty$ can be restricted to many other smaller linear subspaces of full measure. For example, taking any sequence of numbers $\alpha_n > 0$ with $\sum_{n=1}^{\infty} \alpha_n < \infty$, we can restrict γ to the weighted Hilbert space of sequences

$$E := \Big\{(x_n) \in \mathbb{R}^\infty : \sum_{n=1}^{\infty} \alpha_n x_n^2 < \infty\Big\},$$

making this expression the square of the norm. The fact that $\gamma(E) = 1$ follows by the monotone convergence theorem, which shows that

$$\sum_{n=1}^{\infty} \alpha_n x_n^2 < \infty$$

almost everywhere due to convergence of the integrals of the terms (the integral of x_n^2 is 1). Similarly, one can find non-Hilbert full measure Banach spaces of sequences (x_n) with $\sup_n \beta_n |x_n| < \infty$ or $\lim\limits_{n\to\infty} \beta_n |x_n| = 0$ for suitable sequences $\beta_n \to 0$; more precisely, the condition is this:

$$\sum_{n=1}^{\infty} \exp\Big(-\frac{C}{\beta_n^2}\Big) < \infty \quad \forall C > 0.$$

However, there is no minimal linear subspace of full measure. The point is that the intersection of all linear subspaces of positive (equivalently, full) measure is the subspace l^2, which has measure zero, as one can verify directly.

It is known that any Radon Gaussian measure γ has mean $m \in X$, *i.e.*, m is a vector in X such that

$$f(m) = \int_X f(x)\,\gamma(dx) \quad \forall f \in X^*.$$

If $m = 0$, *i.e.*, the measures $\gamma \circ f^{-1}$ for $f \in X^*$ have zero mean, then γ is called *centered*. Any Radon Gaussian measure γ is a shift of a centered Gaussian measure γ_m defined by the formula $\gamma_m(B) := \gamma(B + m)$. Hence for many purposes it suffices to consider only centered Gaussian measures.

For a centered Radon Gaussian measure γ we denote by X_γ^* the closure of X^* in $L^2(\gamma)$. The elements of X_γ^* are called *γ-measurable linear* *functionals*. There is an operator $R_\gamma\colon X^*_\gamma \to X$, called the *covariance operator* of the measure γ, such that

$$f(R_\gamma g) = \int_X f(x)g(x)\,\gamma(dx) \quad \forall f \in X^*,\ g \in X^*_\gamma.$$

Set

$$g := \widehat{h} \quad \text{if } h = R_\gamma g.$$

Then $\widehat{h}$ is called the γ-measurable linear functional generated by h. The following vector equality holds (if X is a Banach space, then it holds in Bochner's sense):

$$R_\gamma g = \int_X g(x)x\,\gamma(dx) \quad \forall g \in X^*_\gamma.$$

For example, if γ is a centered Gaussian measure on a separable Hilbert space X, then there exists a nonnegative nuclear operator K on X for which $Ky = R_\gamma y$ for all $y \in X$, where we identify X^* with X. Then we obtain

$$(Ky, z) = (y, z)_{L^2(\gamma)} \quad \text{and} \quad \widetilde{\gamma}(y) = \exp\bigl(-(Ky, y)/2\bigr).$$

Let us take an orthonormal eigenbasis $\{e_n\}$ of the operator K with eigenvalues $\{k_n\}$. Then γ coincides with the image of the countable power γ_0 of the standard Gaussian measure on $\mathbb{R}^1$ under the mapping

$$\mathbb{R}^\infty \to X, \quad (x_n) \mapsto \sum_{n=1}^{\infty} \sqrt{k_n} x_n e_n.$$

This series converges γ_0-a.e. in X by convergence of the series $\sum_{n=1}^\infty k_n x_n^2$, which follows by convergence of the series of k_n and the fact that the integral of x_n^2 against the measure γ_0 equals 1. Here X^*_γ can be identified with the completion of X with respect to the norm $x \mapsto \|\sqrt{K}x\|_X$, *i.e.*, the embedding $X = X^* \to X^*_\gamma$ is a Hilbert–Schmidt operator.

The space

$$H(\gamma) = R_\gamma(X^*_\gamma)$$

is called the *Cameron–Martin space* of the measure γ. It is a Hilbert space with respect to the inner product

$$(h, k)_H := \int_X \widehat{h}(x)\widehat{k}(x)\,\gamma(dx).$$

The corresponding norm is given by the formula

$$|h|_H := \|\widehat{h}\|_{L^2(\gamma)}.$$

Moreover, it is known that $H(\gamma)$ with the indicated norm is separable and its closed unit ball is compact in the space X. Note that the same norm is given by the formula

$$|h|_H = \sup\{f(h)\colon\ f \in X^*, \|f\|_{L^2(\gamma)} \le 1\}.$$

It should be noted that if $\dim H(\gamma) = \infty$, then $\gamma\big(H(\gamma)\big) = 0$.

In terms of the inner product in H the vector $R_\gamma(l)$ is determined by the identity

$$\big(j_H(f), R_\gamma g\big)_H = f(R_\gamma g) = \int_X fg\, d\gamma, \quad f \in X^*, g \in X_\gamma^*. \tag{2.1}$$

In the above example of a Gaussian measure γ on a Hilbert space we have

$$H(\gamma) = \sqrt{K}(X).$$

Let us observe that $H(\gamma)$ coincides also with the set of all vectors of the form

$$h = \int_X f(x)x\,\gamma(dx), \quad f \in L^2(\gamma).$$

Indeed, letting f_0 be the orthogonal projection of f onto X_γ^* in $L^2(\gamma)$, we see that the integral of the difference $[f(x) - f_0(x)]x$ over X vanishes since the integral of $[f(x) - f_0(x)]l(x)$ vanishes for each $l \in X^*$.

Theorem 2.3. *The mapping $h \mapsto \widehat{h}$ establishes a linear isomorphism between $H(\gamma)$ and X_γ^* preserving the inner product. In addition, $R_\gamma \widehat{h} = h$.*

If $\{e_n\}$ is an orthonormal basis in $H(\gamma)$, then $\{\widehat{e_n}\}$ is an orthonormal basis in X_γ^* and $\widehat{e_n}$ are independent random variables.

One can take an orthonormal basis in X_γ^* consisting of elements $\xi_n \in X^*$. The general form of an element $l \in X_\gamma^*$ is this:

$$l = \sum_{n=1}^{\infty} c_n \xi_n,$$

where the series converges in $L^2(\gamma)$. Since ξ_n are independent Gaussian random variables, this series converges also γ-a.e. The domain of its convergence is a Borel linear subspace L of full measure. One can take a version of l which is linear on all of X in the usual sense; it is called a *proper linear version*. It is easy to show that such a version is automatically continuous on $H(\gamma)$ with the norm $|\cdot|_H$; more precisely,

$$f_0(h) = (R_\gamma f, h)_H = \int_X f\widehat{h}\, d\gamma, \quad h \in H.$$

Conversely, any continuous linear functional l on the Hilbert space $H(\gamma)$ admits a unique extension to a γ-measurable proper linear functional $\widehat{l}$ such that $\widehat{l}$ coincides with l on $H(\gamma)$. For every $h \in H(\gamma)$, such an extension of the functional $x \mapsto (x, h)_H$ is exactly $\widehat{h}$. If $h = \sum_{n=1}^{\infty} c_n e_n$, then $\widehat{h} = \sum_{n=1}^{\infty} c_n \widehat{e_n}$. Two γ-measurable linear functionals are equal almost everywhere precisely when their proper linear versions coincide on $H(\gamma)$.

If a measure γ on $X = \mathbb{R}^\infty$ is the countable power of the standard Gaussian measure on the real line, then X^* can be identified with the space of all sequences of the form $f = (f_1, \ldots, f_n, 0, 0, \ldots)$. Here we have

$$(f, g)_{L^2(\gamma)} = \sum_{i=1}^{\infty} f_i g_i.$$

Hence X_γ^* can be identified with l^2; any element $l = (c_n) \in l^2$ defines an element of $L^2(\gamma)$ by the formula $l(x) := \sum_{n=1}^{\infty} c_n x_n$, where the series converges in $L^2(\gamma)$. Therefore, the Cameron–Martin space $H(\gamma)$ coincides with the space l^2 with its natural inner product. An element l represents a continuous linear functional precisely when only finitely many numbers c_n are nonzero. For the Wiener measure on $C[0, 1]$ the Cameron–Martin space coincides with the class $W_0^{2,1}[0, 1]$ of all absolutely continuous functions h on $[0, 1]$ such that $h(0) = 0$ and $h' \in L^2[0, 1]$; the inner product is given by the formula

$$(h_1, h_2)_H := \int_0^1 h_1'(t) h_2'(t)\, dt.$$

The next classical result, called the *Cameron–Martin formula*, relates measurable linear functionals and vectors in the Cameron–Martin space to the Radon–Nikodym density for shifts of the Gaussian measure.

Theorem 2.4. *The space $H(\gamma)$ is the set of all $h \in X$ such that $\gamma_h \sim \gamma$, where $\gamma_h(B) := \gamma(B + h)$, and the Radon–Nikodym density of the measure γ_h with respect to γ is given by the following Cameron–Martin formula:*

$$d\gamma_h / d\gamma = \exp\big(-\widehat{h} - |h|_H^2/2\big).$$

For every $h \notin H(\gamma)$ we have $\gamma \perp \gamma_h$.

It follows from this formula that for every bounded Borel function f on X we have

$$\int_X f(x+h)\, \gamma(dx) = \int_X f(x) \exp\big(\widehat{h}(x) - |h|_H^2/2\big)\, \gamma(dx).$$

In the case of the standard Gaussian measure on $\mathbb{R}^\infty$ this formula is a straightforward extension of the obvious finite-dimensional expression, one just needs to define $\widehat{h}(x)$ as the sum of a series.

A centered Radon Gaussian measure is uniquely determined by its Cameron–Martin space (with the indicated norm!): if μ and ν are centered Radon Gaussian measures such that $H(\mu) = H(\nu)$ and $|h|_{H(\mu)} = |h|_{H(\nu)}$ for all $h \in H(\mu) = H(\nu)$, then $\mu = \nu$. The Cameron-Martin space is also called the reproducing Hilbert space.

Definition 2.5. A Radon Gaussian measure γ on a locally convex space X is called nondegenerate if for every nonzero functional $f \in X^*$ the measure $\gamma \circ f^{-1}$ is not concentrated at a point.

The nondegeneracy of γ is equivalent to that $\gamma(U) > 0$ for all nonempty open sets $U \subset X$. This is also equivalent to that the Cameron-Martin space $H(\gamma)$ is dense in X. For every degenerate Radon Gaussian measure γ there exists the smallest closed linear subspace $L \subset X$ for which $\gamma(L+m) = 1$, where m is the mean of the measure γ. Moreover, $L+m$ coincides with the topological support of γ. If $m = 0$, then on L the measure γ is nondegenerate.

Let γ be a centered Radon Gaussian measure on a locally convex space X; as usual, one can assume that this is the standard Gaussian measure on $\mathbb{R}^\infty$. *The Ornstein–Uhlenbeck semigroup* is defined by the formula

$$T_t f(x) = \int_X f\big(e^{-t}x - \sqrt{1-e^{-2t}}\, y\big)\, \gamma(dy), \qquad f \in \mathcal{L}^p(\gamma). \tag{2.2}$$

A simple verification of the fact that $\{T_t\}_{t\geq 0}$ is a strongly continuous semigroup on all $L^p(\gamma)$, $1 \leq p < \infty$, can be found in [13]; the semigroup property means that

$$T_{t+s} f = T_s T_s f, \quad t, s \geq 0.$$

An important feature of this semigroup is that the measure γ is invariant for it, that is,

$$\int_X T_t f(x)\, \gamma(dx) = \int_X f(x)\, \gamma(dx).$$

Theorem 2.6. *For every $p \in [1, +\infty)$ and $f \in L^p(\gamma)$ one has*

$$\lim_{t\to 0} \|T_t f - f\|_{L^p(\gamma)} = 0, \qquad \lim_{t\to +\infty} \left\| T_t f - \int f\, d\gamma \right\|_{L^p(\gamma)} = 0$$

and if $1 < p < \infty$, then also $\lim\limits_{t\to 0} T_t f(x) = f(x)$ a.e.

It is also known that in the finite-dimensional case $\lim_{t\to 0} T_t f(x) = f(x)$ a.e. for all $f \in L^1(\gamma)$. It remains an open problem whether this is true in infinite dimensions.

The generator L of the Ornstein–Uhlenbeck semigroup is called the *Ornstein–Uhlenbeck operator* (more precisely, for every $p \in [1, +\infty)$, there is such a generator on the corresponding domain in $L^p(\gamma)$; if p is not explicitly indicated, then usually $p = 2$ is meant). By definition, $Lf = \lim_{t\to 0}(T_t f - f)/t$ if this limit exists in the norm of $L^p(\gamma)$. This operator will be important Section 4. In the case of $\mathbb{R}^\infty$, on smooth functions $f(x) = f(x_1, \ldots, x_n)$ in finitely many variables one can explicitly calculate that

$$Lf(x) = \Delta f(x) - \big(x, \nabla f(x)\big) = \sum_{i=1}^{n} \big[\partial_{x_i}^2 f(x) - x_i \partial_{x_i} f(x)\big].$$

This representation can be also extended to some functions in infinitely many variables. In the general case Lf is the sum of a similar series, but its two parts need converge separately.

In the theory of Gaussian measures an important role is played by the Hermite (or Chebyshev–Hermite) polynomials H_n defined by the equalities

$$H_0 = 1, \quad H_n(t) = \frac{(-1)^n}{\sqrt{n!}} e^{t^2/2} \frac{d^n}{dt^n}\big(e^{-t^2/2}\big), \quad n > 1.$$

They have the following properties:

$$H_n'(t) = \sqrt{n} H_{n-1}(t) = t H_n(t) - \sqrt{n+1} H_{n+1}(t).$$

In addition, the system of functions $\{H_n\}$ is an orthonormal basis in $L^2(\gamma)$, where γ is the standard Gaussian measure on the real line.

For the *standard Gaussian* measure γ_n on $\mathbb{R}^n$ (the product of n copies of the standard Gaussian measure on $\mathbb{R}^1$) an orthonormal basis in $L^2(\gamma_n)$ is formed by the polynomials of the form

$$H_{k_1,\ldots,k_n}(x_1, \ldots, x_n) = H_{k_1}(x_1) \cdots H_{k_n}(x_n), \quad k_i \geq 0.$$

If γ is a centered Radon Gaussian measure on a locally convex space X and $\{l_n\}$ is an orthonormal basis in X_γ^*, then a basis in $L^2(\gamma)$ is formed by the polynomials

$$H_{k_1,\ldots,k_n}(x) = H_{k_1}\big(l_1(x)\big) \cdots H_{k_n}\big(l_n(x)\big), \quad k_i \geq 0, n \in \mathbb{N}.$$

For example, for the countable power of the standard Gaussian measure on the real line such polynomials are $H_{k_1,\ldots,k_n}(x_1, \ldots, x_n)$. It is convenient to arrange polynomials $H_{k_1,\dots,k_n}$ according to their degrees $k_1+\cdots+k_n$. For $k=0,1,\dots$ we denote by $\mathcal{X}_k$ the closed linear subspace of $L^2(\gamma)$ generated by the functions $H_{k_1,\dots,k_n}$ with $k_1+\cdots+k_n=k$. The functions $H_{k_1,\dots,k_n}$ are mutually orthogonal and, for the fixed value $k=k_1+\cdots+k_n$, form an orthonormal basis in $\mathcal{X}_k$.

The one-dimensional space $\mathcal{X}_0$ consists of constants and $\mathcal{X}_1 = X_\gamma^*$. One can show that every element $f \in \mathcal{X}_2$ can be written in the form

$$f=\sum_{n=1}^{\infty}\alpha_n(l_n^2-1),$$

where $\{l_n\}$ is an orthonormal basis in X_γ^* and $\sum_{n=1}^{\infty}\alpha_n^2<\infty$ (*i.e.*, the series for f converges in $L^2(\gamma)$).

The spaces $\mathcal{X}_k$ are mutually orthogonal and their orthogonal sum is the whole $L^2(\gamma)$:

$$L^2(\gamma)=\bigoplus_{k=0}^{\infty}\mathcal{X}_k,$$

which means that, denoting by I_k the operator of orthogonal projection onto $\mathcal{X}_k$, we have an orthogonal decomposition

$$F=\sum_{k=0}^{\infty}I_k(F), \quad F\in L^2(\gamma).$$

One can check that $T_t H_{k_1,\dots,k_n}=e^{-k_1-\cdots-k_n}H_{k_1,\dots,k_n}$, which yields that

$$T_tF=\sum_{k=0}^{\infty}e^{-kt}I_k(F), \quad F\in L^2(\gamma).$$

Given a separable Hilbert space E, one defines similarly the space $\mathcal{X}_k(E)$ of polynomials with values in E as the closure in $L^2(\gamma, E)$ of the liner span of the mappings $f\cdot v$, where $f\in\mathcal{X}_k, v\in E$.

3 Integration by parts and differentiable measures

Suppose that f is a bounded Borel function on a locally convex space X with a centered Radon Gaussian measure γ such that the partial derivative

$$\partial_h f(x)=\lim_{t\to 0}\frac{f(x+th)-f(x)}{t}$$

exists for some vector h in the Cameron-Martin space of γ and is bounded. Applying the Cameron-Martin formula and Lebesgue's dominated convergence theorem, we arrive at the equality

$$\int_X \partial_h f(x)\,\gamma(dx)=\int_X f(x)\widehat{h}(x)\,\gamma(dx),$$

where we also use that the derivative of $t \mapsto e^{t\widehat{h}-t^2|h|_H^2/2}$ at zero is $\widehat{h}$. This simple formula, called the integration by parts formula for the Gaussian measure, plays a very important role in stochastic analysis and is a starting point for far-reaching generalizations connected with differentiabilities of measures in the sense of Fomin [45,46] and in the sense of Skorohod [85].

A measure μ on X is called Skorohod differentiable along a vector h if there exists a measure $d_h\mu$, called the Skorohod derivative of the measure μ along the vector h, such that

$$\lim_{t\to 0}\int_X \frac{f(x-th)-f(x)}{t}\,\mu(dx) = \int_X f(x)d_h\mu(dx) \qquad (1)$$

for every bounded continuous function f on X. If the measure $d_h\mu$ is absolutely continuous with respect to the measure μ, then the measure μ is called Fomin differentiable along the vector h, the Radon–Nikodym density of the measure $d_h\mu$ with respect to μ is denoted by β_h^μ and called the logarithmic derivative of μ along h. The Skorohod differentiability of μ along h is equivalent to the identity

$$\int_X \partial_h f(x)\,\mu(dx) = -\int_X f(x)\,d_h\mu(dx), \quad f\in\mathcal{F}C^\infty.$$

The Fomin differentiability is the equality

$$\int_X \partial_h f(x)\,\mu(dx) = -\int_X f(x)\,\beta_h^\mu(x)\,\mu(dx), \quad f\in\mathcal{F}C^\infty.$$

On the real line the Fomin differentiability is equivalent to the membership of the density in the Sobolev class $W^{1,1}$, and the Skorohod differentiability is the boundedness of variation of the density; the picture is similar also in $\mathbb{R}^n$. A detailed discussion of these types of differentiability of measures can be found in Bogachev [16].

It follows from our previous discussion that for the centered Gaussian measure μ we have

$$\beta_h^\mu = -\widehat{h}, \quad h\in H(\mu).$$

In the case of a probability measure on $\mathbb{R}^\infty$ efficient conditions for both types of differentiability can be expressed in terms of finite-dimensional distributions. The Skorohod differentiability along a vector $h=(h_n)$ is equivalent to the following condition: for every n, the generalized derivative of the projection μ_n on $\mathbb{R}^n$ along the vector $(h_1,\ldots,h_n)$ is a bounded measure and such measures are uniformly bounded. For Fomin's differentiability more is needed: the corresponding logarithmic derivatives

$\beta^{\mu_n}(h_1, \ldots, h_n)$ are uniformly integrable with respect to μ (regarded as functions on $\mathbb{R}^\infty$).

An equivalent characterization is available in terms of conditional measures (see Section 1): Fomin's differentiability of μ along h is equivalent to the following: the conditional measures μ^y on the real line have densities $\varrho^y \in W^{1,1}(\mathbb{R})$ such that the function $y \mapsto \|\partial_t \varrho^y\|_{L^1(\mathbb{R})}$ is μ_Y-integrable, where μ_Y is the projection of μ on a closed hyperplane Y complementing the one-dimensional subspace generated by h. The derivative $d_h\mu$ can be written as

$$d_h\mu(B) = \int_Y \int_{B_y} \partial_t \varrho^y(t)\, dt\, \mu_Y(dy),$$

where $B_y = \{t \in \mathbb{R}\colon (th, y) \in B\}$, $B \in \mathcal{B}(X)$, and X is written as $\mathbb{R}h \times Y$.

It is worth mentioning that for a Gaussian measure, the conditional measures are Gaussian as well (see Bogachev [13, Section 3.10] or [17]).

It is known (see Bogachev [16, Chapter 5]) that the sets $D_C(\mu)$ and $D(\mu)$ of all vectors of differentiability of a nonzero measure μ in the sense of Skorohod and Fomin respectively are Banach spaces with respect to the norm $h \mapsto \|d_h\mu\|$ and that the closed unit ball in $D_C(\mu)$ is compact in X. For any $h \in D(\mu)$ we have $\|d_h\mu\| = \|\beta_h\|_{L^1(\mu)}$. For example, if μ is a Gaussian measure (as before, we need only the countable power of the Gaussian measure on the real line), then the set $D(\mu) = D_C(\mu)$ coincides with the Cameron–Martin space $H(\mu)$ of the measure μ (the set of all vectors the shifts to which give equivalent measures).

We assume further that the measure μ is Fomin differentiable along all vectors in a separable Hilbert space H that is continuously and densely embedded into X (the model example is $l^2 \subset \mathbb{R}^\infty$). Hence the closed graph theorem yields that the natural embedding $H \to D(\mu)$ is continuous and

$$\|d_h\mu\| \le C|h|_H, \quad h \in H$$

for some constant C.

4 Sobolev classes over Gaussian measures

In this section we briefly discuss Sobolev spaces with respect to Gaussian measures. This is a very important analytical tool and one of the mainstreams in modern theory. The reason why such classes are important is that many nonlinear functionals on infinite-dimensional spaces arising in applications have very poor differentiability or even continuity properties from the point of view of the classical analysis (norm continuity,

Fréchet or Gâteaux differentiability), but are Sobolev smooth. This effect is much stronger than in the finite-dimensional case (where it is also notable, *e.g.*, in the theory of partial differential equations), and it was Paul Malliavin [65] who invented special tools (now called the Malliavin calculus) to deal with such problems. It should be noted that important ideas closely connected with Gaussian Sobolev classes were developed already by Gross [51] and the first definition of such classes was given by Frolov [47,48]. Later such classes were studied in Daletskiĭ, Paramonova [35–37], Krée [61,62], Lascar [63], and in many other works.

Similarly to the classical Sobolev spaces (see, *e.g.*, the books Adams, Fournier [1], Lieb, Loss [64], Ziemer [92]), there are essentially three different ways of introducing such spaces: as suitable completions of smooth functions, in terms of integration by parts, and through integral representations. For example, the class $W^{1,1}(\mathbb{R}^d)$ can be defined either as the completion of the class C_0^∞ of smooth compactly supported functions with respect to the Sobolev norm

$$\|f\|_{1,1} = \|f\|_1 + \|\nabla f\|_1,$$

where $\|\cdot\|_1$ denotes the L^1-norm of scalar or vector functions, or the subclass in $L^1(\mathbb{R}^d)$ consisting of the functions whose generalized first order partial derivatives belong to $L^1(\mathbb{R}^d)$, where the generalized partial derivative $\partial_{x_i} f$ is defined by means of the integration by parts formula

$$\int_{\mathbb{R}^d} \varphi \partial_{x_i} f\, dx = -\int_{\mathbb{R}^d} f \partial_{x_i} \varphi\, dx, \qquad \varphi \in C_0^\infty.$$

Using the L^p-norm we arrive at the classes $W^{p,1}(\mathbb{R}^d)$. Similar constructions work in the case of weighted Sobolev classes $W^{p,1}(\varrho)$, where ϱ is a nonnegative locally integrable function, so that in place of Lebesgue measure we use the measure $\mu = \varrho\, dx$. However, in this case some subtleties appear (see, *e.g.* Bogachev [16]). First of all, some conditions on ϱ are needed to ensure the closability of the Sobolev norm, *i.e.*, the property that if a sequence of smooth functions f_j converges to zero in L^p and is fundamental in the Sobolev norm, then it also converges to zero in the Sobolev norm. Next, the use of the integration by parts formula also imposes restrictions on ϱ in the second approach. Finally, these two and other approaches may lead to distinct Sobolev classes unlike the classical case, see Zhikov [90], [91].

We first consider the case of the standard Gaussian measure γ on $\mathbb{R}^d$. The classes $W^{p,1}(\gamma)$, $1 \le p < \infty$, are obtained as the completions of the class $C_0^\infty(\mathbb{R}^d)$ with respect to the Sobolev norms

$$\|f\|_{p,1} := \Bigl(\int |f|^p\,d\gamma\Bigr)^{1/p} + \Bigl(\int |\nabla f|^p\,d\gamma\Bigr)^{1/p}.$$

Similarly one defines the classes $W^{p,1}(\gamma, \mathbb{R}^m)$ of $\mathbb{R}^m$-valued Sobolev mappings. An extension to higher order derivatives is relatively straightforward, but there is a nuance in the choice of the norm on higher order derivatives: for many purposes it turns out to be reasonable to take Hilbert–Schmidt norms (rather than other matrix norms). In particular, the space $W^{p,2}(\gamma)$ is obtained by taking the norm

$$\|f\|_{p,2} := \|f\|_{p,1} + \Bigl(\int \Bigl(\sum_{i,j\le d} |\partial_{x_i}\partial_{x_j} f|^2\Bigr)^{p/2} d\gamma\Bigr)^{1/p}.$$

Continuing inductively we obtain the spaces $W^{p,r}(\gamma), r \in \mathbb{N}$.

The same class $W^{p,r}(\gamma)$ is characterized as follows: it consists of all functions $f \in L^p(\gamma)$ such that f possesses generalized partial derivatives $\partial_{x_{i_1}} \cdots \partial_{x_{i_r}} f$ represented by elements in $L^p(\gamma)$.

The infinite-dimensional case, where γ is a centered Radon Gaussian measure on a locally convex space with the Cameron–Martin space H, is completely analogous, the only difference is that now in place of C_0^∞ we take the class $\mathcal{F}C^\infty$ of all functions on X of the form

$$f(x) = f_0\bigl(l_1(x), \dots, l_n(x)\bigr), \quad l_i \in X^*,\ f_0 \in C_b^\infty(\mathbb{R}^n).$$

Let $\{e_i\}$ be an orthonormal basis in H. Set

$$\partial_h f(x) = \lim_{t\to\infty} \frac{f(x+th) - f(x)}{t}.$$

For all $p \ge 1$ and $r \in \mathbb{N}$, the Sobolev norm $\|\cdot\|_{W^{p,r}}$ is defined by the following formula, where $\partial_i := \partial_{e_i}$:

$$\|f\|_{W^{p,r}} = \sum_{k=0}^{r} \Bigl(\int_X \Bigl[\sum_{i_1,\dots,i_k\ge 1} \bigl(\partial_{i_1}\dots\partial_{i_k} f(x)\bigr)^2\Bigr]^{p/2} \gamma(dx)\Bigr)^{1/p}. \quad (4.1)$$

If $X = \mathbb{R}^\infty$ and $H = l^2$, then $\mathcal{F}C^\infty$ is just the space of functions of finitely many variables of class C_b^∞ and if γ is the standard Gaussian measure on $\mathbb{R}^\infty$, then the Sobolev norms on such functions are the previously defined norms in the finite-dimensional case.

Let $W^{p,r}(\gamma)$ denote the completion of $\mathcal{F}C^\infty$ with respect to the Sobolev norm $\|\cdot\|_{p,r} = \|\cdot\|_{W^{p,r}}$. Note that the same norm can be written as

$$\|f\|_{p,r} = \sum_{k=0}^{r} \|D_H^k f\|_{L^p(\gamma,\mathcal{H}_k)},$$

where $D_H^k f$ stands for the derivative of order k along H and $\mathcal{H}_k$ is the space of Hilbert–Schmidt k-linear forms on H, which can be defined inductively by setting $\mathcal{H}_k = \mathcal{H}(H, \mathcal{H}_{k-1})$, $\mathcal{H}_1 = H$, where $\mathcal{H}(H, E)$ is the space of Hilbert–Schmidt operators between Hilbert spaces H and E equipped with its natural norm defined by

$$\|T\|_{\mathcal{H}}^2 = \sum_{i=1}^{\infty} \|Te_i\|_E^2$$

for an arbitrary orthonormal basis $\{e_i\}$ in H.

After this completion procedure all elements in $W^{p,r}(\gamma)$ acquire Sobolev derivatives $D_H^k f$ of the respective orders. In particular, any $f \in W^{2,1}(\gamma)$ has a Sobolev gradient $D_H f$ along H, which is a limit in $L^2(\gamma, H)$ of the H-gradients of smooth cylindrical functions convergent to f in the norm $\|, \cdot\|_{2,1}$.

For example, in the case of the standard Gaussian measure on $\mathbb{R}^\infty$ the measurable linear functional $f(x) = \sum_{n=1}^{\infty} n^{-1}x_n$ belongs to all classes $W^{p,r}(\gamma)$, since the sums $\sum_{n=1}^{m} n^{-1}x_n$ converge in each norm $\|\cdot\|_{p,r}$; more specifically, $D_H f(x)$ is a constant vector $h = (n^{-1})_{n=1}^{\infty}$ and all higher derivatives vanish. Similarly, the function $f(x) = \sum_{n=1}^{\infty} n^{-2}x_n^2$ belongs to all classes $W^{p,r}(\gamma)$, $D_H f(x) = 2(n^{-2}x_n)_{n=1}^{\infty}$, $D_H^2 f(x)$ is constant and equals the diagonal operator with eigenvalues $2n^{-2}$, higher order derivatives vanish.

In a similar way one defines the Sobolev spaces $W^{p,r}(\gamma, E)$ of mappings with values in a Hilbert space E. The corresponding norms are denoted by the same symbol $\|\cdot\|_{p,r}$.

An equivalent description employs the concept of a Sobolev derivative. Let $p > 1$. We shall say that a function $f \in L^p(\gamma)$ has a generalized (or Sobolev) partial derivative $g \in L^1(\gamma)$ along a vector $h \in H$ if, for every $\varphi \in \mathcal{F}C^\infty$, one has the equality

$$\int_X \partial_h\varphi(x)\, f(x)\,\gamma(dx) = -\int_X \varphi(x)\, g(x)\,\gamma(dx) + \int_X \varphi(x)\, f(x)\,\widehat{h}(x)\,\gamma(dx). \tag{4.2}$$

Set $\partial_h f := g$. Similarly one defines generalized partial derivatives for mappings with values in a separable Hilbert space E.

Definition 4.1. Let $p \in (1, +\infty)$. The class $G^{p,1}(\gamma, E)$ consists of all mappings $f \in L^p(\gamma, E)$ such that there is a mapping $Df \in L^p\big(\gamma, \mathcal{H}(H,E)\big)$ with the property that, for every $h \in H$, the E-valued mapping $x \mapsto Df(x)h$ serves as a generalized partial derivative of f along h.

The classes $G^{p,r}(\gamma, E)$ with $r \in \mathbb{N}$ are defined inductively as follows: the class $D^{p,r+1}(\gamma, E)$ consists of all mappings $f \in G^{p,1}(\gamma, E)$ such that Df belongs to the class $G^{p,r}\big(\gamma, \mathcal{H}_r(H, E)\big)$, defined at the previous inductive step, and the derivative of order $r+1$ is defined by $D_H^{r+1} f = D_H^r D_H f$.

Theorem 4.2. *One has* $G^{p,r}(\gamma, E) = W^{p,r}(\gamma, E)$ *if* $p \in (1, +\infty), r \in \mathbb{N}$.

Remark 4.3. The case $p = 1$ requires a special examination, since in the definition of generalized derivatives we used the fact that $\widehat{h} f \in L^1(\gamma)$, which is true by Hölder's inequality for any $f \in L^p(\gamma)$ with $p > 1$. The space $G^{1,1}(\gamma)$ can be defined as the space of all functions $f \in L^1(\gamma)$ such that $\widehat{h} f \in L^1(\gamma)$ for all $h \in H$ and there is a mapping $D_H f \in L^1(\gamma, H)$ for which the function $(D_H f, h)_H$ serves as a generalized partial derivative along h for each $h \in H$. However, as shall see in the lemma below, the inclusion $\widehat{h} f \in L^1(\gamma)$ is automatically fulfilled if f has a directional partial derivative $\partial_h f \in L^1(\gamma)$ in the sense considered below.

A closely related description focuses on directional properties of functions in the Sobolev classes. We present a typical result for $r = 1$ and $E = \mathbb{R}^1$; extensions to greater r and infinite-dimensional E are straightforward. Let us fix an orthonormal basis $\{e_i\}$ in H.

Theorem 4.4. *A function f in $L^p(\gamma)$, $p \geqq 1$, belongs to $W^{p,1}(\gamma)$ precisely when, for each e_i, it has a version $\widetilde{f}$ such that the functions $t \mapsto \widetilde{f}(x + te_i)$, where $x \in X$, are locally absolutely continuous and, setting*

$$\partial_{e_i} f(x) := \frac{d}{dt} \widetilde{f}(x + te_i)|_{t=0},$$

we obtain a mapping $\nabla f = (\partial_{e_i} f)_{i=1}^{\infty}$ belonging to $L^p(\gamma, H)$. The same is true for the class $G^{p,1}(\gamma)$, so that $W^{p,1}(\gamma) = G^{p,1}(\gamma)$ also for $p = 1$.

It should be added that the partial derivative $\partial_{e_i} f(x)$ exists almost everywhere, since $t \mapsto \widetilde{f}(x + te_i)$ is almost everywhere differentiable on the real line (by a classical result from real analysis), which yields through conditional measures that the derivative at zero exists for almost every fixed x; certainly, for a given x there might be no derivative at zero. The reader is warned that a version $\widetilde{f}$ with the required properties depends in general on e_i, which is suppressed in our notation. This happens already in dimension 2: taking a function $f \in W^{2,1}(\gamma)$ such that every version of it is locally unbounded (it is easy to give an example), we see that f has no version continuous in each variable separately (such a version would have a point of continuity).

Lemma 4.5. *There is a constant C with the following property: if a function f in $L^1(\gamma)$ has a version that is locally absolutely continuous on the lines $x + \mathbb{R}^1 h$ for some $h \in H$ and $\partial_h f \in L^1(\gamma)$, where $\partial_h f$ is defined almost everywhere through the indicated version, then*

$$\int |\widehat{h}||f|\,d\gamma \le C\big(\|f\|_{L^1(\gamma)} + \|\partial_h f\|_{L^1(\gamma)}\big). \tag{4.3}$$

Proof. In the one-dimensional case the assertion is obvious, because the integral of $t|f(t)|$ over $[0,\infty)$ with respect to the standard Gaussian measure is estimated by $C\big(\|f\|_{L^1(\gamma)} + \|f'\|_{L^1(\gamma)}\big)$ with some constant C as follows. Let us deal with a locally absolutely continuous version of f. Then $g'(t) = -tg(t)$ and by the integration by parts formula we have

$$\int_0^R t|f(t)|g(t)\,dt = \int_0^R |f(t)|' g(t)\,dt - |f(t)|g(t)\Big|_0^R,$$

whence, taking into account that $\big||f(t)|'\big| = |f(t)'|$ a.e., we find that

$$\int_{-\infty}^{+\infty} |t||f(t)|g(t)\,dt \le 2\|f'\|_{L^1(\gamma)} + |f(0)|.$$

Let us estimate $f(0)$. We may assume that $f(0) > 0$. Let us take $T > 0$ such that $[0,T]$ has γ-measure $1/4$. Next, we choose $\tau \in [0,T]$ such that $f(\tau) \le 4\|f\|_{L^1(\gamma)}$. Then, letting $C_1^{-1} := \min_{t\in[0,T]} g(t)$, we have

$$f(0) \le f(\tau) + \|f'\|_{L^1[0,\tau]} \le 4\|f\|_{L^1(\gamma)} + C_1\|f'\|_{L^1(\gamma)},$$

so that

$$\int_{-\infty}^{+\infty} |t||f(t)|g(t)\,dt \le C\big(\|f\|_{L^1(\gamma)} + \|f'\|_{L^1(\gamma)}\big), \tag{4.4}$$

where $C = 6 + C_1$ does not depend on f. The general case follows from this special one. Indeed, we can assume that $|h|_H = 1$. Then the conditional measures γ^x on the straight lines $x + \mathbb{R}^1 h$ are standard Gaussian, which yields estimate (4.3). In fact, this can be seen even without conditional measures. The claim reduces to the case where γ is the standard product measure and $h = e_1$. Then it suffices to use Fubini's theorem and (4.4) for the first coordinate and fixed other coordinates. □

Yet another description of Sobolev classes (even with fractional orders of differentiability) employs the Ornstein-Uhlenbeck semigroup $\{T_t\}$. Let $r > 0$. Set

$$V_r f := \Gamma(r/2)^{-1} \int_0^\infty t^{r/2-1} e^{-t} T_t f\, dt, \quad f \in L^p(\gamma),$$

where

$$\Gamma(\alpha) := \int_0^\infty t^{\alpha-1} e^{-t}\, dt.$$

By the same formula we define V_r on $L^p(\gamma, E)$, where E is any separable Hilbert space.

For $p \geq 1$ and $r > 0$ let us consider the space

$$H^{p,r}(\gamma) := V_r\big(L^p(\gamma)\big), \quad \|f\|_{H^{p,r}} = \|V_r^{-1} f\|_{L^p(\gamma)}.$$

It is not difficult to show that this space is complete.

Let us note a common useful property of the classes of any of the three types with $p > 1$: if functions f_n belonging to one of them converge in measure to a function f and $\sup_n \|f_n\|_{p,r} < \infty$, then f belongs to the same class. Another common feature is the reflexivity of these spaces (which follows by the reflexivity of L^p with $1 < p < \infty$).

It is very important that the derivatives in these constructions are taken along the subspace H, so that the geometry of the space X carrying the measure γ is irrelevant. If X itself is a nice space (say, Hilbert or Banach), then smooth functions in the classical Fréchet or Gâteaux sense with appropriate bounds on derivatives become Sobolev differentiable. However, no values of p and r ensure continuity of elements in $W^{p,r}$.

Example 4.6. Let γ be the standard Gaussian measure on $\mathbb{R}^\infty$ restricted to the full measure Hilbert space E of sequence (x_n) with $\sum_{n=1}^\infty n^{-2} x_n^2 < \infty$. Let

$$f(x) = \sum_{n=1}^{\infty} n^{-2/3} x_n.$$

Then the function f has no version continuous on E with its Hilbert norm, but $f \in W^{p,r}(\gamma)$ for all $r \in \mathbb{N}$ and $p \in [1, +\infty)$, moreover, $D_H f$ is a constant vector and $D_H^k f = 0$ if $k \geq 2$.

A similar effect is seen in the case of the stochastic integral

$$f(w) = \int_0^1 \psi(t)\, dw(t),$$

where $\psi \in L^2[0,1]$ has unbounded variation (say, just has no bounded version). Such stochastic integrals regarded as measurable linear functionals on $C[0,1]$ are given by continuous functionals on $C[0,1]$ (are represented as integrals of paths with respect to bounded measures) precisely when ψ is a function of bounded variation as an equivalence class in $L^2[0,1]$, that is, has a modification of bounded variation, see Bogachev [13, Problem 2.12.32].

For integer values of r the spaces $H^{p,r}(\gamma)$ can be compared with the previously defined classes. The following very important result is Meyer's equivalence.

Theorem 4.7. *If $p \in (1,+\infty)$, $r \in \mathbb{N}$, then $H^{p,r}(\gamma) = W^{p,r}(\gamma) = D^{p,r}(\gamma)$ and there exist positive constants $m_{p,r}$ and $M_{p,r}$ such that*

$$m_{p,r}\|D^r_H f\|_{L^p(\gamma,\mathcal{H}_r)} \le \|(I-L)^{r/2} f\|_{L^p(\gamma)} \le$$
$$\le M_{p,r}\Big[\|D^r_H f\|_{L^p(\gamma,\mathcal{H}_r)} + \|f\|_{L^p(\gamma)}\Big]. \tag{4.5}$$

The same is true for E-valued mappings, where E is a separable Hilbert space.

Let us observe that for any function $f \in W^{p,2}(\gamma)$ we have its second derivative $D^2_H f$ and the action Lf of the Ornstein–Uhlenbeck operator on it. In the case of the standard Gaussian measure on $\mathbb{R}^n$ one has

$$Lf(x) = \Delta f(x) - \big(x, \nabla f(x)\big) = \sum_{i=1}^{n} \big[\partial^2_{x_i} f(x) - x_i \partial_{x_i} f(x)\big],$$

where both parts $\Delta f(x) = \operatorname{trace} D^2 f(x)$ and $\big(x, \nabla f(x)\big)$ exist separately. The same holds in the case of $\mathbb{R}^\infty$ for functions of finitely many variables.

However, for general functions $f \in W^{2,2}(\gamma)$ in infinite dimensions this is not true. For example, let us consider a function $f \in \mathcal{X}_2$ given by $f(x) = \sum_{n=1}^\infty n^{-1}(x_n^2 - 1)$. Then

$$D_H f(x) = 2(n^{-1}x_n)_{n=1}^\infty$$

and

$$D^2_H f(x) = A$$

is a constant Hilbert–Schmidt operator defined by the diagonal matrix with the numbers $2n^{-1}$ at the diagonal. We have

$$Lf(x) = 2\sum_{n=1}^\infty n^{-1}(1 - x_n^2),$$

where the series converges in $L^2(\gamma)$ and almost everywhere, but the part "Δf", the trace of the second derivative, which is the series of n^{-1}, does not exist separately.

Let us also note the following estimate (see, *e.g.*, Shigekawa [83, Proposition 4.5] for the proof).

Proposition 4.8. *Let $p > 1$ and $k \in \mathbb{N}$. Then there is a number $C(p,k)$ such that*

$$\|D_H^k f\|_{L^p(\gamma,\mathcal{H}_k)} \le C(p,k)\|D_H^{k+1} f\|_{L^p(\gamma,\mathcal{H}_{k+1})} + C(p,k)\|f\|_p$$

for all $f \in W^{p,k+1}(\gamma)$. An analogous estimate holds for mappings $f \in W^{p,k+1}(\gamma, E)$ with values in a separable Hilbert space E.

A multiplicative inequality for Sobolev norms is ensured by the following result (see, *e.g.*, Shigekawa [83, Proposition 4.10] for the proof).

Proposition 4.9. *Let $\alpha < \beta < \kappa$. Then there is a number $C(\alpha, \beta, \kappa) > 0$ such that*

$$\|(I-L)^\beta f\|_p \le C(\alpha,\beta,\kappa)\|(I-L)^\alpha f\|_p^{\frac{\kappa-\beta}{\kappa-\alpha}} \|(I-L)^\kappa f\|_p^{\frac{\beta-\alpha}{\kappa-\alpha}}$$

for all $f \in H^{p,\kappa}(\gamma)$.

For example, as a special case of this estimate (in fact, obtained as a step of the proof) one can obtain that

$$\|(I-L)f\|_p \le 2\|f\|_p^{1/2}\|(I-L)^2 f\|_p^{1/2},$$

which can be written as

$$\|f\|_{p,2} \le 2\|f\|_p^{1/2}\|f\|_{p,4}^{1/2}.$$

With a suitable number c we also have

$$\|f\|_{p,1} \le c\|f\|_p^{1/2}\|f\|_{p,2}^{1/2}.$$

Sobolev functions satisfy certain vector integration by parts formulas.

Theorem 4.10. *Suppose that $v \in W^{p,1}(\gamma, H)$, where $p > 1$. Then there is a function $\delta v \in L^p(\gamma)$, called the divergence of v, such that*

$$\int_X (D_H\varphi, v)_H \, d\gamma = \int_X \delta v \varphi \, d\gamma, \quad \varphi \in W^{q,1}(\gamma),\ q = p/(p-1). \quad (4.6)$$

If $v = D_H f$, where $f \in W^{p,2}(\gamma)$, then $\delta v = Lf$.

The function δv is called the divergence of the vector field v. It plays an important role in stochastic analysis. For a constant vector field $v = h \in H$ we have $\delta v = \widehat{h}$. For vector fields v on $\mathbb{R}^\infty$ of the simplest form

$$v(x) = u(x_1, \ldots, x_n)e_k$$

the divergence is also easily evaluated:

$$\delta v(x) = (D_H u(x_1, \ldots, x_n), e_k)_H + u(x_1, \ldots, x_n)x_k.$$

Similarly, in a slightly more general case of the field

$$v = u_1 h_1 + \cdots + u_n h_n, \quad u_i \in W^{p,1}(\gamma),\ h_i \in H,$$

we have

$$\delta v = \sum_{i=1}^{n} \Big[(D_H u_i, h_i)_H + u_i \widehat{h_i} \Big].$$

The previous theorem says essentially that the L^p-norm of this function can be controlled through the Sobolev norm of v; this is quite easy for $p = 2$, but requires some work in the general case.

In fact, the following result is true (see, *e.g.*, Shigekawa [83, Theorem 4.17] for the proof).

Theorem 4.11. *The divergence operator δ extends to a continuous linear operator*

$$\delta \colon\ W^{p,r+1}(\gamma, H) \to W^{p,r}(\gamma).$$

An analogous result is true for mappings with values in the space of Hilbert–Schmidt operators between H and a separable Hilbert space E, in which case the divergence takes values in E.

5 Inequalities and embeddings

In the theory of Sobolev classes on $\mathbb{R}^n$, a very important role is played by various inequalities related to embeddings of Sobolev classes into other functional classes such as L^p-spaces. For example, every function f of class $W^{1,1}(\mathbb{R}^n)$ belongs not only to $L^1(\mathbb{R}^n)$, but also to $L^{n/(n-1)}(\mathbb{R}^n)$. There are also local embeddings of this sort. For example, any function in $W^{p,1}(\mathbb{R}^n)$ with $p > n$ has a continuous version. The Sobolev classes we discuss are analogs of weighted Sobolev classes on $\mathbb{R}^n$. For the latter even on the real line the usual global embeddings fail. For example, the function $f(x) = \varphi(x)e^{-x}e^{x^2/4}$, where φ is smooth, $\varphi(x) = 0$ if $x \le 0$ and $\varphi(x) = 1$ if $x \ge 1$, belongs to all classes $W^{2,r}(\gamma)$ for the standard Gaussian measure on the real line, but is in no class $L^p(\gamma)$ with $p > 2$.

Nevertheless, local embeddings hold for reasonable weights (say, positive continuous).

In the infinite-dimensional case, there are no even local embeddings. In particular, no continuity is ensured by the membership even in all $W^{p,r}(\gamma)$, $p < \infty$, $r \in \mathbb{N}$. This is seen in the simplest example of a measurable linear functional $f(x) = \sum_{n=1}^{\infty} n^{-1}x_n$ on $\mathbb{R}^\infty$ with the standard Gaussian measure γ. Indeed, let us show that there is no function g continuous on $\mathbb{R}^\infty$ and equal f a.e. (the fact that f itself is not continuous, is obvious, since every continuous linear function on $\mathbb{R}^\infty$ depends on finitely many variables). Otherwise there is a neighborhood of zero V such that $|f(x)| \le M$ a.e. in V for some M. Hence there exist k and c such that $|f(x)| \le M$ a.e. on the set $S = \{|x_i| < c, i = 1, \dots, k\}$. There is N such that $\left|\sum_{n=k+2}^{\infty} n^{-1}x_n\right| \le N$ with probability at least $1/2$. Since the set $\{x\colon x_{k+1} > ck + N + M + 1\}$ has positive measure, we arrive at the contradiction: there is a positive measure set of points $x \in S$ such that $f(x) > M$.

Moreover, by using the previous example, one can find a function f with compact support in $\mathbb{R}^\infty$ belonging to all classes $W^{p,r}(\gamma)$ for the standard Gaussian measure and having no continuous version. Also a function f with compact support can be constructed such that $f \in W^{2,k}(\gamma)$ for all k, but f does not belong to the union of $L^p(\gamma)$ with $p > 2$. Such a function can be constructed in the form $f = \sum_{n=1}^{\infty} f_n$, where f_n is a function of x_n of class C_0^∞. Nevertheless, there are some dimension-free inequalities that extend to the infinite-dimensional case.

Two important inequalities central for Gaussian analysis are presented in the next theorem.

Theorem 5.1. *Suppose that γ is a centered Radon Gaussian measure on a locally convex space X. Then, for any $f \in W^{2,1}(\gamma)$, one has the logarithmic Sobolev inequality*

$$\int_X f^2 \log|f|\,d\gamma \le \int_X |D_H f|_H^2 d\gamma + \frac{1}{2}\Big(\int_X f^2 d\gamma\Big)\log\Big(\int_X f^2 d\gamma\Big). \quad (5.1)$$

In addition, there holds the Poincaré inequality

$$\int_X \Big(f - \int_X f\,d\gamma\Big)^2 d\gamma \le \int_X |D_H f|_H^2\,d\gamma. \quad (5.2)$$

Moreover, if $p \ge 1$, then

$$\int_X \Big|f - \int_X f\,d\gamma\Big|^p d\gamma \le (\pi/2)^p M_p \int_X |D_H f|_H^p\,d\gamma, \quad (5.3)$$

where M_p is the moment of order p of the standard Gaussian measure on the real line.

Several authors contributed in discovering these inequalities in different form; Nash's paper [68] is the earliest one I know where the Poincaré inequality is explicitly given in the stated form with gradients (certainly, when written in terms of the Hermite expansions it becomes trivial); Stam [86] proved the logarithmic Sobolev inequality, later different derivations via hypercontractivity were found, see Federbush [43] and Nelson [69]; the paper of Gross [52] (where (5.1) was proved explicitly with gradients) became a starting point of intensive research related to characterizations and applications of logarithmic Sobolev inequalities (see references in Blanchere *et al.* [11], Bogachev [16], Lieb, Loss [64], and also the recent paper Cianchi, Pick [31]).

As an application of the logarithmic Sobolev inequality let us consider the following situation that often arises in stochastic analysis. Suppose that $\nu = \varrho \cdot \gamma$ is a probability measure, where γ is the standard Gaussian measure on $\mathbb{R}^n$ or on $\mathbb{R}^\infty$. Its entropy (or the entropy of ϱ) is defined by

$$\mathrm{Ent}_\gamma \varrho := \int \varrho \log \varrho \, d\gamma,$$

whenever $\varrho \log \varrho$ is integrable; otherwise we set $\mathrm{Ent}_\gamma \varrho := +\infty$. Since the function $t \mapsto t \log t$ is convex and the logarithm of the integral of ϱ vanishes, we have by Jensen's inequality that

$$\mathrm{Ent}_\gamma \varrho \geq 0.$$

Upper bounds for entropy are often of interest in applications. Suppose that we have $\sqrt{\varrho} \in W^{2,1}(\gamma)$. Then the Sobolev inequality yields the estimate

$$\mathrm{Ent}_\gamma \varrho \leq \frac{1}{2} \mathrm{I}(\varrho), \quad \mathrm{I}(\varrho) := \int \frac{|D_H \varrho|^2}{\varrho} \, d\gamma,$$

where $|D_H \varrho|^2/\varrho = 0$ on the set $\{\varrho = 0\}$ and ∇ is the .

To justify this we consider the standard Gaussian measure on $\mathbb{R}^n$ and note that $\varrho \in W^{1,1}(\gamma)$ and that $\nabla\sqrt{\varrho} = 2^{-1}\varrho^{-1/2}\nabla\varrho$ with the above convention. Indeed, the integrability of ϱ and $|\nabla\varrho|^2/\varrho$ yield the integrability of $|\nabla\varrho|$ by the Cauchy inequality. We can calculate the derivatives pointwise by using the corresponding versions.

The logarithmic Sobolev inequality is a certain weak replacement for missing analogs of the classical Sobolev inequalities in $\mathbb{R}^d$ which improve the initial integrability of a function on the basis of the integrability of its derivative. This is used, *e.g.*, in the study of invariant measures of infinite-dimensional diffusions (see Bogachev, Röckner [23]).

The logarithmic Sobolev inequality is known to be equivalent to the so-called hypercontractivity of the Ornstein–Uhlenbeck semigroup.

Theorem 5.2. *The Ornstein–Uhlenbeck semigroup* $\{T_t\}_{t\geq 0}$ *is hypercontractive,* i.e., *whenever* $p > 1$, $q > 1$, *one has*

$$\|T_t f\|_q \leq \|f\|_p$$

for all $t > 0$ *such that* $e^{2t} \geq (q-1)/(p-1)$.

Applying this theorem we obtain a number of important results for polynomials.

Corollary 5.3. *Let* $p \geq 2$. *Then the operator* $I_n\colon f \mapsto I_n(f)$ *from* $L^2(\gamma)$ *to* $L^p(\gamma)$ *is continuous and*

$$\|I_n(f)\|_p \leq (p-1)^{n/2}\|f\|_2. \tag{5.4}$$

In addition, for every $p \in (1,\infty)$, *the operators* I_n *are continuous on* $L^p(\gamma)$ *and*

$$\|I_n\|_{\mathcal{L}(L^p(\gamma))} \leq (M-1)^{n/2}, \tag{5.5}$$

where $M = \max\bigl(p, p(p-1)^{-1}\bigr)$.

Corollary 5.4. *Let* $f \in \mathcal{X}_d$. *For any* $\alpha \in (0, d/(2e))$, *there holds the inequality*

$$\gamma\Bigl(x\colon\ |f(x)| \geq t\|f\|_2\Bigr) \leq c(\alpha,d)\exp(-\alpha t^{2/d}),$$

where $c(\alpha,d) = \exp\alpha + \dfrac{d}{d-2e\alpha}$.

Corollary 5.5. *The spaces* $\mathcal{X}_d$ *are closed with respect to convergence in measure. Moreover, any sequence from* $\bigoplus_{k=0}^{d}\mathcal{X}_k$ *that converges in measure, is convergent in* $L^p(\gamma)$ *for every* $p \in [1,\infty)$. *The same is true for the spaces* $\mathcal{X}_d(E)$ *of mappings with values in any separable Hilbert space* E.

Corollary 5.6. *The norms from* $L^p(\gamma)$, $p \in [1,\infty)$, *are equivalent on every* $\mathcal{X}_n$. *In addition, for every* $p > 0$, *the topology on* $\mathcal{X}_n$ *induced by the metric from* $L^p(\gamma)$ *coincides with the topology of convergence in measure. Finally, if* $q > p > 1$, *one has*

$$\|f\|_p \leq \|f\|_q \leq \left(\frac{q-1}{p-1}\right)^{n/2}\|f\|_p \quad \forall f \in \mathcal{X}_n. \tag{5.6}$$

It should be noted that the classes $W^{p,r}(\gamma)$ can be also defined by completing the set of measurable polynomials with the respect to the Sobolev norm.
Embedding inequalities in the case $p=1$ are studied in Shigekawa [82].

Riesz transforms and other operators related to Gaussian L^p- and Sobolev spaces are studied in Aimar, Forzani, Scotto [2], Brandolini, Chiacchio, Trombetti [26], Sjögren, Soria [84]. Besov-type Gaussian spaces, a recent topic in this area, are studied in Pineda, Urbina [76] and Nikitin [70–72].

6 Sobolev classes over differentiable measures

We now turn to general Fomin differentiable measures. Let μ be a Radon probability measure on a locally convex space X and let H be a separable Hilbert space continuously and densely embedded into X; as above, one can think that $X = \mathbb{R}^\infty$ and $H = l^2$. Sobolev classes over μ can be introduced in the same three ways as in the Gaussian case: as completions, via integration by parts, and using semigroups. We only give these definitions, referring the reader to Chapter 8 of the book [16], where a detailed discussion with proofs is given.

Let $\{e_n\}$ be an orthonormal basis in H. For $p \geq 1$ and $r \in N$ the Sobolev norm $\|\cdot\|_{p,r}$ is defined by the formula

$$\|f\|_{p,r} := \sum_{k=0}^{r}\Big(\int_X\Big[\sum_{i_1,\dots,i_k=1}^{\infty}\big(\partial_{e_{i_1}}\cdots\partial_{e_{i_k}} f\big)^2\Big]^{p/2}\mu(dx)\Big)^{1/p}. \qquad (6.1)$$

The same norm can be written as

$$\|f\|_{p,r} = \sum_{k=0}^{r}\|D_H^k f\|_{L^p(\mu,\mathcal{H}_k)}.$$

For example, for $r = 1$ we obtain

$$\|f\|_{p,1} = \|f\|_{L^p(\mu)} + \|D_H f\|_{L^p(\mu,H)}.$$

If $f \in \mathcal{F}C^\infty$, then $\|f\|_{p,r} < \infty$ since $\|D_H^k f\|_{\mathcal{H}_k} \in C_b(X)$. Denote by $W^{p,r}(\mu)$ or by $W_H^{p,r}(\mu)$ the completion of $\mathcal{F}C^\infty$ with respect to the norm $\|\cdot\|_{p,r}$.

The next definition introduces Sobolev functions by means of versions possessing directional derivatives with suitable integrability conditions.

Let E be one more separable Hilbert space.

Definition 6.1. Let $F\colon X \to E$ be μ-measurable. The mapping F is called absolutely ray continuous if, for every $h \in H$, the mapping F has a modification F_h such that for every $x \in X$ the mapping $t \mapsto F_h(x+th)$ is absolutely continuous on bounded intervals.

Definition 6.2. A μ-measurable mapping $F\colon X \to E$ is stochastically Gâteaux differentiable if there exists a measurable mapping $D_H F\colon X \to \mathcal{H}(H, E)$ such that for every $h \in H$ we have

$$\frac{F(x+th) - F(x)}{t} - D_H F(x)(h) \underset{t\to 0}{\longrightarrow} 0 \quad \text{in measure } \mu.$$

The derivative of the nth order $D^n_H F$ is defined inductively as $D_H(D^{n-1}_H F)$. An alternative notation is $\nabla^n_H F$.

Definition 6.3. Let $1 \le p < \infty$. The space $D^{p,1}(\mu, E)$ is defined as the class of all mappings $f \in L^p(\mu, E)$ such that f is ray absolutely continuous, stochastically Gâteaux differentiable and $D_H f \in L^p\big(\mu, \mathcal{H}(H, E)\big)$. The space $D^{p,1}(\mu, E)$ (another notation is $D^{p,1}_H(\mu, E)$) is equipped with the norm

$$\|f\|^0_{p,1,E} := \|f\|_{L^p(\mu)} + \|D_H f\|_{L^p(\mu,\mathcal{H}(H,E))}.$$

For $r = 2, 3, \dots$ we define the classes $D^{p,r}(\mu, E) = D^{p,r}_H(\mu, E)$ inductively:

$$D^{p,r}(\mu, E) := \Big\{ f \in D^{p,r-1}(\mu, E)\colon D_H f \in D^{p,r-1}\big(\mu, \mathcal{H}(H, E)\big)\Big\}.$$

The corresponding norms are defined by the equalities

$$\|f\|^0_{p,r,E} := \|f\|_{L^p(\mu)} + \|D_H f\|_{L^p(\mu,\mathcal{H}(H,E))} + \dots + \|D^r_H f\|_{L^p(\mu,\mathcal{H}_r(H,E))}.$$

We set $D^{p,r}(\mu) := D^{p,r}_H(\mu) := D^{p,r}(\mu, \mathbb{R})$.

Let us turn to the definition with generalized derivatives.

Let $j\colon X^* \to H$ be the adjoint embedding for the embedding $H \to X$, *i.e.*, we have

$$(j(l), h)_H = l(h) \quad \text{for all } l \in X^*,\ h \in H.$$

Since H is dense in X, we see that $j(X^*)$ is dense in H.

Definition 6.4. We shall say that a function $f \in \mathcal{L}^1(\mu)$ has a generalized partial derivative $g = \partial_h f \in L^1(\mu)$ along a vector $h \in H$ if $f\beta^\mu_h \in \mathcal{L}^1(\mu)$ and for every $\varphi \in \mathcal{F}C^\infty$ one has

$$\begin{aligned}\int_X \partial_h\varphi(x) f(x)\,\mu(dx) = &-\int_X g(x)\varphi(x)\,\mu(dx)\\ &-\int_X f(x)\varphi(x)\beta^\mu_h(x)\,\mu(dx).\end{aligned} \tag{6.2}$$

It is clear that a generalized partial derivative is uniquely determined as an element of $L^1(\mu)$.

Definition 6.5. Let $G^{p,1}(\mu) = G_H^{p,1}(\mu)$ be the class of all real functions $f \in L^p(\mu)$ possessing generalized partial derivatives along all vectors in $j(X^*)$ and having finite norms

$$\|f\|_{p,1} := \|f\|_{L^p(\mu)} + \|D_H f\|_{L^p(\mu,H)} < \infty,$$

where $D_H f$ is defined as follows: there is a mapping $T\colon X \to H$ such that for every $l \in X^*$ we have $\langle l, T(x)\rangle = \partial_{j(l)} f(x)$ a.e. Then we set $D_H f := \nabla_H f := T$. The space $G^{p,1}(\mu)$ is equipped with the norm $\|\cdot\|_{p,1}$. Similarly we define the class $G^{p,1}(\mu, E) = G_H^{p,1}(\mu, E)$ of mappings with values in a Hilbert space E. Hence one can inductively introduce the classes $G^{p,r}(\mu) = G_H^{p,r}(\mu)$ of functions $f \in L^p(\mu)$ with $D_H^k f \in L^p(\mu, \mathcal{H}_k)$ whenever $k \le r$ equipped with the norms

$$\|f\|_{p,r} := \|f\|_{L^p(\mu)} + \sum_{k=1}^{r} \|D_H^k f\|_{L^p(\mu,\mathcal{H}_k)}.$$

In our model example $X = \mathbb{R}^\infty$ and $H = l^2$, the inclusion $f \in G^{p,1}(\mu)$ means that $f \in L^p(\mu)$ has generalized partial derivatives $\partial_{e_n} f$ such that $D_H f := (\partial_{e_n} f)_{n=1}^\infty \in L^p(\mu, H)$, where $|D_H f|_H^2 = \sum_{n=1}^\infty |\partial_{e_n} f|^2$. Certainly, the inclusions $f\beta_{j(l)}^\mu \in L^1(\mu)$ for all $l \in X^*$ are implicitly meant (here this reduces to $f\beta_{e_n}^\mu \in L^1(\mu)$ for all n).

Proposition 6.6. *Let $p \ge 1$ and $\beta_h^\mu \in L^{p/(p-1)}(\mu)$ for all $h \in j(X^*)$. Then the spaces $G^{p,1}(\mu, E)$ with the respective norms are complete.*

Remark 6.7. Let $p \ge 1$ and $\beta_h^\mu \in L^{p/(p-1)}(\mu)$ for all $h \in H$. If $f \in G^{p,1}(\mu)$, then f has generalized partial derivatives $\partial_h f \in L^p(\mu)$ for all $h \in H$, not only for the elements of $j(X^*)$, as required by the definition. Indeed, let $\{e_n\}$ be an orthonormal basis in H contained in $j(X^*)$; in the case of $\mathbb{R}^\infty$ just the usual basis. Since $\beta_h^\mu \in L^{p/(p-1)}(\mu)$, we obtain a linear mapping $H \to L^{p/(p-1)}(\mu)$. It is readily seen that its graph is closed. Therefore, letting $h_n = \sum_{i=1}^n (h, e_i)_H e_i$, we obtain convergence $\beta_{h_n}^\mu \to \beta_h^\mu$ in $L^{p/(p-1)}(\mu)$. Therefore, (6.2) remains valid for h once it holds for each h_n.

Finally, there is an exact analog of the definition involving a symmetric semigroup (which has been used to obtain fractional classes $H^{p,r}$ in the Gaussian case).

Let $\{e_n\}$ be an orthonormal basis in H such that $e_n = j(l_n)$, $l_n \in X^*$; it is again wise to assume that we deal with the space $X = \mathbb{R}^\infty$ and $H = l^2$, in which case we take the standard basis. Suppose that

$$\beta^{\mu}_{e_n} \in L^2(\mu) \quad \forall n \in \mathbb{N}.$$

Then we obtain the operator

$$Lf = \sum_{i=1}^{n} [\partial^2_{e_i} f + \beta^{\mu}_{e_i} \partial_{e_i} f]$$

acting on functions f of the form $f(x) = f_0(l_1(x), \dots, l_n(x))$, $f_0 \in C_b^\infty(\mathbb{R}^n)$; that is, on smooth cylindrical functions in the case of $\mathbb{R}^\infty$. This operator is densely defined and symmetric in $L^2(\mu)$:

$$\begin{aligned}\int_X g(x)Lf(x)\,\mu(dx) &= -\sum_{i=1}^{n} \int_X \partial_{e_i} g(x) \partial_{e_i} f(x)\,\mu(dx) \\ &= -\int_X (D_H f, D_H g)_H \, d\mu\end{aligned}$$

due to the integration by parts formula. In addition, $(Lf, f)_2 \le 0$. Therefore, the operator L has a non-positive selfadjoint extension, namely, we take its Friedrichs extension denoted by the same symbol L (see Reed, Simon [80, Section X.3]). The domain of definition of this extension will be denoted by $D(L)$. The bounded operator $(I - L)^{-1}$ is a selfadjoint nonnegative contraction on $L^2(\mu)$. Hence we can also define the powers $(I - L)^{-r/2}$ on $L^2(\mu)$ and obtain the Sobolev spaces

$$H^{2,r}(\mu) := (I - L)^{-r/2}\big(L^2(\mu)\big), \quad r > 0.$$

However, we can get even more: one can show that the operators $(I-L)^{\frac{-r}{2}}$ extend as nonnegative contractions on the spaces $L^p(\mu)$, $1 \le p < \infty$, so that the Sobolev classes

$$H^{p,r}(\mu) := (I - L)^{-r/2}\big(L^p(\mu)\big), \quad r > 0, \ p \ge 1,$$

arise. Unlike the Gaussian case, their relation (for natural values of r) to the previously defined classes has not been clarified. Moreover, the exact relations between the other classes remain unclear except for rather special cases. One of them is considered in part (iii) of the next theorem.

Theorem 6.8. (i) *If $\beta^{\mu}_h \in L^{p/(p-1)}(\mu)$ for all $h \in j(X^*)$, then we have*

$$W^{p,r}(\mu) \subset G^{p,r}(\mu) \quad \textit{and} \quad D^{p,r}(\mu) \subset G^{p,r}(\mu)$$

for all $p \in [1, \infty)$, $r \in \mathbb{N}$.

(ii) *Suppose that for every* $h \in j(X^*)$ *the corresponding conditional measures* μ^y *on the real line have continuous positive densities (for example, this holds if there exist* $c_h > 0$ *such that* $\exp\big(c_h \beta_h^\mu\big) \in L^1(\mu)$*). Then*

$$D^{p,1}(\mu) = G^{p,1}(\mu).$$

(iii) *Suppose that* $X = \mathbb{R}^\infty$, *condition* (ii) *is fulfilled and that*

$$\sup_n \Big\| \sum_{i=1}^n \beta_{e_i}^\mu e_i - \mathbb{E}_n \sum_{i=1}^n \beta_{e_i}^\mu e_i \Big\|_{L^p(\mu,l^2)} < \infty, \tag{6.3}$$

where $\mathbb{E}_n$ *is the conditional expectation with respect to the* σ*-algebra generated by the coordinates* $x_1, \ldots, x_n$. *Then*

$$W^{p,1}(\mu) = D^{p,1}(\mu) = G^{p,1}(\mu) \quad \textit{for all } p > 1.$$

Condition (6.3) here is very restrictive (but it holds for product-measures and some measures absolutely continuous with respect to product-measures). It would be interesting to find more general conditions ensuring the coincidence of the classes $W^{p,1}(\mu)$, $D^{p,1}(\mu)$, and $G^{p,1}(\mu)$; also their relation to $H^{p,1}(\mu)$ is open in the general case.

The interpolation approach to fractional Sobolev classes was suggested in Watanabe [89] and further employed by several authors, see, *e.g.*, Airault, Bogachev, Lescot [3], Bogachev [16], Nikitin [70–72].

7 The class BV: the Gaussian case

In a particular way one introduces the space $BV(\gamma)$ of functions of bounded variation, containing $W^{1,1}(\gamma)$, see Fukushima [49], Fukushima, Hino [50], Hino [55–57], Ambrosio, Miranda, Maniglia, Pallara [8,9], Bogachev, Pilipenko, Rebrova [20], Bogachev, Rebrova [22], Bogachev, Shaposhnikov [24], and Röckner, Zhu, Zhu [81]. In the case of $\mathbb{R}^\infty$ with the standard Gaussian measure it consists of functions $f \in L^1(\gamma)$ such that $x_n f \in L^1(\gamma)$ for all n and there is an H-valued measure Λf of bounded variation for which the scalar measures $(\Lambda f, e_n)_H$ satisfy the identity

$$\int \varphi(x)\,(\Lambda f, e_n)_H(dx) = -\int [f(x)\partial_{e_n}\varphi(x) - x_n f(x)\varphi(x)]\,\gamma(dx)$$

for all $\varphi \in \mathcal{F}C_b^\infty$. In the general case the definition is the same, we just take for $\{e_n\}$ an orthonormal basis in H and use the functionals $\widehat{e_n}$ in place of the coordinate functions x_n. So the general definition reads as follows.

Definition 7.1. The class $BV(\gamma)$ consists of all functions $f \in L^1(\gamma)$ such that $f\widehat{h} \in L^1(\gamma)$ for all $h \in H$ and there is an H-valued measure Λf of bounded variation satisfying the identity

$$\int_X \varphi(x)\,(\Lambda f, h)_H(dx) = -\int [f(x)\partial_h\varphi(x) - \widehat{h}(x) f(x)\varphi(x)]\,\gamma(dx) \quad (7.1)$$

for all $\varphi \in \mathcal{F}C^\infty$ and all $h \in H$.

If $f \in W^{1,1}(\gamma)$, then we take $\Lambda f = \nabla f \cdot \gamma$ and see that $f \in BV(\gamma)$, since $x_i f \in L^1(\gamma)$ by Lemma 4.5.

The following fact is known (see, *e.g.*, Ambrosio, Miranda, Maniglia, Pallara [9]).

Theorem 7.2. *A function $f \in L^1(\gamma)$ belongs to $BV(\gamma)$ precisely when*

$$\sup_{t>0} \|T_t f\|_{1,1} < \infty.$$

This is also equivalent to the existence of a sequence of functions $f_n \in \mathcal{F}C^\infty$ convergent to f in $L^1(\gamma)$ and bounded in $W^{1,1}(\gamma)$.

We now consider a broader class of functions of bounded variation: its definition is based on vector measures of semibounded variation in place of measures of bounded variation.

Definition 7.3. The class $SBV(\gamma)$ consists of all functions $f \in L^1(\gamma)$ such that $f\widehat{h} \in L^1(\gamma)$ for all $h \in H$ and there is an H-valued measure Λf of semibounded variation satisfying (7.1).

In the finite-dimensional case the classes $BV(\gamma)$ and $SBV(\gamma)$ coincide as sets, but their norms are different. Already for smooth functions f, where Λf is given by a vector density ∇f with respect to γ, the BV-norm may be much larger, since it involves the integral of $|\nabla f|$, while the SBV-norm deals with the integrals of $|\partial_h f|$ with $|h| \le 1$ (see the example below where this is shown explicitly).

Lemma 7.4. *There is a number $C > 0$ such that*

$$\|\widehat{h} f\|_{L^1(\gamma)} \le C\big(\|f\|_1 + V(\Lambda f)\big), \quad f \in SBV(\gamma),\ |h|_H \le 1.$$

Proof. This can be derived by using Lemma 4.5 (in fact, by Theorem 7.2, it suffices to consider $f \in W^{1,1}(\gamma)$) or by a similar reasoning for functions of class SBV. □

Proposition 7.5. *The space $BV(\gamma)$ is Banach with the norm*

$$\|f\|_{BV} = \|f\|_1 + \|\Lambda f\|.$$

The space $SBV(\gamma)$ is Banach with the norm

$$\|f\|_{SBV} = \|f\|_1 + V(\Lambda f).$$

Proof. Let $\{f_n\}$ be a Cauchy sequence in $BV(\gamma)$. Then it converges in $L^1(\gamma)$ to some function $f \in L^1(\gamma)$. In addition, the measures Λf_n converge in variation to some H-valued measure Λ. Finally, for each $h \in H$ the sequence of functions $\widehat{h} f_n$ is fundamental in $L^1(\gamma)$ by the estimate in the lemma above, hence it converges in $L^1(\gamma)$ to $\widehat{h} f$. Therefore, we can set $\Lambda f := \Lambda$ and obtain (7.1), *i.e.*, $f \in BV(\gamma)$. By construction, $f_n \to f$ in $BV(\gamma)$. The proof of the second assertion is the same. □

Example 7.6. In the infinite-dimensional case $BV(\gamma) \neq SBV(\gamma)$. For the proof we have to find a function $f \in SBV(\gamma)$ for which the variation of Λf is infinite. It suffices to verify that on every $\mathbb{R}^k$ we can find a smooth function f_k such that the ratio of the variation of the measure Λf_k and its semivariation tends to infinity as $k \to \infty$. As indicated above, $\Lambda f_k = \nabla f_k \cdot \gamma_k$, where γ_k is the standard Gaussian measure on $\mathbb{R}^k$. In addition, $\|\Lambda f_k\|$ is the integral of $|\nabla f_k|$ with respect to the measure γ_k, $V(\Lambda f_k)$ is the maximum of the integrals of $|\partial_h f_k|$ with respect to the measure γ_k taken over h in the unit ball. Let us take f_k of the form $g(|x|^2)$, where g is a smooth nonzero function with support in $[0, 1]$. Then $V(\Lambda f_k)$ is the integral of $|\partial_{x_1} f_k|$ with respect to the measure γ_k. Passing to spherical coordinates we obtain that $V(\Lambda f_k)/\|\Lambda f_k\|$ is the ratio of the integrals of $\cos\varphi \sin^{k-2}\varphi$ and $\sin^{k-2}\varphi$ over $[0, \pi/2]$. It is straightforward to verify that this ratio tends to zero, since the integral of the function $\sin^{k-2}\varphi$ over $[0, t_k]$, where $t_k = \arccos k^{-1/4}$, is estimated by $C\exp(-k^{1/2}/2)$, the integral of $\sin^{k-2}\varphi$ over $[0, \pi/2]$ is greater than the integral of $\cos\varphi \sin^{k-2}\varphi$, which equals $(k-1)^{-1}$, and on $[t_k, \pi/2]$ one has $\cos\varphi \leq k^{-1/4}$.

8 The class BV: the general case

In this section we consider BV-functions over general Fomin differentiable measures. Suppose that μ is a Radon probability measure on a locally convex space X that is Fomin differentiable (*i.e.*, has a logarithmic derivative) along all vectors from a continuously embedded Hilbert space H. For simplicity, one can assume that $X = \mathbb{R}^\infty$ and $H = l^2$. As in the Gaussian case, there are two options in the definition of functions of bounded variation: based on vector measures of bounded and semibounded variation. The proofs of the results of this section can be found in Bogachev, Rebrova [22] or in the more general case of domains in the last section and in the papers Bogachev, Pilipenko, Rebrova [20], Bogachev, Pilipenko, Shaposhnikov [21].

Recall that β_h^μ is the logarithmic derivative of μ along h.

For the subsequent discussion the following facts may be useful. For any measure μ differentiable along H we obtain an H-valued measure $D\mu$ defined by the equality

$$(D\mu(B), h)_H = d_h\mu(B).$$

If H is infinite-dimensional and μ is not zero, then the measure $D\mu$ has unbounded variation (see Proposition 7.3.2 in Bogachev [16]). However, if $D\mu$ is regarded as an X-valued measure and X is a Banach space, then under broad assumptions this X-valued measure has bounded variation (*e.g.*, if the embedding $H \to X$ is absolutely summing, see Chapter 7 in Bogachev [15]). If μ is a centered Gaussian measure and H is its Cameron–Martin space, then $D\mu$ as an X-valued measure has vector density $-x$ with respect to μ. Note that the vector measure with density $-x$ with respect to the standard Gaussian measure on $\mathbb{R}^n$ has variation $\int |x|\gamma(x)$, which goes to $+\infty$ as $n \to +\infty$, but its semivariation is independent of n.

Set

$$M_H(\mu) = \{f \in L^1(\mu)\colon\ f\beta_h^\mu \in L^1(\mu)\ \forall\, h \in H\}.$$

Theorem 8.1. *The set $M_H(\mu)$ is a Banach space with the norm*

$$\|f\|_M := \|f\|_{L^1(\mu)} + \sup_{|h|_H\le 1} \|f\beta_h^\mu\|_{L^1(\mu)}.$$

Proof. For each $f \in M_H(\mu)$ the quantity $\|f\|_M$ is finite, since the mapping $h \mapsto f\beta_h^\mu$ from H to $L^1(\mu)$ has a closed graph. Indeed, if $h_n \to h$ in H and $f\beta_{h_n}^\mu \to g$ in $L^1(\mu)$, then $g = f\beta_h^\mu$, because by the continuity of the embedding $H \to DD(\mu)$ (which follows by the closed graph theorem) we have $\beta_{h_n}^\mu \to \beta_h^\mu$ in $L^1(\mu)$, hence in measure. Thus, the operator $h \mapsto f\beta_h^\mu$ is bounded, so $\|f\|_M < \infty$. Now, if $\{f_n\}$ is a Cauchy sequence in this norm, it converges to a function f in $L^1(\mu)$. Clearly, $\|f\|_M < \infty$. Moreover, given $\varepsilon > 0$, we take N such that $\|f_n - f_k\|_M \le \varepsilon$ for all $n, k \ge N$ and by Fatou's theorem conclude that $\|f_n - f\|_M \le \varepsilon$ for all $n \ge N$. □

Definition 8.2. Let

$$SV(\mu) = \{f \in L^1(\mu)\colon \text{ the Skorohod derivative } d_h(f\cdot\mu) \text{ exists for all } h \in H\},$$

$$SBV(\mu) = SV(\mu) \cap M_H(\mu).$$

In other words, the class $SBV(\mu)$ consists of all functions $f \in L^1(\mu)$ for which

$$\sup_{|h|\le 1} |f\beta_h|_{L^1(\mu)} < \infty$$

and there exists an H-valued measure Λf of bounded semivariation such that the Skorohod derivative $d_h(f\mu)$ exists and equals $(\Lambda f, h)_H + f\beta_h\mu$ for each $h \in H$.

It is important to note that the measure $(\Lambda f, h)_H$ can be singular with respect to μ (say, have atoms in the one-dimensional case), but it also admits a disintegration

$$(\Lambda f, h)_H = (\Lambda f, h)_H^{y,h}\,\mu_Y(dy)$$

with some measures $(\Lambda f, h)_H^{y,h}$ on the straight lines $y+\mathbb{R}h$, where $y \in Y$ and Y is a closed hyperplane complementing $\mathbb{R}h$. Indeed, we have

$$(\Lambda f, h)_H = d_h(f\mu) - f\beta_h\mu,$$

where the projections of $|d_h(f\mu)|$ and $|f\beta_h|\mu$ on Y are absolutely continuous with respect to the projection of μ, since the projection of the measure $|d_h(f\mu)|$ is absolutely continuous with respect to the projection of $|f|\mu$. The latter follows from the fact that the projection on Y of the Skorohod derivative $d_h\sigma$ of a nonnegative measure σ on X is absolutely continuous with respect to the projection of σ (although $d_h\sigma$ itself need not be absolutely continuous with respect to σ unlike the case of Fomin's derivative), because for any Borel set $B \subset Y$ with $\sigma_Y(B) = 0$ we have

$$\begin{aligned} d_h\sigma(\mathbb{R}h \times B) &= \int_Y d_h\sigma^y(\mathbb{R}h \times B)\,\sigma_Y(dy) \\ &= \int_B d_h\sigma^y(\mathbb{R}h \times B)\,\sigma_Y(dy) = 0. \end{aligned}$$

Theorem 8.3. (i) *The set $SV(\mu)$ is a Banach space with the norm*

$$\|f\|_{SV} := \|f\|_{L^1(\mu)} + \sup_{|h|_H \le 1} \|d_h(f\cdot\mu)\|,$$

and for every function $f \in SV(\mu)$ there is an H-valued measure $D(f\cdot\mu)$ of bounded semivariation such that $d_h(f\cdot\mu) = (D(f\cdot\mu), h)_H$ for all $h \in H$.

(ii) *The set $SBV(\mu)$ is a Banach space with the norm*

$$\|f\|_{SBV} := \|f\|_M + \|f\|_{SV},$$

and for every function $f \in SBV(\mu)$ there is an H-valued measure Λf of bounded semivariation such that $d_h(f\cdot\mu) = (\Lambda f, h)_H + f\cdot d_h\mu$ for all $h \in H$.

Definition 8.4. Let $BV(\mu)$ be the class of all functions $f \in SBV(\mu)$ such that the H-valued measure Λf has bounded variation.

Theorem 8.5. *The set $BV(\mu)$ is a Banach space with the norm*

$$\|f\|_{BV} := \|f\|_{SBV} + \|\Lambda f\|.$$

For an interval $J \subset \mathbb{R}$ (possibly unbounded) let $BV_{\rm loc}(J)$ denote the class of all functions on J having bounded variation on every compact interval in J.

Lemma 8.6. *A function $f \in M_H(\mu)$ belongs to $SBV(\mu)$ precisely when there is an H-valued measure Λf of bounded semivariation such that for every $h \in H$ for μ-almost every x the function $t \mapsto f(x+th)$ belongs to $BV_{\rm loc}(\mathbb{R})$ and its generalized derivative is*

$$(\Lambda f, h)_H^{x,h} / \varrho^{x,h}(t),$$

where $\varrho^{x,h}$ is the density of the conditional measure $\mu^{x,h}$ for μ on the straight line $x + \mathbb{R}h$.

A similar assertion is true for $BV(\mu)$, where the measure Λ must have bounded variation.

Example 8.7. Let $X = H = \mathbb{R}^n$, let μ be a probability measure with a smooth density ϱ, and let f be a smooth function such that f, $|\nabla f|$ and $f|\nabla\varrho|/\varrho$ are integrable with respect to μ. Then $d_h\mu$ is a measure with density $\partial_h\varrho$, $d_h(f\cdot\mu)$ is a measure with density $\partial_h(f\varrho) = f\partial_h\varrho + \varrho\partial_h f$, whence it follows that Λf is a measure with vector density $(\nabla f)\varrho$ with respect to Lebesgue measure, *i.e.*, with density ∇f with respect to the measure μ. If the function $f \in L^1(\mu)$ with $f|\nabla\varrho| \in L^1(\mathbb{R}^n)$ is not smooth, but belongs to the class $BV_{\rm loc}(\mathbb{R}^n)$ of locally integrable functions whose generalized first order derivatives are locally bounded measures, then f belongs to $BV(\mu)$ and $\Lambda f = \varrho \cdot Df$ provided that the latter measure is bounded. It is clear that in the finite-dimensional case the classes SBV and BV coincide as sets and their norms are equivalent, but these norms are different.

Theorem 8.8. *The spaces $BV(\mu)$ and $SBV(\mu)$ possess the following property: if a sequence of functions f_n is norm bounded in it and converges almost everywhere to a function f, then f belongs to the same class and the norm of f does not exceed the precise upper bound of the norms of the functions f_n.*

Functions of bounded variation on spaces with convex measures are considered in Ambrosio, Da Prato, Goldys, Pallara [4].

9 Sobolev functions on domains and their extensions

Let us proceed to domains. In this section, we introduce Sobolev classes on domains, and BV-functions will be discussed in the next section.

Given a set $U \subset X$, the symbol $L^p(U,\mu)$ will denote the space of equivalence classes of all μ-measurable functions f on U for which the functions $|f|^p$ are integrable with respect to the measure μ on U.

We first consider the Gaussian case and then briefly discuss the case of a general differentiable measure.

Let $V \subset X$. If the sets $(V - x) \cap H$ are open in H for all $x \in V$, then V is called H-open (this property is equivalent to the fact that $V - x$ contains a ball from H for every $x \in V$, and is weaker than openness of V in X), and if all such sets are convex, then V is called H-convex. The latter property is weaker than the usual convexity. Obviously, for any H-convex and H-open set V all nonempty sets $V_{x,h}$ are open intervals (possibly unbounded), where

$$V_{x,h} := V \cap (x + \mathbb{R}h).$$

Example 9.1. Any open convex set is H-open and H-convex. However, the convex ellipsoid

$$U = \Big\{x \in \mathbb{R}^\infty \colon \sum_{n=1}^{\infty} n^{-2}x_n^2 < 1\Big\}$$

is not open in $\mathbb{R}^\infty$, but is H-open, where $H = l^2$. This ellipsoid has positive measure with respect to the standard Gaussian product-measure γ. The set

$$Z = \Big\{x \in \mathbb{R}^\infty \colon \lim_{N\to\infty} N^{-1}\sum_{n=1}^{N} x_n^2 = 1\Big\}$$

is Borel and has full measure with respect to γ (by the law of large numbers). It is not convex: if $x \in Z$, then $-x \in Z$, but $0 \notin Z$. It is clear that Z has no interior (its intersection with the set of finite sequences is empty). However, Z is H-open and H-convex, since for every $x \in Z$ we have $(Z - x) \cap H = H$, that is, for every $h \in H$ we have $x + h \in Z$. Indeed,

$$N^{-1}\sum_{n=1}^{N}(x_n + h_n)^2 - N^{-1}\sum_{n=1}^{N} x_n^2 = N^{-1}\sum_{n=1}^{N}(h_n^2 + 2x_nh_n),$$

which tends to zero as $N \to \infty$, because $N^{-1}\sum_{n=1}^{N} h_n^2 \to 0$ for each $h \in H$ and $|N^{-1}\sum_{n=1}^{N} x_nh_n| \to 0$ by the Cauchy inequality.

Suppose now that we are given a Borel or γ-measurable set $V \subset X$ of positive γ-measure such that its intersection with every straight line of the form $x + \mathbb{R}^1 e_n$ is a convex set V_{x,e_n}.

There are several natural ways of introducing Sobolev classes on V. The first one is considering the class $W^{p,1}(V,\gamma)$ equal to the completion of $\mathcal{F}\mathcal{C}^\infty$ with respect to the Sobolev norm $\|\cdot\|_{p,1,V}$ with the order of integrability p, evaluated with respect to the restriction of γ to V. This class is contained in the class $D^{p,1}(V,\gamma)$ consisting of all functions f on V belonging to $L^p(V,\gamma)$ and having versions of the type indicated above, but with the difference that now the absolute continuity is required only on the closed intervals belonging to the sections V_{x,e_n}. The class $D^{p,1}(V,\gamma)$ is naturally equipped with the Sobolev norm $\|\cdot\|_{p,1,V}$ defined by the restriction of γ to V:

$$\|f\|_{p,1,V} = \left(\int_V |f|^p \, d\gamma\right)^{1/p} + \left(\int_V |\nabla f|^p \, d\gamma\right)^{1/p}.$$

In the paper Hino [57] the Sobolev class $D^{2,1}(V,\gamma)$ was used (denoted there by $W^{1,2}(V)$). In the finite-dimensional case for convex V both classes coincide, the infinite-dimensional situation is less studied, but for H-convex H-open sets one has

$$W^{2,1}(V,\gamma) = D^{2,1}(V,\gamma),$$

which follows from [57], where it is shown that $D^{2,1}(V,\gamma)$ contains a dense set of functions possessing extensions of class $W^{2,1}(\gamma)$ and for this reason belonging to $W^{2,1}(V,\gamma)$. It is readily verified that the spaces $W^{p,1}(V,\gamma)$ and $D^{p,1}(V,\gamma)$ with the Sobolev norm are Banach.

Note that one can introduce more narrow Sobolev classes on V that admit extensions. For example, in the space $W^{p,1}(V,\gamma)$ one can take the closure $W_0^{p,1}(V,\gamma)$ of the set of functions from $W^{p,1}(\gamma)$ with compact support in V; the functions from $W_0^{p,1}(V,\gamma)$ extended by zero outside of V belong to $W^{p,1}(\gamma)$. For certain very simple sets V, say, half-spaces, it is easy to define explicitly an extension operator.

Example 9.2. Let $V = \{x\colon \widehat{h}(x) > 0\}$, where $|h|_H = 1$. Then any function $f \in W^{p,1}(V,\gamma)$ has an extension of class $W^{p,1}(\gamma)$ defined by

$$\widetilde{f}(x) = f(x - 2\widehat{h}(x)h) \quad \text{if } \widehat{h}(x) < 0.$$

For example, if $X = \mathbb{R}^\infty$ and $h = e_1$, then $V = \{x_1 > 0\}$ and

$$\widetilde{f}(x_1, x_2, \ldots) = f(-x_1, x_2, \ldots) \quad \text{whenever } x_1 < 0.$$

Taking a sequence of functions $f_j \in \mathcal{F}C^\infty$ whose restrictions to V converge to f in the norm of $W^{p,1}(\gamma, V)$, we see that the functions $\widetilde{f}_j$ redefined on the set $\{\widehat{h} \le 0\}$ by $\widetilde{f}_j(x) = f_j(x - 2\widehat{h}(x)h)$ belong to $W^{p,1}(\gamma)$ and converge in the Sobolev norm. In particular, $\|\widetilde{f}_j\|_{L^p(\gamma)} = 2^{1/p}\|f_j\|_{L^p(V,\gamma)}$, $\|D_H\widetilde{f}_j\|_{L^p(\gamma)} = 2^{1/p}\|D_H f_j\|_{L^p(V,\gamma)}$.

It is not clear whether there are essentially infinite-dimensional domains V for which all Sobolev functions have extensions.

However, functions with bounded derivatives extend from any H-convex domains.

Proposition 9.3. *Let V be H-convex and let $f \in W^{2,1}(\gamma, V)$ be such that*

$$|D_H f(x)| \le C \quad \text{a.e. in } V.$$

Then there is a function $g \in W^{p,1}(\gamma)$ for all $p < \infty$ such that $g|_V = f|_V$ a.e. on V and $|g(x+h) - g(x)| \le C|h|_H$ for all $x \in X$ and $h \in H$.

Lemma 9.4. *If the set V is such that for some $p \ge 1$ every function $f \in W^{p,1}(V, \gamma)$ has an extension $g \in W^{p,1}(\gamma)$, then there exists an extension $g_f \in W^{p,1}(\gamma)$ such that $\|g_f\|_{p,1} \le C\|f\|_{p,1,V}$ with some common constant C.*

The following theorem is a negative result about extensions. For notational simplicity, we formulate it for the Gaussian product-measure on $\mathbb{R}^\infty$. A complete proof is given in Bogachev, Pilipenko, Shaposhnikov [21].

Theorem 9.5. *The space $\mathbb{R}^\infty$ contains a convex Borel H-open set K of positive γ-measure with the following property: for every $p \in [1, +\infty)$ there is a function in the class $W^{p,1}(K, \gamma)$ having no extensions to a function of class $W^{p,1}(\gamma)$. One can also find a convex compact set K with the same property.*

Remark 9.6. In the proof of this theorem in [21] a certain Hilbert space L of full measure is taken such that, passing to the restriction of the measure γ to L, we obtain a convex and open in L set K of positive measure, on which for every $p \in [1, +\infty)$ there is a function in the class $W^{p,1}(K, \gamma)$ without restrictions to a function in $W^{p,1}(\gamma)$. It is clear that the same example can be realized also on a larger weighted Hilbert space of sequences in which K will be precompact. Hence it is possible to combine H-openness of K with its relative compactness in a Hilbert space (clearly, our set K in $\mathbb{R}^\infty$ is relatively compact).

In addition, if we take for γ the classical Wiener measure on the space $C[0,1]$ or $L^2[0,1]$ and embed this space into $\mathbb{R}^\infty$ by means of the mapping $x \mapsto n(x, e_n)_{L^2}$, where $\{e_n\}$ is the orthonormal basis in $L^2[0,1]$

formed by the eigenfunctions of the covariance operator of the Wiener measure, that is, $e_n(t) = c_n \sin((\pi n - \pi/2)t)$, $n \in \mathbb{N}$, with the eigenvalues $\lambda_n = (\pi n - \pi/2)^{-2}$, then the image of γ will coincide with the standard Gaussian product-measure and the space L mentioned above will coincide with the image of $L^2[0,1]$ under the embedding, hence our convex set K will be open in the corresponding Hilbert space.

For any centered Radon Gaussian measure γ on a locally convex space X with the infinite-dimensional Cameron-Martin space H, the results presented above yield existence of an H-open convex Borel set V of positive γ-measure and, for every $p \in [1, +\infty)$, a function f in $W^{p,1}(V, \gamma)$ without extensions to functions in the class $W^{p,1}(\gamma)$. It would be interesting to construct an example of a function in the intersection of all $W^{p,1}(V, \gamma)$ without extensions of class $W^{1,1}(\gamma)$. Apparently, there are bounded functions with such a property.

Certainly, it is natural to ask about such examples on a ball in a Hilbert space. However, we have no such examples.

10 BV functions on domains and their extensions

In this section we follow the paper Bogachev, Pilipenko, Shaposhnikov [21].

We assume below that the measure μ on a locally convex space X is Fomin differentiable along all vectors in a separable Hilbert space H continuously and densely embedded into X (again the model example is $l^2 \subset \mathbb{R}^\infty$) and that for every fixed $h \in H$ the continuous versions of the conditional densities on the straight lines $x + \mathbb{R}h$ are positive (a sufficient condition for this is the integrability of $\exp|\beta_h^\mu|$ with respect to μ). Below these densities are denoted by $\varrho^{x,h}$ without indicating μ.

Let $U \subset X$ be a Borel set that is H-convex and H-open, that is (see Section 9), all sets $(U - x) \cap H$ are convex and open in H. For example, this can be a set that is convex and open in X.

For any H-convex and H-open set U the one-dimensional sections

$$U_{x,h} := U \cap (x + \mathbb{R}h),$$

are open intervals on the straight lines $x + \mathbb{R}h$. We shall often identify these intervals with the intervals

$$J_{x,h} := \{t \in \mathbb{R}\colon\ x + th \in U\}.$$

In particular, when speaking about functions on intervals $U_{x,h}$ we shall mean sometimes functions of the real argument on $J_{x,h}$.

Let $M_H(U,\mu)$ denote the class of all functions $f \in L^1(U,\mu)$ such that

$$\|f\|_M := \|f\|_{L^1(U,\mu)} + \sup_{|h|\le 1} \int_U |f(x)|\,|\beta_h(x)|\,\mu(dx) < \infty.$$

The corresponding space of equivalence classes will be denoted by the same symbol.

This is the exact analog of the class $M_H(\mu)$ in Section 8.

Lemma 10.1. *The set $M_H(U,\mu)$ is a Banach space with the norm*

$$\|f\|_M := \|f\|_{L^1(U,\mu)} + \sup_{|h|_H\le 1} \|f\beta_h^\mu\|_{L^1(U,\mu)}.$$

Proof. Let us observe that the operator $h \mapsto f\beta_h$ from H to $L^1(U,\mu)$ is linear and has a closed graph. Indeed, suppose that $h_n \to h$ in H and $f\beta_{h_n} \to g$ in $L^1(U,\mu)$. By the continuity of the embedding $H \to D(\mu)$ we have $\beta_{h_n} \to \beta_h$ in $L^1(\mu)$, whence it follows that $f\beta_{h_n} \to f\beta_h$ in measure on U, hence $g = f\beta_h$. Therefore, for every $f \in M_H(U,\mu)$ the quantity $\|f\|_M$ is finite. Obviously, it is a norm. Let $\{f_n\}$ be a Cauchy sequence in $M_H(U,\mu)$. Then $\{f_n\}$ converges in $L^1(U,\mu)$ to some function f. By Fatou's theorem $f \in M_H(U,\mu)$. In addition, f is a limit of $\{f_n\}$ with respect to the norm in $M_H(\mu)$: if $\|f_n - f_k\|_M \le \varepsilon$ for all $n, k \ge n_1$, then $\|f_n - f\|_M \le \varepsilon$ for all $n \ge n_1$. □

Definition 10.2. We shall say that $f \in M_H(U,\mu)$ belongs to the class $SBV_H(U,\mu)$ if the function $t \mapsto f(x+th)\varrho^{x,h}(t)$ belongs to the class $BV_{\mathrm{loc}}(U_{x,h})$ for every fixed $h \in H$ for almost all x and there exists an H-valued measure $\Lambda_U f$ on U of bounded semivariation such that, for every $h \in H$, the measure $(\Lambda_U f, h)_H$ admits the representation

$$(\Lambda_U f, h)_H = (\Lambda_U f, h)_H^{x,h,\mu}\,\mu(dx),$$

where the measures $(\Lambda_U f, h)_H^{x,h,\mu}$ on the straight lines $x + \mathbb{R}h$ possess the property that

$$(\Lambda_U f, h)_H^{x,h} + f(x+th)\partial_t \varrho^{x,h}(t)$$

is the generalized derivative of the function $t \mapsto f(x+th)\varrho^{x,h}(t)$ on $U_{x,h}$.

The class $BV_H(U,\mu)$ consists of all $f \in SBV_H(U,\mu)$ such that the measure $\Lambda_U f$ has bounded variation.

In other words, an analog of the characterization from Lemma 8.6 is now taken as a definition.

As we have warned above, the sections $U_{x,h}$ in this definition are identified with intervals $J_{x,h}$ of the real line.

Note that the defining relation for $(\Lambda_U f, h)_H^{x,h}$ can be stated in terms of the functions $t \mapsto f(x+th)$ (not multiplied by conditional densities): the generalized derivatives of these functions must be (as in Lemma 8.6)

$$\varrho^{x,h}(t)^{-1}(\Lambda_U f, h)_H^{x,h}.$$

An equivalent description of functions in $SBV_H(U,\mu)$ can be given in the form of integration by parts if in place of the class $\mathcal{F}C^\infty$ we use appropriate classes of test functions for every $h \in H$.

For any fixed vector $h \in H$ we choose a closed hyperplane Y complementing $\mathbb{R}h$ and consider the class $\mathcal{D}_h$ of all bounded functions φ on X with the following properties: φ is measurable with respect to all Borel measures, for each $y \in Y$ the function $t \mapsto \varphi(y+th)$ is infinitely differentiable and has compact support in the interval

$$J_{y,h} = \{t\colon\ y+th \in U\},$$

and the functions $\partial_h^n \varphi$ are bounded for all $n \geq 1$. Here $\partial_h^n \varphi(y+th)$ is the derivative of order n at the point t for the function $t \mapsto \varphi(y+th)$.

Note that $\psi\varphi \in \mathcal{D}_h$ for all $\varphi \in \mathcal{D}_h$ and $\psi \in \mathcal{F}C^\infty$.

Lemma 10.3. *A function $f \in M_H(U,\mu)$ belongs to $SBV_H(U,\mu)$ precisely when there exists an H-valued measure $\Lambda_U f$ on U of bounded semivariation such that, for every $h \in H$ and all $\varphi \in \mathcal{D}_h$, one has the equality*

$$\int_X \partial_h\varphi(x) f(x)\,\mu(dx) = -\int_X \varphi(x)\,(\Lambda_U f, h)_H(dx) - \int_X \varphi(x) f(x)\beta_h(x)\,\mu(dx).$$

A similar assertion with variation in place of semivariation is true for the class $BV_H(U,\mu)$.

Proof. If $f \in SBV_H(U,\mu)$, then the indicated equality follows from the definition and the integration by parts formula for conditional measures. Let us prove the converse assertion. Let us fix $k \in \mathbb{N}$. The set Y_k of all points $y \in Y$ such that the length of the interval $J_{y,h}$ is not less than $8/k$ is measurable with respect to every Radon measure (see Bogachev [15, Theorem 7.14.49]). In addition, it is not difficult to show that there exists a function $g_k \in \mathcal{D}_h$ measurable with respect to every Borel measure and possessing the following properties: $0 \leq g_k \leq 1$, $g_k(y) = 0$ if the length of $J_{y,h}$ is less than $8/k$, $g_k(y+th) = 0$ if $t \notin J_{y,h}$ or if $t \in J_{y,h}$ and the distance from t to an endpoint of $J_{y,h}$ is less than $1/k$, $g_k(y+th) = 1$ if $t \in J_{y,h}$ and the distance from t to an endpoint of $J_{y,h}$ is not less than $2/k$. It follows from our hypothesis that for all $\psi \in \mathcal{FC}^\infty$ we have the equality

$$\begin{aligned}\int_X \partial_h\psi(x)g_k(x)f(x)\,\mu(dx) = &-\int_X \psi(x)g_k(x)\,(\Lambda_U f, h)_H(dx)\\ &-\int_X \psi(x)g_k(x)f(x)\beta_h(x)\,\mu(dx)\\ &-\int_X \psi(x)\partial_h g_k(x)f(x)\,\mu(dx).\end{aligned}$$

Therefore, the measure $fg_k\mu$ is Skorohod differentiable and

$$d_h(fg_k\mu) = g_k(\Lambda_U f, h)_H + fg_k\beta_h\mu + \partial_h g_k f\mu.$$

Using the disintegration for $fg_k\mu$ and letting $k \to \infty$, we obtain the disintegration for $f\mu$ required by the definition. □

Lemma 10.4. *If $f \in SBV_H(U,\mu)$ and $\psi \in C_b^1(\mathbb{R})$, then*

$$\psi(f) \in SBV_H(U,\mu)$$

and for any $h \in H$ one has

$$(\Lambda_U\psi(f), h)_H = (\Lambda_U\psi(f), h)_H^{x,h,\mu}\,\mu(dx)$$

with

$$(\Lambda_U\psi(f), h)_H^{x,h,\mu} = \psi'(f)(x+th)(\Lambda_U f, h),$$

where $\psi'(f)(x+th)$ is redefined at the points of jumps of the function $t \mapsto f(x+th)$ by the expression

$$\frac{\psi(f(x+th+)) - \psi(f(x+th-))}{f(x+th+) - f(x+th-)}.$$

Moreover,

$$V(\Lambda_U\psi(f)) \leq LV(\Lambda_U f),$$

where $L = \sup_{u\neq v}\frac{|\psi(u)-\psi(v)|}{|u-v|}$ is the Lipschitz constant of ψ.

A similar assertion is true for $BV_H(U,\mu)$.

Proof. It suffices to use conditional measures and apply the chain rule for BV-functions in the one-dimensional case, see, *e.g.*, Ambrosio, Fusco, Pallara [6, page 188]. □

Theorem 10.5. *The set $SBV_H(U,\mu)$ is a Banach space with the norm*

$$\|f\|_{SBV} := \|f\|_M + V(\Lambda_U f).$$

The set $BV_H(U,\mu)$ is a Banach space with the norm

$$\|f\|_{BV} := \|f\|_M + \mathrm{Var}(\Lambda_U f).$$

Proof. Since the space $M_H(U,\mu)$ is complete, every Cauchy sequence $\{f_n\}$ in the space $SBV_H(U,\mu)$ converges in the M-norm to some function $f \in L^1(U,\mu)$. The sequence of measures $\Lambda_U f_n$ is Cauchy in semivariation, hence converges in the norm V to some measure ν of bounded semivariation. Applying Lemma 10.3 it is easy to show that $f \in SBV_H(U,\mu)$ and $\nu = \Lambda_U f$ is the corresponding H-valued measure. Since

$$\|f_n - f\|_M + V(\Lambda_U f_n - \Lambda_U f) \to 0,$$

it follows that f is a limit of $\{f_n\}$ in the norm of the space $SBV_H(U,\mu)$. The proof of completeness of the space $BV_H(U,\mu)$ is similar. □

Theorem 10.6. *The classes $SBV_H(U,\mu)$ and $BV_H(U,\mu)$ have the following property: if a sequence of functions $\{f_n\}$ is norm bounded in it and converges almost everywhere to a function f, then f belongs to the same class, and the norm of f does not exceed the precise upper bound of the norms of the functions f_n.*

Moreover, for every fixed $h \in H$, the measures $(\Lambda_U f_n, h)_H$ converge to $(\Lambda_U f, h)_H$ in the weak topology generated by the duality with $\mathcal{D}_h$ (in the case $U = X$ also with respect to the duality with $\mathcal{F}C^\infty$).

Proof. These assertions are true on the real line, since our assumption about the conditional densities means that the density of μ is positive, so $\Lambda f_n = \varrho f_n'$, where f_n' is the generalized derivative, and these measures converge to $\varrho f'$ in the sense of distributions.

In the general case suppose first that $\{f_n\}$ is uniformly bounded. Let us fix $h \in H$. Since the measures $(\Lambda_U f_n, h)_H$ are uniformly bounded and

$$(\Lambda_U f_n, h)_H = (\Lambda_U f_n, h)_H^{x,h,\mu}\ \mu(dx),$$

by Fatou's theorem the function

$$\liminf_{n\to\infty} \|(\Lambda_U f_n, h)_H^{x,h,\mu}\|$$

is μ-integrable. In particular, it is finite almost everywhere, hence the restriction of the function f to almost every straight line $x+\mathbb{R}h$ is in BV_{loc}. Moreover, we obtain finite measures $(\Lambda_U f, h)_H^{x,h,\mu}$ on these straight lines such that the measure

$$(\Lambda_U f, h)_H := (\Lambda_U f, h)_H^{x,h,\mu}\ \mu(dx)$$

is finite for every $h \in H$. By the Pettis theorem (see Dunford, Schwartz [39, Chapter IV, §10]) we obtain an H-valued measure $\Lambda_U f$. It meets the requirements in Definition 10.2. For any $\varphi \in \mathcal{D}_h$ by the Lebesgue dominated convergence theorem we have

$$\int_X \partial_h\varphi(x) f(x)\mu(dx) = \lim_{n\to\infty} \int_X \partial_h\varphi(x) f_n(x)\mu(dx),$$

$$\int_X \varphi(x) f(x)\beta_h(x)\mu(dx) = \lim_{n\to\infty} \int_X \varphi(x) f_n(x)\beta_h(x)\mu(dx).$$

Therefore, the integrals of φ with respect to the measures $(\Lambda_U f_n, h)_H$ converge. Moreover, the limit is the integral of φ with respect to $(\Lambda_U f, h)_H$, which follows by the one-dimensional case applied to the conditional measures. This completes the proof in the case of SBV and bounded $\{f_n\}$.

In the case of BV it is necessary to show that $\Lambda_U f$ has bounded variation. The measures Λf_n possess H-valued vector densities R_n with respect to some common nonnegative measure ν and the sequence of functions $|R_n|_H$ is bounded in $L^1(\nu)$. It suffices to show that for every Borel mapping v such that $|v|_H \le 1$ and $v = \sum_{i=1}^k v_i h_i$, where $h_i \in H$ are constant and orthonormal, we have the estimate

$$\sum_{i=1}^k \int_X v_i(x)\,(\Lambda_U f, h_i)_H(dx) \le \sup_n \operatorname{Var}(\Lambda_U f_n).$$

It is readily seen that it is enough to do this for v with functions v_i such that $v_i \in \mathcal{D}_{h_i}$. Indeed, by using convolutions we reduce the general case to the case where the function v_i has bounded derivatives of any order along the vector h_i. Next, we approximate such functions in $L^1(|(\Lambda_U f, h_i)_H|)$ by their products with functions $w_{i,n} \in \mathcal{D}_{h_i}$ with the following properties: $0 \le w_{i,n} \le 1$, $w_{i,n}(y + th_i) = 0$ whenever the length $\delta_{y,h,i}$ of U_{y,h_i} is less than $4/n$ and otherwise $w_{i,n} = 1$ on the inner interval of length $\delta_{y,h_i} - 2/n$ with the same center as U_{y,h_i}.

For such functions we have

$$\sum_{i=1}^{k} \int_X v_i(x)\, (\Lambda_U f, h_i)_H(dx) = \lim_{n\to\infty} \sum_{i=1}^{k} \int_X v_i(x)\, (\Lambda_U f_n, h_i)_H(dx)$$
$$= \lim_{n\to\infty} \int_X (v(x), R_n(x))_H\, \nu(dx)$$
$$\le \sup_n \int_X |R_n(x)|_H\, \nu(dx)$$
$$= \sup_n \operatorname{Var} \Lambda_U f_n.$$

Thus, the case of a uniformly bounded $\{f_n\}$ is considered.

Let us now proceed to the general case where $\{f_n\}$ is not uniformly bounded. Take a smooth increasing function ψ on the real line such that $\psi(t) = t$ if $|t| \le 1$, $\psi(t) = 2\,\mathrm{sign}\, t$ if $|t| \ge 3$, $|\psi(t)| \le |t|$, and $|\psi'(t)| \le 1$. Let $\psi_m(t) = \psi(t/m)$. According to Lemma 10.4, for every fixed m the functions $\psi_m(f_n)$ belong to the respective (SBV or BV) class and their norms are uniformly bounded in n and m. Hence the function $\psi_m(f)$ belongs to the same class and its norm does not exceed the supremum of the norms of $\{f_n\}$. We shall deal with a Borel version of f, so the functions $\psi_m(f)$ are also Borel. The Borel sets $B_m = \{|f| < m\}$ are increasing to X and $|\psi_m(f)| \le |f|$. Clearly, $f \in M_H(U, \mu)$ and $\|f\|_M = \lim_{m\to\infty} \|\psi_m(f)\|_M$. Since $\psi_{m+1}(f)$ coincides with $\psi_m(f)$ on the set B_m, we obtain that the conditional measures $(\Lambda_U \psi_m(f), h)_H^{x,h}$ on the straight lines $x + \mathbb{R}h$ have a finite setwise limit for almost every x, and the measures $\sigma^{x,h}$ obtained in the limit give rise to bounded measures $\sigma^{x,h}\, \mu(dx)$, which can be taken for $(\Lambda_U f, h)_H$. In the case of $BV_H(U, \mu)$ we have additionally that the resulting vector measure Λf is of bounded variation. □

Corollary 10.7. *If $f \in SBV_H(U, \mu)$ and ψ is a Lipschitzian function on the real line, then $\psi(f) \in SBV_H(U, \mu)$ and*

$$\|\psi(f)\|_{SBV} \le C\|f\|_{SBV},$$

where C is a Lipschitz constant for ψ.

The same is true in the case of $BV_H(U, \mu)$.

Proof. For smooth ψ this has already been noted. The general case follows by approximation and the above theorem. □

Theorem 10.8. *Suppose that $I_U \in SBV_H(\mu)$. Then for every function*

$$f \in SBV_H(U, \mu) \cap L^\infty(U, \mu)$$

its extension by zero outside of U gives a function in the class $SBV_H(\mu)$.

In the opposite direction, the restriction to U of every function in $SBV_H(\mu)$, *not necessarily bounded, gives a function in* $SBV_H(U,\mu)$.

If $I_U \in BV_H(\mu)$, *then the analogous assertions are true for the class* $BV_H(U,\mu)$.

Proof. Let $f \in SBV_H(U,\mu)\cap L^\infty(U,\mu)$. We may assume that $|f| \le 1$. Let us fix $h \in H$. Then we can find a version of f whose restrictions to the straight lines $x+\mathbb{R}h$ have locally bounded variation. Let a_x be an end-point of the interval $U_{x,h}$ (if it exists). Then the considered version of f has a limit at a_x (left or right, respectively), bounded by 1 in the absolute value. Defined by zero outside of U, the function f remains a function of locally bounded variation on all these straight lines, but at the end-points of $U_{x,h}$ its generalized derivative may gain Dirac measures with coefficients bounded by 1 in the absolute value. However, such Dirac measures (with the coefficient 1 at the left end and the coefficient -1 at the right end) are already present in the derivative of the restriction of I_U. Thus, after adding these point measures to $(\Lambda_U f, h)_H^{x,h,\mu}$, we obtain a measure that differs from $(\Lambda_U f, h)_H$ by some measure with semivariation not exceeding $\|(\Lambda I_U, h)_H\|$, hence is also of bounded semivariation. Therefore, Lemma 8.6 gives the inclusion of the extension to $SBV_H(\mu)$.

The fact that $f|_U \in SBV_H(U,\mu)$ for any $f \in SBV_H(\mu)$ follows by Lemma 10.3, since the restriction of Λf to U serves as $\Lambda_U f$.

In the case of $BV_H(\mu)$ we also use the fact that any H-valued measure of bounded variation is given by a Bochner integrable vector density with respect to a suitable scalar measure. □

Proposition 10.9. *Let U be a Borel convex set. Then* $I_U \in SBV_H(\mu)$.

Proof. Let us fix $h \in H$. If $U_{x,h}$ is not empty and not the whole straight line, the generalized derivative $\sigma^{x,h}$ of the function $t \mapsto I_U(x+th)$ is either the difference of two Dirac's measures at the endpoints of $U_{x,h}$ or Dirac's measure (with the sign plus or minus) at the single endpoint (if $U_{x,h}$ is a ray). Let us define $(\Lambda I_U, h)_H^{x,h}$ by

$$(\Lambda I_U, h)_H^{x,h} := \sigma^{x,h}\varrho^{x,h}.$$

We obtain a bounded measure $(\Lambda I_U, h)_H^{x,h}\,\mu(dx)$ (indeed, $\|\sigma^{x,h}\| \le 2$, $0 \le I_{U_{x,h}} \le 1$, $\|\varrho^{x,h}\|$ and $\|\partial_t\varrho^{x,h}\|$ are μ-integrable), which defines an H-valued measure ΛI_U with the properties mentioned in Lemma 8.6, which yields the desired conclusion. □

It should be noted that even for a bounded convex Borel set U in a separable Hilbert space the indicator function I_U does not always belong to $BV_H(\gamma)$; explicit examples are given in Caselles, Lunardi, Miranda, Novaga [27]. On the other hand, the indicator of any open convex set in the Gaussian case belongs to $BV_H(\gamma)$ (see [27]).

It is worth noting that it is also of interest to consider restrictions of Sobolev functions to finite-dimensional subspaces and to infinite-dimensional surfaces, see Airault, Bogachev, Lescot [3], Celada, Lunardi [29].

Finally, note that a number of other interesting topics related to Sobolev and BV classes on infinite-dimensional spaces have not been mentioned in this survey. In particular, Cruzeiro [32] initiated the study of the continuity equation and related transformations of measures in infinite dimensions, and this direction is further developed by many authors, see Ambrosio, Figalli [5], Bogachev, Mayer-Wolf [19], Kolesnikov, Röckner [60], Peters [75], Üstünel, Zakai [87]; Sobolev classes can be useful in the study of the Monge–Kantorovich problem in infinite dimensions, see Feyel, Üstünel [44] and the subsequent papers Bogachev, Kolesnikov [18], Cavalletti [28], Fang, Shao, Sturm [42]. Differential equations for Sobolev functions on Hilbert spaces are considered in Da Prato [33]. Additional references on diverse aspects of Sobolev classes can be found in Bogachev [16] and [17].

References

[1] R.A. ADAMS and J. J. F. FOURNIER, "Sobolev Spaces", 2nd ed., Academic Press, New York, 2003.

[2] H. AIMAR, L. FORZANI and R. SCOTTO, *On Riesz transforms and maximal functions in the context of Gaussian harmonic analysis*, Trans. Amer. Math. Soc. **359** (2007), n. 5., 2137–2154.

[3] H. AIRAULT, V. I. BOGACHEV and P. LESCOT, *Finite-dimensional sections of functions from fractional Sobolev classes on infinite-dimensional spaces*, Dokl. Ross. Akad. Nauk. 2003. **391** (2003), n. 3, 320–323 (in Russian); English transl.: Dokl. Math. **68** (2003), n. 1, 71–74.

[4] L. AMBROSIO, G. DA PRATO, B. GOLDYS and D. PALLARA, *Bounded variation with respect to a log-concave measure*, Comm. Partial Differerential Equations 2012. **37** (2012), n. 12, 2272–2290.

[5] L. AMBROSIO and A. FIGALLI, *On flows associated to Sobolev vector fields in Wiener spaces: an approach à la DiPerna–Lions*, J. Funct. Anal. 2009. **256** (2009), n. 1, 179–214.

[6] L. AMBROSIO, N. FUSCO and D. PALLARA, "Functions of bounded Variation and Free Discontinuity Problems", Clarendon Press, Oxford University Press, New York, 2000.

[7] L. AMBROSIO and S. DI MARINO, *Equivalent definitions of BV space and of total variation on metric measure spaces*. J. Funct. Anal. 2014. **266** (2014), 4150–4188.

[8] L. AMBROSIO, M. MIRANDA, S. MANIGLIA and D. PALLARA, *Towards a theory of BV functions in abstract Wiener spaces*. Physica D: Nonlin. Phenom. 2010. **239** (2010), n. 15, 1458–1469.

[9] L. AMBROSIO, M. MIRANDA (JR.), S. MANIGLIA and D. PALLARA, *BV functions in abstract Wiener spaces*, J. Funct. Anal. 2010. **258** (2010), n. 3, 785–813.

[10] L. AMBROSIO and P. TILLI, "Topics on Analysis in Metric Spaces", Oxford University Press, Oxford, 2004.

[11] S. BLANCHERE, D. CHAFAI, F. FOUGERES, I. GENTIL, F. MALRIEN, C. ROBERTO and G. SCHEFFER, "Sur les inégalités de Sobolev logarithmiques. Panoramas et Synthèses", Soc. Math. France, 2000.

[12] V. I. BOGACHEV, *Differentiable measures and the Malliavin calculus*. J. Math. Sci. (New York) **87** (1997), n. 4, 3577–3731.

[13] V. I. BOGACHEV, "Gaussian Measures", Amer. Math. Soc., Providence, Rhode Island, 1998.

[14] V. I. BOGACHEV, *Extensions of H-Lipschitzian mappings with infinite-dimensional range*, Infin. Dim. Anal., Quantum Probab. Relat. Top. **2** (1999), n. 3, 1–14.

[15] V. I. BOGACHEV, "Measure Theory", Vol. 1, 2, Springer, Berlin, 2007.

[16] V. I. BOGACHEV, "Differentiable Measures and the Malliavin Calculus", Amer. Math. Soc., Providence, Rhode Island, 2010.

[17] V. I. BOGACHEV, *Gaussian measures on infinite-dimensional spaces*, In: "Real and Stochastic Analysis, Current Trends", M. M. Rao (ed.), World Sci., Singapore, 2014, 1–83.

[18] V. I. BOGACHEV and A. V. KOLESNIKOV, *The Monge-Kantorovich problem: achievements, connections, and perspectives*, Russian Math. Surveys. **67** (2012), n. 5, 3–110.

[19] V. I. BOGACHEV and E. MAYER-WOLF, *Absolutely continuous flows generated by Sobolev class vector fields in finite and infinite dimensions*, J. Funct. Anal. **167** (1999), n. 1, 1–68.

[20] V. I. BOGACHEV, A. YU. PILIPENKO and E. A. REBROVA, *Classes of funtions of bounded variation on infinite-dimensional domains*. Dokl. Russian Acad. Sci. **451** (2013), n. 2., 127–131 (in Russian); English transl.: Dokl. Math. **88** (2013), n. 1, 391–395.

[21] V. I. BOGACHEV, A. YU. PILIPENKO and A. V. SHAPOSHNIKOV, *Sobolev functions on infinite-dimensional domains*, J. Math. Anal. Appl. **419** (2014), 1023–1044.

[22] V. I. BOGACHEV and E. A. REBROVA, *Functions of bounded variation on infinite-dimensional spaces with measures*, Dokl. Russian Acad. Sci. **449** (2013), n. 2. 131–135 (in Russian); English transl.: Dokl. Math. **87** (2013), n. 2, 144–147.

[23] V. I. BOGACHEV and M. RÖCKNER, *Regularity of invariant measures on finite and infinite dimensional spaces and applications*, J. Funct. Anal. **133** (1995), n. 1, 168–223.

[24] V. I. BOGACHEV and A. V. SHAPOSHNIKOV, *On extensions of Sobolev functions on the Wiener space*, Dokl. Ross. Akad. Nauk. **448** (2013), n. 4, 379–383 (in Russian); English transl.: Dokl. Math. **87** (2013), n. 1, 58–61.

[25] N. BOULEAU and F. HIRSCH, "Dirichlet Forms and Analysis on Wiener Space", De Gruyter, Berlin – New York, 1991.

[26] B. BRANDOLINI, F. CHIACCHIO and C. TROMBETTI, *Hardy type inequalities and Gaussian measure*, Commun. Pure Appl. Anal. **6** (2007), n. 2, 411–428.

[27] V. CASELLES, A. LUNARDI, M. MIRANDA (JUN.) and M. NOVAGA, *Perimeter of sublevel sets in infinite dimensional spaces*, Adv. Calc. Var. **5** (2012), n. 1, 59–76.

[28] F. CAVALLETTI, *The Monge problem in Wiener space*, Calc. Var. Partial Diff. Equ. **45** (2012), n. 1-2, 101–124.

[29] P. CELADA and A. LUNARDI, *Traces of Sobolev functions on regular surfaces in infinite dimensions*, J. Funct. Anal. **266** (2014), 1948–1987.

[30] J. CHEEGER, *Differentiability of Lipschitz functions on metric measure spaces*, Geom. Funct. Anal. **9** (1999), 428–517.

[31] A. CIANCHI and L. PICK, *Optimal Gaussian Sobolev embeddings*, J. Funct. Anal. **256** (2009), n. 11, 3588–3642.

[32] A. B. CRUZEIRO, *Équations différentielles sur l'espace de Wiener et formules de Cameron–Martin non-linéaires*, J. Funct. Anal. 1983. **54** (1983), n. 2, 206–227.

[33] G. DA PRATO, "Kolmogorov Equations for Stochastic PDEs", Birkhäuser Verlag, Basel, 2004.

[34] G. DA PRATO, "Introduction to Stochastic Analysis and Malliavin Calculus", Scuola Normale Superiore, Pisa, 2007.

[35] YU. L. DALETSKIĬ and S. N. PARAMONOVA, *Stochastic integrals with respect to a normally distributed additive set function*, Dokl. Akad. Nauk SSSR. **208** (1973), n. 3. 512–515 (in Russian); English transl.: Sov. Math. Dokl. **14** (1973), 96–100.

[36] YU. L. DALETSKIĬ and S. N. PARAMONOVA, *On a formula from the theory of Gaussian measures and on the estimation of stochastic integrals*, Teor. Verojatn. i Primen. **19** (1974), n. 4, 844–849 (in

Russian); English. transl.: Theory Probab. Appl. 1 **19** (1974), n. 4, 812–817.

[37] YU. L. DALETSKIĬ and S. N. PARAMONOVA, *Integration by parts with respect to measures in function spaces. I, II*. Teor. Verojatn. Mat. Stat. **17** (1977), 51–60; **18** (1978), 37–45 (in Russian); English transl.: Theory Probab. Math. Stat. 1979. **17** (1979), 55–65; **18**, 39–46.

[38] J. DIESTEL and J. J. UHL, "Vector Masures" Amer. Math. Soc., Providence, 1977.

[39] N. DUNFORD and J. T. SCHWARTZ, "Linear Operators, I. General Theory", Interscience, New York, 1958.

[40] C. EVANS and R. F. GARIEPY, "Measure Theory and Fine Properties of Functions", CRC Press, Boca Raton – London, 1992.

[41] S. FANG, "Introduction to Malliavin Calculus", Beijing, 2004.

[42] S. FANG, J. SHAO and K.-TH. STURM, *Wasserstein space over the Wiener space*, Probab. Theory Related Fields. **146** (2010), n. 3-4, 535–565.

[43] P. FEDERBUSH, *Partially alternate derivation of a result of Nelson*, J. Math. Phys. **10** (1969), 50–52.

[44] D. FEYEL and A.S. ÜSTÜNEL, *Monge-Kantorovitch measure transportation and Monge-Ampère equation on Wiener space*, Probab. Theory Related Fields **128** (2004), n. 3, 347–385.

[45] S. V. FOMIN, *Differentiable measures in linear spaces*, Proc. Int. Congress of Mathematicians, Sec. 5, Izdat. Moskov. Univ., Moscow, 1966 (in Russian), 78–79.

[46] S. V. FOMIN, *Differentiable measures in linear spaces*, Uspehi Matem. Nauk. 1968. **23** (1968), n. 1, 221–222 (in Russian).

[47] N. N. FROLOV, *Embedding theorems for spaces of functions of countably many variables, I*, Proceedings Math. Inst. of Voronezh Univ., Voronezh University 1970. n. 1, 205–218 (in Russian).

[48] N. N. FROLOV, *Embedding theorems for spaces of functions of countably many variables and their applications to the Dirichlet problem*, Dokl. Akad. Nauk SSSR **203** (1972), n. 1, 39–42 (in Russian); English transl.: Soviet Math. **13** (1972), n. 2, 346–349.

[49] M. FUKUSHIMA, *BV functions and distorted Ornstein–Uhlenbeck processes over the abstract Wiener space*, J. Funct. Anal. **174** (2000), 227–249.

[50] M. FUKUSHIMA and M. HINO, *On the space of BV functions and a related stochastic calculus in infinite dimensions*, J. Funct. Anal. 2001. **183** (2001), n. 1, 245–268.

[51] L. GROSS, *Potential theory on Hilbert space*, J. Funct. Anal. **1** (1967), n. 2, 123–181.

[52] L. GROSS, *Logarithmic Sobolev inequalities*, Amer. J. Math. **97** (1975), n. 4, 1061–1083.

[53] P. HAJŁASZ and P. KOSKELA, "Sobolev met Poincaré", Mem. Amer. Math. Soc., 2000, Vol. 145, n. 688.

[54] J. HEINONEN, "Lectures on Analysis on Metric Spaces", Springer, New York, 2001.

[55] M. HINO, *Integral representation of linear functionals on vector lattices and its application to BV functions on Wiener space*, In: "Stochastic Analysis and Related Topics in Kyoto", Adv. Stud. Pure Math., Vol. 41, Math. Soc. Japan, Tokyo, 2004, 121–140.

[56] M. HINO, *Sets of finite perimeter and the Hausdorff–Gauss measure on the Wiener space*. J. Funct. Anal. **258** (1010), n. 5, 1656–1681.

[57] M. HINO, *Dirichlet spaces on H-convex sets in Wiener space*, Bull. Sci. Math. **135** (2011), 667–683. Erratum: ibid. **137** (2013), 688–689.

[58] S. JANSON, "Gaussian Hilbert Spaces", Cambridge Univ. Press, Cambridge, 1997.

[59] S. KEITH, *A differentiable structure for metric measure spaces*, Adv. Math. **183** (2004), n. 2, 271–315.

[60] A. KOLESNIKOV and M. RÖCKNER, *On continuity equations in infinite dimensions with non-Gaussian reference measure*, J. Funct. Anal. **266** (2014), n. 7, 4490–4537.

[61] M. KRÉE, *Propriété de trace en dimension infinie, d'espaces du type Sobolev*, C. R. Acad. Sci., Sér. A **297** (1974), 157–164.

[62] M. KRÉE, *Propriété de trace en dimension infinie, d'espaces du type Sobolev*, Bull. Soc. Math. France. **105** (1977), n. 2, 141–163.

[63] B. LASCAR, *Propriétés locales d'espaces de type Sobolev en dimension infinie*. Comm. Partial Differential Equations **1** (1976), n. 6, 561–584.

[64] E. H. LIEB and M. LOSS, "Analysis", Amer. Math. Soc., Providence, Rhode Island, 2001.

[65] P. MALLIAVIN, *Stochastic calculus of variation and hypoelliptic operators*, Proc. Intern. Symp. on Stoch. Differential Equations (Res. Inst. Math. Sci., Kyoto Univ., Kyoto, 1976), Wiley, New York – Chichester – Brisbane, 1978, 195–263.

[66] P. MALLIAVIN, "Stochastic Analysis", Springer-Verlag, Berlin, 1997.

[67] P. MALLIAVIN and A. THALMAIER, "Stochastic Calculus of Variations in Mathematical Finance", Springer-Verlag, Berlin, 2006.

[68] J. NASH, *Continuity of solutions of parabolic and elliptic equations*. Amer. J. Math. **80** (1958), 931–954.

[69] E. NELSON, *The free Markov field*, J. Funct. Anal. **12** (1973), 211–227.

[70] E. V. NIKITIN, *Fractional order Sobolev clsses on infinite-dimensional spaces*, Dokl. Russian Acad. Sci. **452** (2013), n. 2, 130–135 (in Russan); English tranls.: Dokl. Math. **88** (2013), n. 2, 518–523.

[71] E. V. NIKITIN, *Besov classes on infinite-dimensional spaces*, Matem. Zametki **93** (2013), n. 6, 951–953 (in Russan); English tranls.: Math. Notes **93** (2013), n. 6, 936–939.

[72] E. V. NIKITIN, *Comparison of two definitions of Besov classes on infinite–dimensional spaces*, Matem. Zametki. **95** (2014), n. 1, 150–153 (in Russan); English tranls.: Math. Notes **95** (2014), n. 1, 133–135.

[73] I. NOURDIN and G. PECCATI, "Normal Approximations Using Malliavin Calculus: from Stein's Method to Universality", Cambridge University Press, Cambridge, 2012.

[74] D. NUALART, "The Malliavin Calculus and Related Topics", 2nd ed. Springer-Verlag, Berlin, 2006.

[75] G. PETERS, *Anticipating flows on the Wiener space generated by vector fields of low regularity*, J. Funct. Anal. **142** (1996), n. 1. 129–192.

[76] E. PINEDA and W. URBINA, *Some results on Gaussian Besov–Lipschitz spaces and Gaussian Triebel—Lizorkin spaces*, J. Approx. Theory **161** (2009), n. 2, 529–564.

[77] YU. G. RESHETNYAK, *Sobolev classes of functions with values in a metric space*, Sibirsk. Mat. Zh. **38** (1997), n. 3, 657–675 (in Russian); English transl.: Siberian Math. J. **38** (1997), n. 3, 567–583.

[78] YU. G. RESHETNYAK, *Sobolev classes of functions with values in a metric space. II*, Sibirsk. Mat. Zh. **45** (2004), n. 4, 855–870 (in Russian); English transl.: Siberian Math. J. **45** (2004), n. 4, 709–721.

[79] YU. G. RESHETNYAK, *On the theory of Sobolev classes of functions with values in a metric space*, Sibirsk. Mat. Zh. **47** (2006), n. 1, 146–168 (in Russian); English transl.: Siberian Math. J. **47** (2006), n. 1, 117–134.

[80] M. REED and B. SIMON, "Methods of Modern Mathematical Physics", Vol. II, Academic Pres, New York – London, 1975.

[81] M. RÖCKNER, R.-CH. ZHU and X.-CH. ZHU, *The stochastic reflection problem on an infinite dimensional convex set and BV functions in a Gelfand triple*. Ann. Probab. **40** (2012), n. 4, 1759–1794.

[82] I. SHIGEKAWA, *The Meyer inequality for the Ornstein–Uhlenbeck operator in L^1 and probabilistic proof of Stein's L^p multiplier the-*

orem, Trends in probability and related analysis (Taipei, 1996), World Sci. Publ., River Edge, New Jersey, 1997, 273–288.

[83] I. SHIGEKAWA, "Stochastic Analysis", Amer. Math. Soc., Providence, Rhode Island, 2004.

[84] P. SJÖGREN and F. SORIA, *Sharp estimates for the non-centered maximal operator associated to Gaussian and other radial measures*, Adv. Math. **181** (2004), n. 2, 251–275.

[85] A. V. SKOROHOD, "Integration in Hilbert Space", Springer-Verlag, Berlin – New York, 1974.

[86] A. J. STAM, *Some inequalities satisfied by the quantities of information of Fisher and Shannon*, Information and Control **2** (1959), 101–112.

[87] A. S. ÜSTÜNEL and M. ZAKAI, "Transformation of Measure on Wiener Space", Springer-Verlag, Berlin, 2000.

[88] S. K. VODOP'YANOV, *Topological and geometric properties of mappings with an integrable Jacobian in Sobolev classes. I*, Sibirsk. Mat. Zh. **41** (2000), n. 1, 23–48 (in Russian); English transl.: Siberian Math. J. **41** (2000), n. 1, 19–39.

[89] S. WATANABE, *Fractional order Sobolev spaces on Wiener space*, Probab. Theory Relat. Fields **95** (1993), 175–198.

[90] V. V. ZHIKOV, *Weighted Sobolev spaces*, Matem. Sbornik **189** (1998), n. 8, 27–58 (in Russian); English transl.: Sbornik Math. **189** (1998), n. 8, 1139–1170.

[91] V. V. ZHIKOV, *On the density of smooth functions in a weighted Sobolev space*, Dokl. Russian Acad. Sci. **453** (2013), n. 3, 247–251 (in Russian); English transl.: Dokl. Math. **88** (2013), n. 3, 669–673.

[92] W. ZIEMER, "Weakly Differentiable Functions", Springer-Verlag, New York – Berlin, 1989.

Isoperimetric problem and minimal surfaces in the Heisenberg group

Roberto Monti

Contents

This text is an extended version of the lecture notes of the course *Isoperimetric problem and minimal surfaces in the Heisenberg group* given at the *ERC-School on Geometric Measure Theory and Real Analysis*, held in Pisa between 30th September and 4th October 2013.

1 Introduction to the Heisenberg group $\mathbb{H}^n$

1.1 Algebraic structure

The $2n+1$-dimensional Heisenberg group is the manifold $\mathbb{H}^n = \mathbb{C}^n \times \mathbb{R}$, $n \in \mathbb{N}$, endowed with the group product

$$(z,t)\cdot(\zeta,\tau) = \big(z+\zeta, t+\tau+2\,\mathrm{Im}\langle z,\bar{\zeta}\rangle\big), \tag{1.1}$$

where $t, \tau \in \mathbb{R}$, $z, \zeta \in \mathbb{C}^n$ and $\langle z, \bar{\zeta}\rangle = z_1\bar{\zeta}_1 + \ldots + z_n\bar{\zeta}_n$. The Heisenberg group is a noncommutative Lie group. The identity element is $0 = (0,0) \in \mathbb{H}^n$. The inverse element of (z,t) is $(-z,-t)$. The center of the group is $Z = \{(z,t) \in \mathbb{H}^n : z = 0\}$. We denote elements of $\mathbb{H}^n$ by $p = (z,t) \in \mathbb{C}^n \times \mathbb{R}$.

The *left translation* by $p \in \mathbb{H}^n$ is the mapping $L_p : \mathbb{H}^n \to \mathbb{H}^n$

$$L_p(q) = p \cdot q.$$

Left translations are linear mappings in $\mathbb{H}^n = \mathbb{R}^{2n+1}$. For any $\lambda > 0$, the mapping $\delta_\lambda : \mathbb{H}^n \to \mathbb{H}^n$

$$\delta_\lambda(z,t) = (\lambda z, \lambda^2 t), \tag{1.2}$$

is called *dilation*. Dilations are linear mappings and form a 1-parameter group $(\delta_\lambda)_{\lambda>0}$ of automorphisms of $\mathbb{H}^n$.

We denote by $|E|$ the Lebesgue measure of a Lebesgue measurable set $E \subset \mathbb{H}^n = \mathbb{R}^{2n+1}$. The differential dL_p of any left translation is an upper triangular matrix with 1 along the principal diagonal. It follows that $\det dL_p = 1$ on $\mathbb{H}^n$ for any $p \in \mathbb{H}^n$ and, as a consequence,

$$|L_p E| = |E|, \quad \text{for any } p \in \mathbb{H}^n \text{ and for any } E \subset \mathbb{H}^n.$$

Lebesgue measure is the Haar measure of the Heisenberg group. Moreover, we have $\det \delta_\lambda = \lambda^Q$, where the integer

$$Q = 2n + 2 \tag{1.3}$$

is called *homogeneous dimension* of $\mathbb{H}^n$. As a consequence, we have

$$|\delta_\lambda E| = \lambda^Q |E|.$$

We introduce the Lie algebra of left invariant vector fields of $\mathbb{H}^n$. A C^∞ vector field X in $\mathbb{H}^n$ is left invariant if for any function $f \in C^\infty(\mathbb{H}^n)$ and for any $p \in \mathbb{H}^n$ there holds

$$X(f \circ L_p) = (Xf) \circ L_p.$$

Equivalently, X is left invariant if $X(p) = dL_pX(0)$, where dL_p is the differential of the left translation by p. Left invariant vector fields with the bracket form a nilpotent Lie algebra $\mathfrak{h}_n$, called Heisenberg Lie algebra. The algebra $\mathfrak{h}_n$ is spanned by the vector fields

$$X_j = \frac{\partial}{\partial x_j} + 2y_j\frac{\partial}{\partial t}, \quad Y_j = \frac{\partial}{\partial y_j} - 2x_j\frac{\partial}{\partial t}, \quad \text{and} \quad T = \frac{\partial}{\partial t}, \tag{1.4}$$

with $j = 1, \ldots, n$. In other words, any left invariant vector field is a linear combination with real coefficients of the vector fields (1.4). We are using the notation $p = (z, t)$ and $z = x + iy$ with $x, y, \in \mathbb{R}^n$. The vector fields (1.4) are determined by the relations

$$X_j(p) = dL_pX_j(0) = dL_p\frac{\partial}{\partial x_j},$$

$$Y_j(p) = dL_pY_j(0) = dL_p\frac{\partial}{\partial y_j},$$

$$T(p) = dL_pT(0) = dL_p\frac{\partial}{\partial t}.$$

The distribution of $2n$-dimensional planes H_p spanned by the vector fields X_j and Y_j, $j = 1, \ldots, n$, is called *horizontal distribution*:

$$H_p = \text{span}\{X_j(p), Y_j(p) \; : \; j = 1, \ldots, n\}. \tag{1.5}$$

The horizontal distribution is nonintegrable. In fact, for any $j = 1, \ldots, n$ there holds

$$[X_j, Y_j] = -4T \neq 0. \tag{1.6}$$

All other commutators vanish. The horizontal distribution is bracket generating of step 2.

When $n = 1$, we write $X = X_1$ and $Y = Y_1$.

1.2 Metric structure

We introduce the Carnot-Carathéodory metric of $\mathbb{H}^n$ and we describe the geodesics of this metric. In $\mathbb{H}^1$, these curves are important in the structure of H-minimal surfaces and surfaces with constant H-curvature.

A Lipschitz curve $\gamma : [0, 1] \to \mathbb{H}^n$ is horizontal if $\dot{\gamma}(t) \in H_{\gamma(t)}$ for a.e. $t \in [0, 1]$. Equivalently, γ is horizontal if there exist functions $h_j \in L^\infty([0, 1])$, $j = 1, \ldots, 2n$, such that

$$\dot{\gamma} = \sum_{j=1}^{n} h_jX_j(\gamma) + h_{n+j}Y_j(\gamma), \quad \text{a.e. on } [0, 1]. \tag{1.7}$$

The coefficients h_j are unique, and by the structure of the vector fields X_j and Y_j they satisfy $h_j = \dot{\gamma}_j$, where $\gamma = (\gamma_1, \dots, \gamma_{2n+1})$ are the coordinates of γ given by the identification $\mathbb{H}^n = \mathbb{R}^{2n+1}$. We call the Lipschitz curve $\kappa : [0,1] \to \mathbb{R}^{2n}$, $\kappa = (\gamma_1, \dots, \gamma_{2n})$, *horizontal projection* of γ.

The vertical component of γ is determined by the horizontality condition (1.7). Namely, we have

$$\dot{\gamma}_{2n+1} = 2\sum_{j=1}^{n} h_j \gamma_{n+j} - h_{n+j}\gamma_j = 2\sum_{j=1}^{n} \dot{\kappa}_j \kappa_{n+j} - \dot{\kappa}_{n+j}\kappa_j,$$

and, by integrating, we obtain for any $t \in [0,1]$

$$\gamma_{2n+1}(t) = \gamma_{2n+1}(0) + 2\sum_{j=1}^{n} \int_0^t (\dot{\kappa}_j \kappa_{n+j} - \dot{\kappa}_{n+j}\kappa_j)ds. \tag{1.8}$$

If κ is a given Lipschitz curve in $\mathbb{R}^{2n}$, the curve γ with $(\gamma_1, \dots, \gamma_{2n}) = \kappa$ and γ_{2n+1} as in (1.8) is called a *horizontal lift* of κ and we write $\gamma = \mathrm{Lift}(\kappa)$. The horizontal lift is unique modulo the initial value $\gamma_{2n+1}(0)$.

Now we define the Carnot-Carathéodory metric of $\mathbb{H}^n$. For any pair of points $p, q \in \mathbb{H}^n$, there exists a horizontal curve $\gamma : [0,1] \to \mathbb{H}^n$ such that $\gamma(0) = p$ and $\gamma(1) = q$. This follows from the nonintegrability condition (1.6) and it can be checked via a direct computation. The basic observation is that for any $t \in \mathbb{R}$

$$\exp(-tY_j)\exp(-tX_j)\exp(tY_j)\exp(tX_j)(0,0) = (0, -4t^2),$$

where $\exp(tV)(p)$ is the flow of the vector field V at time t starting from p.

We fix on the horizontal distribution H_p the positive quadratic form $g(p; \cdot)$ making the vector fields $X_1, \dots, X_n, Y_1, \dots, X_n$ orthonormal at every point $p \in \mathbb{H}^n$. Since the vector fields are left invariant, the quadratic form is left invariant. We use the quadratic form g to define the length of a horizontal curve $\gamma : [0,1] \to \mathbb{H}^n$ with horizontal projection κ:

$$L(\gamma) = \int_0^1 g(\gamma; \dot{\gamma})^{1/2} dt = \int_0^1 |\dot{\kappa}| dt,$$

where $|\dot{\kappa}|$ is the Euclidean norm in $\mathbb{R}^{2n}$ of $\dot{\kappa}$. For any couple of points $p, q \in \mathbb{H}^n$, we define

$$d(p,q) = \inf\Big\{L(\gamma) : \gamma : [0,1] \to \mathbb{H}^n \text{ is horizontal}, \gamma(0) = p \text{ and } \gamma(1) = q\Big\}. \tag{1.9}$$

We already observed that the above set is nonempty for any $p, q \in \mathbb{H}^n$, and thus $0 \le d(p,q) < \infty$.

The function $d : \mathbb{H}^n \times \mathbb{H}^n \to [0,\infty)$ is a distance on $\mathbb{H}^n$, called *Carnot-Carathéodory distance*. It can be proved that for any compact set $K \subset \mathbb{H}^n = \mathbb{R}^{2n+1}$ there exists a constant $0 < C_K < \infty$ such that

$$d(p,q) \ge C_K |p-q| \tag{1.10}$$

for all $p, q \in K$, where $|p-q|$ is the Euclidean distance between the points. In particular, we have $d(p,q) \ne 0$ if $p \ne q$. The distance d is left invariant and 1-homogeneous. Namely, for any $p, q, w \in \mathbb{H}^n$ and $\lambda > 0$ there holds:

i) $d(w \cdot p, w \cdot q) = d(p,q)$;
ii) $d(\delta_\lambda(p), \delta_\lambda(q)) = d(p,q)$.

Statement i) follows from the fact that $L(w \cdot \gamma) = L(\gamma)$ for any horizontal curve γ and for any $w \in \mathbb{H}^n$, because the quadratic form g is left invariant. Analogously, ii) follows from $L(\delta_\lambda(\gamma)) = \lambda L(\gamma)$, that is a consequence of the identities

$$X_j(f \circ \delta_\lambda) = \lambda (X_j f) \circ \delta_\lambda, \quad Y_j(f \circ \delta_\lambda) = \lambda (Y_j f) \circ \delta_\lambda,$$

holding for any $f \in C^\infty(\mathbb{H}^n)$ and $\lambda > 0$.

For any $p \in \mathbb{H}^n$ and $r > 0$, we define the Carnot-Carathéodory ball

$$B_r(p) = \{q \in \mathbb{H}^n : d(p,q) < r\}.$$

We also let $B_r = B_r(0)$. The size of Carnot-Carathéodory balls can be described by means of anisotropic homogeneous norms. For any $p = (z,t) \in \mathbb{H}^n$ let

$$\|p\|_\infty = \max\{|z|, |t|^{1/2}\}. \tag{1.11}$$

The "box norm" $\|\cdot\|_\infty$ has the following properties:

i) $\|\delta_\lambda(p)\|_\infty = \lambda \|p\|_\infty$, for all $p \in \mathbb{H}^n$ and $\lambda > 0$;
ii) $\|p \cdot q\|_\infty \le \|p\|_\infty + \|q\|_\infty$, for all $p, q \in \mathbb{H}^n$.

By ii), the function $\varrho : \mathbb{H}^n \times \mathbb{H}^n \to [0,\infty)$,

$$\varrho(p,q) = \|p^{-1} \cdot q\|_\infty, \tag{1.12}$$

satisfies the triangle inequality and is a distance on $\mathbb{H}^n$. By an elementary argument based on continuity, compactness, and homogeneity, there exists an absolute constant $C > 0$ such that

$$C^{-1} d(p,q) \le \varrho(p,q) \le C d(p,q)$$

for all $p, q \in \mathbb{H}^n$. The distance functions d and ϱ are equivalent.

All the previous observations are still valid when the "box norm" $\|\cdot\|_\infty$ is replaced with the Koranyi norm $\|p\| = \big(|z|^4 + t^2\big)^{1/4}$.

The metric space $(\mathbb{H}^n, d)$ is complete and locally compact. By the definition of d, it is also a length space. Then, a standard application of Ascoli-Arzelà theorem shows that it is a geodesic space, namely for all $p, q \in \mathbb{H}^n$ there exists a horizontal curve $\gamma : [0,1] \to \mathbb{H}^n$ such that $\gamma(0) = q$, $\gamma(1) = p$, and $L(\gamma) = d(p,q)$. The curve γ is called *geodesic* or *length minimizing curve* joining q to p.

We classify geodesics in $\mathbb{H}^1$ starting from the initial point 0. Let $\Phi : [0, 2\pi] \times [-2\pi, 2\pi] \to \mathbb{H}^1$ be the mapping

$$\Phi(\psi, \varphi) = \Big(\frac{e^{i\psi}(e^{i\varphi} - 1)}{\varphi}, 2\frac{\varphi - \sin\varphi}{\varphi^2}\Big). \tag{1.13}$$

When $\varphi = 0$, the formula is determined by analytic continuation and we have $\Phi(\psi, 0) = (ie^{i\psi}, 0)$. The set $S = \Phi([0, 2\pi] \times [-2\pi, 2\pi]) \subset \mathbb{H}^1$ is homeomorphic to a 2-dimensional sphere. It is a C^∞ surface at points $(z,t) \in S$ such that $z \neq 0$, i.e., $(z,t) = \Phi(\psi, \varphi)$ with $|\varphi| \neq 2\pi$. The antipodal points $(0, \pm 1/\pi) \in S$ are obtained for $\varphi = \pm 2\pi$ and are Lipschitz points. We will show that S is the unitary Carnot-Carathéodory sphere of $\mathbb{H}^1$ centered at 0, $S = \partial B_1(0)$.

Theorem 1.1. *For any* $\psi \in [0, 2\pi]$ *and* $\varphi \in [-2\pi, 2\pi]$, *the curve* $\gamma_{\psi,\varphi} : [0,1] \to \mathbb{H}^1$

$$\gamma_{\psi,\varphi}(s) = \Big(\frac{e^{i\psi}(e^{i\varphi s} - 1)}{\varphi}, 2\frac{\varphi s - \sin\varphi s}{\varphi^2}\Big), \quad s \in [0,1], \tag{1.14}$$

is length minimizing. When $|\varphi| < 2\pi$, $\gamma_{\psi,\varphi}$ *is the unique length minimizing curve from* 0 *to* $\Phi(\psi, \varphi)$. *When* $\varphi = \pm 2\pi$, *for every* $\psi \in [0, 2\pi]$ *the curve* $\gamma_{\psi,\varphi}$ *is length minimizing from* 0 *to* $(0, \pm 1/\pi)$.

Proof. Let $(z_0, t_0) \in \mathbb{H}^1$ be any point and introduce the family of admissible curves

$$\mathscr{A} = \Big\{\kappa \in \mathrm{Lip}([0,1]; \mathbb{R}^2) : \kappa(0) = 0, \kappa(1) = z_0\Big\}.$$

The end-point mapping relative to the third coordinate End : $\mathscr{A} \to \mathbb{R}$ is

$$\mathrm{End}(\kappa) = 2\int_0^1 \big(\kappa_2\dot\kappa_1 - \kappa_1\dot\kappa_2\big)ds = 2\int_0^1 \mathrm{Im}(\kappa\dot{\bar\kappa})ds,$$

where $\kappa\dot{\bar\kappa}$ is a complex product.

The geodesic γ joining 0 to (z_0, t_0) is the horizontal lift of the curve κ in the plane that solves the problem

$$\min\left\{\int_0^1 |\dot\kappa| ds \,:\, \kappa \in \mathscr{A} \text{ and } \mathrm{End}(\kappa) = t_0\right\}. \tag{1.15}$$

Let κ be a minimizer for problem (1.15). We compute the first variation of the length functional at the curve κ with constraint $\mathrm{End}(\kappa) = t_0$. For $\tau \in \mathbb{R}$ and $\vartheta \in C_c^\infty\big((0,1);\mathbb{R}^2\big)$ the curve $\kappa^\tau = \kappa + \tau\vartheta$ satisfies

$$\begin{aligned}
\frac{d}{d\tau}\mathrm{End}(\kappa^\tau)\Big|_{\tau=0} &= 2\frac{d}{d\tau}\int_0^1 \big((\kappa_2+\tau\vartheta_2)(\dot\kappa_1+\tau\dot\vartheta_1) \\
&\qquad - (\kappa_1+\tau\vartheta_1)(\dot\kappa_2+\tau\dot\vartheta_2)\big)ds\Big|_{\tau=0} \\
&= 2\int_0^1 \big(\vartheta_2\dot\kappa_1 + \kappa_2\dot\vartheta_1 - \kappa_1\dot\vartheta_2 - \vartheta_1\dot\kappa_2\big)ds \\
&= 4\int_0^1 \big(\vartheta_2\dot\kappa_1 - \vartheta_1\dot\kappa_2\big)ds.
\end{aligned}$$

We have $|\dot\kappa| \neq 0$ a.e., and thus there exists $\vartheta \in C_c^\infty\big((0,1);\mathbb{R}^2\big)$ such that

$$\frac{d}{d\tau}\mathrm{End}(\kappa^\tau)\Big|_{\tau=0} \neq 0. \tag{1.16}$$

Fix a function ϑ satisfying (1.16) and let $\eta \in C_c^\infty\big((0,1);\mathbb{R}^2\big)$ be an arbitrary vector valued function. The curve $\kappa+\tau\vartheta+\epsilon\eta$ belongs to $\mathscr{A}$. Define the function in the plane $E:\mathbb{R}^2\to\mathbb{R}$

$$E(\epsilon,\tau) = \mathrm{End}(\kappa+\tau\vartheta+\epsilon\eta).$$

This function is C^1-smooth and $H := \partial E(0,0)/\partial\tau \neq 0$, by (1.16). By the implicit function theorem, there exist $\epsilon_0 > 0$ and a function $\tau \in C^1(-\epsilon_0,\epsilon_0)$ such that $E(\epsilon,\tau(\epsilon)) = E(0,0) = t_0$ for all $\epsilon \in (-\epsilon_0,\epsilon_0)$. Moreover, we have

$$\begin{aligned}
\tau'(0) &= -\Big(\frac{\partial E(0,0)}{\partial\tau}\Big)^{-1}\Big(\frac{\partial E(0,0)}{\partial\epsilon}\Big) \\
&= -\frac{1}{H}\int_0^1 (\eta_2\dot\kappa_1 - \eta_1\dot\kappa_2)ds = -\frac{1}{H}\int_0^1 \langle\dot\kappa^\perp,\eta\rangle ds,
\end{aligned} \tag{1.17}$$

where $\dot\kappa^\perp = (-\dot\kappa_2,\dot\kappa_1)$, or equivalently, in the complex notation $\dot\kappa^\perp = i\dot\kappa$.

Since κ is a solution to the minimum problem (1.15) and $\kappa + \tau(\epsilon)\vartheta + \epsilon\eta \in \mathscr{A}$ with $\mathrm{End}(\kappa + \tau(\epsilon)\vartheta + \epsilon\eta) = t_0$, then we have

$$1 = \int_0^1 |\dot\kappa| ds \le \int_0^1 |\dot\kappa + \tau(\epsilon)\dot\vartheta + \epsilon\dot\eta| ds = L(\epsilon),$$

and thus $L'(0) = 0$. We can without loss of generality assume that κ is parameterized by arc-length, i.e., $|\dot\kappa| = 1$. The equation $L'(0) = 0$ gives (we also use (1.17))

$$\begin{aligned} 0 &= \tau'(0) \int_0^1 \langle \dot\kappa, \dot\vartheta \rangle ds + \int_0^1 \langle \dot\kappa, \dot\eta \rangle ds \\ &= \varphi \int_0^1 \langle \dot\kappa^\perp, \eta \rangle ds + \int_0^1 \langle \dot\kappa, \dot\eta \rangle ds, \end{aligned}$$

where $\varphi \in \mathbb{R}$ is the constant

$$\varphi = -\frac{1}{H} \int_0^1 \langle \dot\kappa, \dot\vartheta \rangle ds.$$

Eventually, for any test function $\eta \in C_c^\infty\big((0,1); \mathbb{R}^2\big)$ we have

$$\int_0^1 \big\{ \langle \dot\kappa, \dot\eta \rangle + \varphi \langle \dot\kappa^\perp, \eta \rangle \big\} ds = 0,$$

and a standard argument implies that κ is in $C^\infty([0,1]; \mathbb{R}^2)$ and it solves the differential equation $\ddot\kappa = \varphi\dot\kappa^\perp = i\varphi\dot\kappa$. Then we have $\dot\kappa(s) = i e^{i\psi} e^{i\varphi s}$, $s \in \mathbb{R}$, for some $\psi \in [0, 2\pi]$. Integrating with $\kappa(0) = 0$, we find

$$\kappa(s) = \frac{e^{i\psi}(e^{i\varphi s} - 1)}{\varphi}, \quad s \in \mathbb{R}.$$

The vertical coordinate of the horizontal lift γ of κ is

$$\gamma_3(s) = 2 \int_0^s \mathrm{Im}(\kappa(\sigma)\overline{\dot\kappa(\sigma)}) d\sigma = 2\frac{\varphi s - \sin\varphi s}{\varphi^2},$$

and thus for any $\psi \in [0, \pi]$ and $\varphi \in \mathbb{R}$ we get the curve

$$\gamma_{\psi,\varphi}(s) = \Big(\frac{e^{i\psi}(e^{i\varphi s} - 1)}{\varphi}, 2\frac{\varphi s - \sin\varphi s}{\varphi^2} \Big), \quad s \in \mathbb{R}. \tag{1.18}$$

When $\varphi = 0$, γ reduces to the line $\gamma(s) = (i e^{i\psi} s, 0)$.

The curve $\gamma_{\psi,\varphi}$ is length minimizing on the interval $0 \le s \le 2\pi/|\varphi|$ and, after $s = 2\pi/|\varphi|$, it ceases to be length minimizing. We prove this claim in the case $\varphi = 2\pi$ by a geometric argument. For $s = 1$ we have

$$\gamma_{\psi,2\pi}(1) = \big(0, 1/\pi\big).$$

At the point $(0, 1/\pi) \in \mathbb{C} \times \mathbb{R}$, the surface $S = \Phi([0, 2\pi] \times [-2\pi, 2\pi])$ introduced in (1.13) has a conical point directed downwards. By this, we mean that near $(0, 1/\pi)$ the surfaces S stays above the cone $t = 1/\pi + \delta|z|$ for some $\delta > 0$. Then for any $\epsilon > 0$ small enough there exist $0 < \lambda < 1$ and $(\bar{\psi}, \bar{\varphi}) \in [0, 2\pi] \times [-2\pi, 2\pi]$ such that $\gamma_{\psi,2\pi}(1+\epsilon) = \delta_\lambda \Phi(\bar{\psi}, \bar{\varphi})$. Since $d(\Phi(\bar{\psi}, \bar{\varphi}), 0) \le 1$ (a posteriori we have equality, here), we deduce that

$$d(\gamma_{\psi,2\pi}(1+\epsilon), 0) = \lambda d(\Phi(\bar{\psi}, \bar{\varphi}), 0) \le \lambda < 1.$$

Since the length of $\gamma_{\psi,2\pi}$ on the interval $[0, 1+\epsilon]$ is $1+\epsilon$, we see that the curve is not length minimizing.

For any point $(z_0, t_0) \in \mathbb{H}^1$ with $z_0 \neq 0$, the system of equations

$$\frac{e^{i\psi}(e^{i\varphi s} - 1)}{\varphi} = z_0, \qquad 2\frac{\varphi s - \sin\varphi s}{\varphi^2} = t_0, \tag{1.19}$$

has unique solutions $s \ge 0$, $\psi \in [0, 2\pi)$, and $\varphi \in \mathbb{R}$ subject to the constraint $s|\varphi| < 2\pi$ (we omit details). Thus $\gamma_{\psi,\varphi}$ is the unique length minimizing curve from 0 to (z_0, t_0) and $s = d((z_0, t_0), 0)$. □

Remark 1.2. The Heisenberg isoperimetric problem is related to the classical Dido problem, that asks to bound a region of the half plane with a curve with minimal length, where the boundary of the half plane (the coast) is a free length.

Let $\gamma : [0, 1] \to \mathbb{H}^1$ be a horizontal curve such that $\gamma(0) = 0$ and let $\kappa : [0, 1] \to \mathbb{R}^2$ be its horizontal projection. By formula (1.8), the third coordinate of γ at time $t \in [0, 1]$ is

$$\gamma_3(t) = 2\int_0^t (\kappa_2\dot{\kappa}_1 - \kappa_1\dot{\kappa}_2)ds = 2\int_{\kappa|[0,t]} ydx - xdy.$$

Let $E_t \subset \mathbb{R}^2$ be the region of the plane bounded by the curve κ restricted to $[0, t]$ and by the line segment joining $\kappa(t)$ to 0. Assume that the concatenation of κ and of the line segment bounds E_t counterclockwise. Then by Stokes' theorem we have

$$\gamma_3(t) = -4\int_{E_t} dx \wedge dy = -4|E_t|.$$

If the orientation is clockwise, $-4|E_t|$ is replaced by $4|E_t|$. If the orientation is different in subregions of E_t, there are area cancellations.

So the minimum problem (1.15) consists in finding the shortest curve in the plane enclosing an amount of area given by the t_0 coordinate of the final point (z_0, t_0). In the Heisenberg isoperimetric problem, the point z_0 is also fixed, differently from Dido problem.

2 Heisenberg perimeter and other equivalent measures

2.1 H-perimeter

We introduce the notion of H-perimeter for a set $E \subset \mathbb{H}^n$. We preliminarily need the definition of H-divergence of a vector valued function $\varphi \in C^1(\mathbb{H}^n; \mathbb{R}^{2n})$.

Let V be a smooth vector field in $\mathbb{H}^n = \mathbb{R}^{2n+1}$. We may express V using both the basis X_j, Y_j, T and the standard basis of vector fields of $\mathbb{R}^{2n+1}$:

$$
\begin{aligned}
V &= \sum_{j=1}^{n} (\varphi_j X_j + \varphi_{n+j} Y_j) + \varphi_{2n+1} T \\
&= \sum_{j=1}^{n} \left\{ \varphi_j \frac{\partial}{\partial x_j} + \varphi_{n+j} \frac{\partial}{\partial y_j} + (2y_j \varphi_j - 2x_j \varphi_{n+j}) \frac{\partial}{\partial t} \right\} + \varphi_{2n+1} \frac{\partial}{\partial t},
\end{aligned} \tag{2.1}
$$

where $\varphi_j, \varphi_{n+j}, \varphi_{2n+1} \in C^\infty(\mathbb{H}^n)$ are smooth functions. The standard divergence of V is

$$
\begin{aligned}
\operatorname{div} V &= \sum_{j=1}^{n} \left\{ \frac{\partial \varphi_j}{\partial x_j} + \frac{\partial \varphi_{n+j}}{\partial y_j} + \left(2y_j \frac{\partial \varphi_j}{\partial t} - 2x_j \frac{\partial \varphi_{n+j}}{\partial t} \right) \right\} \\
&\quad + \frac{\partial \varphi_{2n+1}}{\partial t} \\
&= \sum_{j=1}^{n} (X_j \varphi_j + Y_j \varphi_{n+j}) + T\varphi_{2n+1}.
\end{aligned} \tag{2.2}
$$

The vector field V is said to be horizontal if $V(p) \in H_p$ for all $p \in \mathbb{H}^n$. Namely, a vector field V as in (2.1) is horizontal when $\varphi_{2n+1} = 0$. These observations suggest the following definition.

Let $A \subset \mathbb{H}^n$ be an open set. We define the horizontal divergence of a vector valued mapping $\varphi \in C^1(A; \mathbb{R}^{2n})$ as

$$
\operatorname{div}_H \varphi = \sum_{j=1}^{n} (X_j \varphi_j + Y_j \varphi_{n+j}). \tag{2.3}
$$

By (2.2), $\mathrm{div}_H\varphi = \mathrm{div}\, V$ is the standard divergence of the horizontal vector field V with coordinates $\varphi = (\varphi_1, \dots, \varphi_{2n})$ in the basis $X_1, \dots, X_n$, $Y_1, \dots, Y_n$. If $\|\cdot\|$ is the norm on H_p that makes $X_1, \dots, X_n, Y_1, \dots, Y_n$ orthonormal, then we have

$$\|V(p)\| = |\varphi(p)|,$$

where $|\cdot|$ is the standard norm on $\mathbb{R}^{2n}$.

The following definition is the starting point of the fundamental paper [27] (see also [33]).

Definition 2.1 (*H*-**perimeter**). The H-perimeter in an open set $A \subset \mathbb{H}^n$ of a Lebesgue measurable set $E \subset \mathbb{H}^n$ is

$$P(E; A) = \sup\left\{ \int_E \mathrm{div}_H\varphi\, dzdt \ :\ \varphi \in C_c^1(A; \mathbb{R}^{2n}),\ \|\varphi\|_\infty \le 1 \right\}. \quad (2.4)$$

Above, we let

$$\|\varphi\|_\infty = \sup_{p\in A} |\varphi(p)|.$$

If $P(E; A) < \infty$, we say that E has finite H-perimeter in A. If $P(E; A') < \infty$ for any open set $A' \Subset A$, we say that E has locally finite H-perimeter in A.

H-perimeter has the following invariance properties.

Proposition 2.2. *Let $E \subset \mathbb{H}^n$ be a set with finite H-perimeter in an open set $A \subset \mathbb{H}^n$. Then for any $p \in \mathbb{H}^n$ and for any $\lambda > 0$ we have:*

i) $P(L_pE; L_pA) = P(E; A)$;
ii) $P(\delta_\lambda E; \delta_\lambda A) = \lambda^{Q-1} P(E; A)$.

Proof. Statement i) follows from the fact that the vector fields X_j and Y_j are left invariant, and thus

$$(\mathrm{div}_H\varphi) \circ L_p = \mathrm{div}_H(\varphi \circ L_p).$$

We prove ii) in the case $A = \mathbb{H}^n$. First notice that for any $\varphi \in C_c^1(\mathbb{H}^n; \mathbb{R}^{2n})$ we have

$$\mathrm{div}_H(\varphi \circ \delta_\lambda) = \lambda (\mathrm{div}_H\varphi) \circ \delta_\lambda,$$

and thus

$$\int_{\delta_\lambda E} \mathrm{div}_H\varphi\, dzdt = \lambda^Q \int_E (\mathrm{div}_H\varphi) \circ \delta_\lambda\, dzdt = \lambda^{Q-1} \int_E \mathrm{div}_H(\varphi \circ \delta_\lambda) dzdt.$$

The claim easily follows. □

Let $E \subset \mathbb{H}^n$ be a set with locally finite H-perimeter in an open set $A \subset \mathbb{H}^n$. The linear functional $T : C_c^1(A; \mathbb{R}^{2n}) \to \mathbb{R}$

$$T(\varphi) = \int_E \mathrm{div}_H \varphi(z,t)\, dzdt$$

is locally bounded in $C_c(A; \mathbb{R}^{2n})$. Namely, for any open set $A' \Subset A$ we have

$$T(\varphi) \le \|\varphi\|_\infty P(E; A') \tag{2.5}$$

for all $\varphi \in C_c^1(A'; \mathbb{R}^{2n})$. By density, T can be extended to a bounded linear operator on $C_c(A'; \mathbb{R}^{2n})$ satisfying the same bound (2.5). Thus, by Riesz' representation theorem we deduce the following proposition.

Proposition 2.3. *Let $E \subset \mathbb{H}^n$ be a set with locally finite H-perimeter in the open set $A \subset \mathbb{H}^n$. There exist a positive Radon measure μ_E on A and a μ_E-measurable function $\nu_E : A \to \mathbb{R}^{2n}$ such that:*

1) $|\nu_E| = 1$ *μ_E-a.e. on A.*
2) *The following generalized Gauss-Green formula holds*

$$\int_E \mathrm{div}_H \varphi\, dzdt = -\int_A \langle \varphi, \nu_E \rangle d\mu_E \tag{2.6}$$

for all $\varphi \in C_c^1(A; \mathbb{R}^{2n})$.

Above, $\langle \cdot, \cdot \rangle$ is the standard scalar product in $\mathbb{R}^{2n}$.

Definition 2.4 (Horizontal normal). The measure μ_E is called *H-perimeter measure* and the function ν_E is called *measure theoretic inner horizontal normal* of E.

We shall refer to ν_E simply as to the *horizontal normal*. In Section 3, we describe geometrically ν_E in the smooth case (see (3.3)). In Proposition 2.10 below, we shall see that the vertical hyperplane in $\mathbb{H}^n$ orthogonal to $\nu_E(p)$ is the "tangent plane" to ∂E at points of the reduced boundary.

Remark 2.5. By Proposition 2.3, the open sets mapping $A' \mapsto P(E; A')$, with $A' \Subset A$ open, extends to the Radon measure μ_E. We show that for any open set $A' \Subset A$ we have $\mu_E(A') = P(E; A')$.

The inequality $P(E; A') \le \mu_E(A')$ follows from the sup-definition (2.4) of H-perimeter. The opposite inequality can be proved by a standard approximation argument. By Lusin's theorem, for any $\epsilon > 0$ there exists a compact set $K \subset A'$ such that $\mu_E(A' \setminus K) < \epsilon$ and $\nu_E : K \to \mathbb{R}^{2n}$ is continuous. By Titze's theorem, there exists $\psi \in C_c(A'; \mathbb{R}^{2n})$ such that $\psi = \nu_E$ on K and $\|\psi\|_\infty \le 1$. Finally, by mollification there exists

$\varphi \in C_c^\infty(A'; \mathbb{R}^{2n})$ such that $\|\varphi - \psi\|_\infty < \epsilon$ and $\|\varphi\|_\infty \leq 1$. Then we have

$$P(E; A') \geq \int_E \mathrm{div}_H \varphi dz dt = -\int_{A'} \langle \varphi, \nu_E \rangle d\mu_E \geq (1-\epsilon)\mu_E(A') - 2\epsilon,$$

and the claim follows.

In the sequel, we need a metric structure on $\mathbb{H}^n$. For most purposes, the Carnot-Carathéodory metric would be fine. In some cases, however, as in the characterization (2.13) of H-perimeter by means of spherical Hausdorff measures, the structure of Carnot-Carathéodory balls is less manageable. For this reason, we closely follow [27] and we use the metric ϱ introduced in (1.12) via the "box-norm" $\|\cdot\|_\infty$ in (1.11). We denote the open ball in ϱ centered at $p \in \mathbb{H}^n$ and with radius $r > 0$ in the following way

$$U(p,r) = \{q \in \mathbb{H}^n : \|p^{-1} \cdot q\|_\infty < r\}. \tag{2.7}$$

We also let $U_r(p) = U(p,r)$ and $U_r = U_r(0)$.

Definition 2.6 (Measure theoretic boundary). The *measure theoretic boundary* of a measurable set $E \subset \mathbb{H}^n$ is the set

$$\partial E = \{p \in \mathbb{H}^n : |E \cap U_r(p)| > 0 \text{ and } |U_r(p) \setminus E| > 0 \text{ for all } r > 0\}.$$

The measure theoretic boundary is a subset of the topological boundary. The definition does not depend on the specific balls $U_r(p)$. We may also consider the set of points with density $1/2$:

$$E_{1/2} = \left\{p \in \mathbb{H}^n : \lim_{r\to 0} \frac{|E \cap U_r(p)|}{|U_r(p)|} = \frac{1}{2}\right\}.$$

We clearly have $E_{1/2} \subset \partial E$. The definition of $E_{1/2}$ is sensitive to the choice of the metric.

The perimeter measure μ_E is concentrated in a subset of $E_{1/2}$ called reduced boundary. The following definition is introduced and studied in [27].

Definition 2.7 (Reduced boundary). The *reduced boundary* of a set $E \subset \mathbb{H}^n$ with locally finite H-perimeter is the set $\partial^* E$ of all points $p \in \mathbb{H}^n$ such that the following three conditions hold:

(1) $\mu_E(U_r(p)) > 0$ for all $r > 0$.
(2) We have

$$\lim_{r\to 0} \fint_{U_r(p)} \nu_E \, d\mu_E = \nu_E(p).$$

(3) There holds $|\nu_E(p)| = 1$.

As usual $\fint$, stands for the averaged integral. The definition of reduced boundary is sensitive to the metric. It also depends on the representative of ν_E.

The proof of the Euclidean model of Proposition 2.8 below relies upon Lebesgue-Besicovitch differentiation theorem for Radon measures in $\mathbb{R}^n$. In $\mathbb{H}^n$ with metrics equivalent to the Carnot-Carathéodory distance, however, Besicovitch's covering theorem fails (see [36] and [65]). This problem is bypassed in [27] using an asymptotic doubling property established, in a general context, in [1].

Proposition 2.8. *Let $E \subset \mathbb{H}^n$ be a set with locally finite H-perimeter. Then the perimeter measure μ_E is concentrated on $\partial^* E$. Namely, we have $\mu_E(\mathbb{H}^n \setminus \partial^* E) = 0$.*

Proof. By [1], Theorem 4.3, there exists a constant $\tau(n) > 0$ such that for μ_E-a.e. $p \in \mathbb{H}^n$ there holds

$$\tau(n) \le \liminf_{r\to 0} \frac{\mu_E(U_r(p))}{r^{Q-1}} \le \limsup_{r\to 0} \frac{\mu_E(U_r(p))}{r^{Q-1}} < \infty.$$

As a consequence, we have the following asymptotic doubling formula

$$\limsup_{r\to 0} \frac{\mu_E(U_{2r}(p))}{\mu_E(U_r(p))} < \infty, \tag{2.8}$$

for μ_E-a.e. $p \in \mathbb{H}^n$. Thus, by Theorems 2.8.17 and 2.9.8 in [25], for any function $f \in L^1_{\mathrm{loc}}(\mathbb{H}^n; \mu_E)$ there holds

$$\lim_{r\to 0} \fint_{U_r(p)} f \, d\mu_E = f(p)$$

for μ_E-a.e. $p \in \mathbb{H}^n$.

Assume that $p \in \mathbb{H}^n \setminus \partial^* E$. There are three possibilities:

1) We have $\mu_E(U_r(p)) = 0$ for some $r > 0$. The set of points with this property has null μ_E measure.
2) We have

$$\lim_{r\to 0} \fint_{U_r(p)} \nu_E \, d\mu_E \neq \nu_E(p).$$

By the above argument with $f = \nu_E$, the set of such points has null μ_E measure.

3) We have $|\nu_E(p)| \neq 1$. By Proposition 2.3, the set of such points has null μ_E measure.

This ends the proof. □

Definition 2.9 (Vertical plane). For any $\nu \in \mathbb{R}^{2n}$ with $|\nu| = 1$, we call the set

$$H_\nu = \{(z,t) \in \mathbb{H}^n : \langle \nu, z\rangle \geq 0,\ t \in \mathbb{R}\}$$

the *vertical half-space* through $0 \in \mathbb{H}^n$ with inner normal ν. The boundary of H_ν, the set

$$\partial H_\nu = \{(z,t) \in \mathbb{H}^n : \langle \nu, z\rangle = 0,\ t \in \mathbb{R}\},$$

is called *vertical plane* orthogonal to ν passing through $0 \in \mathbb{H}^n$.

At points $p \in \partial^* E$, the set E blows up to the vertical half space H_ν with $\nu = \nu_E(p)$. In this sense, the boundary of H_ν is the anisotropic tangent space of $\partial^* E$ at p. The problem of the characterization of blow-ups in Carnot groups is still open. In general, it is known that in the blow-up of blow-ups there are vertical hyperplanes (see [3]). Hereafter, we let $E_\lambda = \delta_\lambda E$ for $\lambda > 0$.

Theorem 2.10 (Blow-up). *Let $E \subset \mathbb{H}^n$ be a set with finite H-perimeter, assume that $0 \in \partial^* E$ and let $\nu = \nu_E(0)$. Then we have*

$$\lim_{\lambda\to\infty} \chi_{E_\lambda} = \chi_{H_\nu}, \tag{2.9}$$

where the limit is in $L^1_{\mathrm{loc}}(\mathbb{H}^n)$. Moreover, for a.e. $r > 0$ we have

$$\lim_{\lambda\to\infty} P(E_\lambda; U_r) = P(H_\nu; U_r) = c_n r^{Q-1}, \tag{2.10}$$

where $c_n = P(H_\nu; U_1) > 0$ is an absolute constant.

Proof. Let $\varphi \in C^1_c(\mathbb{H}^n; \mathbb{R}^{2n})$ be a test vector valued function. For a.e. $r > 0$, we have the following integration by parts formula

$$\int_{E\cap U_r} \mathrm{div}_H \varphi\, dzdt = -\int_{U_r} \langle \varphi, \nu_E\rangle d\mu_E - \int_{\partial U_r \cap E} \langle \varphi, \nu_{U_r}\rangle d\mu_{U_r}. \tag{2.11}$$

This formula can be proved in the following way. Let $(f_j)_{j\in\mathbb{N}}$ be a sequence of functions $f_j \in C^\infty(\mathbb{H}^n)$ such that $f_j \to \chi_E$, as $j \to \infty$, in $L^1_{\mathrm{loc}}(\mathbb{H}^n)$ and $\nabla_H f_j dzdt \rightharpoonup \nu_E d\mu_E$ in the weak sense of Radon measures. We are denoting by

$$\nabla_H f = (X_1 f, \ldots, X_n f, Y_1 f, \ldots, Y_n f)$$

the horizontal gradient of a function f.

The set U_r supports the standard divergence theorem and therefore we have

$$\int_{U_r} f_j \mathrm{div}_H \varphi\, dzdt = -\int_{U_r} \langle \varphi, \nabla_H f_j\rangle dzdt - \int_{\partial U_r} f_j \langle \varphi, \nu_{U_r}\rangle d\mu_{U_r}. \tag{2.12}$$

We can assume that, for a.e. $r > 0$, $f_j \to \chi_E$ in $L^1(\partial U_r)$ and $\mu_E(\partial U_r) = 0$. Letting $j \to \infty$ in (2.12) we obtain (2.11).

Let $\varphi \in C_c^1(\mathbb{H}^n; \mathbb{R}^{2n})$ be such that $\varphi(z,t) = \nu_E(0)$ for all $(z,t) \in U_r$. From (2.11), we have

$$0 = -\int_{U_r} \langle \nu_E(0), \nu_E \rangle d\mu_E - \int_{\partial U_r \cap E} \langle \nu_E(0), \nu_{U_r} \rangle d\mu_{U_r}.$$

Using $|\nu_E(0)| = |\nu_{U_r}| = 1$ a.e. and Proposition 2.2, we have

$$\begin{aligned}\int_{U_r} \langle \nu_E(0), \nu_E \rangle d\mu_E &= -\int_{\partial U_r \cap E} \langle \nu_E(0), \nu_{U_r} \rangle d\mu_{U_r} \le P(U_r; \mathbb{H}^n) \\ &= r^{Q-1} P(U_1; \mathbb{H}^n).\end{aligned}$$

Since $0 \in \partial^* E$, there holds

$$\int_{U_r} \langle \nu_E(0), \nu_E \rangle d\mu_E = (1 + o(1)) P(E; U_r),$$

where $o(1) \to 0$ as $r \to 0$. Using these estimates, we conclude that for any $\lambda \ge 1$ we have

$$P(E_\lambda; U_r) = \lambda^{Q-1} P(E; U_{r/\lambda}) \le 2P(U_1; \mathbb{H}^n) r^{Q-1}.$$

The family of sets $(E_\lambda)_{\lambda > 1}$ has locally uniformly bounded perimeter. By the compactness theorem for BV_H functions (see [33]), there exists a set $F \subset \mathbb{H}^n$ with locally finite perimeter and a sequence $\lambda_j \to \infty$ such that $E_{\lambda_j} \to F$ in the $L^1_{\mathrm{loc}}(\mathbb{H}^n)$ convergence of characteristic functions. From the Gauss-Green formula (2.6), it follows that

$$\nu_{E_{\lambda_j}} \mu_{E_{\lambda_j}} \rightharpoonup \nu_F \mu_F, \quad \text{as } j \to \infty,$$

in the sense of the weak convergence of Radon measures.

Starting from the identity

$$\fint_{U_r} \nu_{E_{\lambda_j}} d\mu_{E_{\lambda_j}} = \fint_{U_{r/\lambda_j}} \nu_E d\mu_E,$$

using $0 \in \partial^* E$, and choosing $r > 0$ such that $\mu_F(\partial U_r) = 0$ – this holds for a.e. $r > 0$, – letting $j \to \infty$ we find

$$\fint_{U_r} \langle \nu_F, \nu_E(0) \rangle d\mu_F = 1.$$

This implies that $\nu_F = \nu_E(0)$ μ_F-a.e. in $\mathbb{H}^n$, because $r > 0$ is otherwise arbitrary. By the characterization of sets with constant horizontal normal (see Remark 5.7 below), we have $F = H_\nu$ with $\nu = \nu(0)$. We are omitting the proof that $0 \in \partial F$. The limit $F = H_\nu$ is thus independent of the sequence $(\lambda_j)_{j\in\mathbb{N}}$ and this observation concludes the proof of (2.9).

We prove (2.10). From

$$\fint_{U_r} \langle \nu, \nu_{E_\lambda}\rangle d\mu_{E_\lambda} = \fint_{U_{r/\lambda}} \langle \nu, \nu_E\rangle d\mu_E = 1 + o(1), \quad \text{as} \quad \lambda \to \infty,$$

we deduce that

$$P(E_\lambda; U_r) = (1 + o(1)) \int_{U_r} \langle \nu, \nu_{E_\lambda}\rangle d\mu_{E_\lambda}.$$

Letting $\lambda \to \infty$, using the weak convergence $\nu_{E_\lambda} d\mu_{E_\lambda} \to \nu_F d\mu_F$ and choosing $r > 0$ with $\mu_F(\partial U_r) = 0$, we get the claim. □

2.2 Equivalent notions for H-perimeter

In this section, we describe some characterizations of H-perimeter related to the metric structure of $\mathbb{H}^n$.

2.2.1 Hausdorff measures The Heisenberg perimeter has a representation in terms of spherical Hausdorff measures. We use the metric ϱ in (1.12). The diameter of a set $K \subset \mathbb{H}^n$ is

$$\operatorname{diam} K = \sup_{p,q\in K} \varrho(p,q).$$

If U_r is a ball in the distance ϱ with radius r, then we have $\operatorname{diam} U_r = 2r$. Let $E \subset \mathbb{H}^n$ be a set. For any $s \geq 0$ and $\delta > 0$ define the premeasures

$$\mathscr{H}_\varrho^{s,\delta}(E) = \inf\Big\{\sum_{i\in\mathbb{N}} (\operatorname{diam} K_i)^s : E \subset \bigcup_{i\in\mathbb{N}} K_i,\ K_i \subset \mathbb{H}^n,\ \operatorname{diam} K_i < \delta\Big\},$$

$$\mathscr{S}_\varrho^{s,\delta}(E) = \inf\Big\{\sum_{i\in\mathbb{N}} (\operatorname{diam} U_i)^s : E \subset \bigcup_{i\in\mathbb{N}} U_i, U_i \ \varrho\text{-balls in } \mathbb{H}^n, \operatorname{diam} U_i < \delta\Big\},$$

Letting $\delta \to 0$, we define

$$\mathscr{H}_\varrho^{s}(E) = \sup_{\delta>0} \mathscr{H}_\varrho^{s,\delta}(E) = \lim_{\delta\to 0} \mathscr{H}_\varrho^{s,\delta}(E),$$
$$\mathscr{S}_\varrho^{s}(E) = \sup_{\delta>0} \mathscr{S}_\varrho^{s,\delta}(E) = \lim_{\delta\to 0} \mathscr{S}_\varrho^{s,\delta}(E).$$

By Carathèodory's construction, $E \mapsto \mathscr{H}_\varrho^s(E)$ and $E \mapsto \mathscr{S}_\varrho^s(E)$ are Borel measures in $\mathbb{H}^n$. The measure $\mathscr{H}_\varrho^s$ is called s-dimensional Hausdorff measure. The measure $\mathscr{S}_\varrho^s$ is called s-dimensional spherical Hausdorff measure. These measures are equivalent, in the sense that for any

$E \subset \mathbb{H}^n$ there holds

$$\mathscr{H}_\varrho^s(E) \leq \mathscr{S}_\varrho^s(E) \leq 2^s \mathscr{H}_\varrho^s(E).$$

The measures $\mathscr{H}_\varrho^Q(E)$ and $\mathscr{S}_\varrho^Q$ are Haar measures in $\mathbb{H}^n$ and therefore they coincide with the Lebesgue measure, up to a multiplicative constant factor. The natural dimension to measure hypersurfaces, as the boundary of smooth sets, is $s = Q - 1$.

The following theorem is proved in [27], Theorem 7.1 part (iii). The proof relies on Federer's differentiation theorems, Theorem 2.10.17 and Theorem 2.10.19 part (3) of [25]. Extensions of this result are based on general differentiation theorems for measures, see [41]. Formula (2.14) for the geometric constant c_n in (2.13) depends on the shape (convexity and symmetries) of the metric unit ball U_1, [42].

Theorem 2.11 (Franchi-Serapioni-Serra Cassano). *For any set $E \subset \mathbb{H}^n$ with locally finite H-perimeter we have*

$$\mu_E = c_n \mathscr{S}_\varrho^{Q-1} \llcorner \partial^* E, \tag{2.13}$$

where μ_E is the perimeter measure of E, $\mathscr{S}_\varrho^{Q-1} \llcorner \partial^* E$ is the restriction of $\mathscr{S}_\varrho^{Q-1}$ to the reduced boundary $\partial^* E$, and the constant $c_n > 0$ is given by

$$c_n = P(H_\nu; U_1). \tag{2.14}$$

Remark 2.12. It is not known whether in (2.13) the spherical measure $\mathscr{S}_\varrho^{Q-1}$ can be replaced by the Hausdorff measure $\mathscr{H}_\varrho^{Q-1}$, even when $\partial^* E$ is a smooth set. In $\mathbb{R}^n$ with the standard perimeter, the identity $\mathscr{S}^{n-1} \llcorner \partial^* E = \mathscr{H}^{n-1} \llcorner \partial^* E$ follows from Besicovitch's covering theorem, that fails to hold in the Heisenberg group, see [36] and [65].

2.2.2 Minkowski content and H-perimeter In the description of H-perimeter in terms of Minkowski content, the correct choice of the metric is the Carnot-Carathéodory distance d on $\mathbb{H}^n$.

The Carnot-Carathéodory distance from a closed set $K \subset \mathbb{H}^n$ is the function

$$\text{dist}_K(p) = \min_{q \in K} d(p, q), \qquad p \in \mathbb{H}^n.$$

For $r > 0$, the r-tubular neighborhood of K is the set

$$I_r(K) = \{ p \in \mathbb{H}^n : \text{dist}_K(p) < r \}.$$

The *upper* and *lower Minkowski content* of K in an open set $A \subset \mathbb{H}^n$ are, respectively,

$$\mathcal{M}^+(K; A) = \limsup_{r \to 0} \frac{|I_r(K) \cap A|}{2r},$$
$$\mathcal{M}^-(K; A) = \liminf_{r \to 0} \frac{|I_r(K) \cap A|}{2r}.$$

Above, $|\cdot|$ stands for Lebesgue measure. If $\mathcal{M}^+(K; A) = \mathcal{M}^-(K; A)$, the common value is called *Minkowski content* of K in A and it is denoted by $\mathcal{M}(K; A)$.

Below, $\mathcal{H}^{2n}$ is the standard $2n$-dimensional Hausdorff measure in $\mathbb{H}^n = \mathbb{R}^{2n+1}$.

Theorem 2.13 (Monti-Serra Cassano). *Let $A \subset \mathbb{H}^n$ be an open set and let $E \subset \mathbb{H}^n$ be a bounded set with C^2 boundary such that $\mathcal{H}^{2n}(\partial E \cap \partial A) = 0$. Then we have*

$$P(E; A) = \mathcal{M}(\partial E; A). \tag{2.15}$$

This result is proved in [54], in a general framework. It is an open problem to prove formula (2.15) for sets E with less regular boundary. The tools used in the proof in [54] are the eikonal equation for the Carnot-Carathéodory distance and the coarea formula. Assume $A = \mathbb{H}^n$. We have

$$|I_r(\partial E)| = \int_{I_r(\partial E)} |\nabla_H \mathrm{dist}_{\partial E}(z, t)| dz dt,$$

because $|\nabla_H \mathrm{dist}_K(z, t)| = 1$ a.e. in $\mathbb{H}^n$. By the coarea formula in the sub-Riemannian setting, we have

$$\int_{I_r(\partial E)} |\nabla_H \mathrm{dist}_{\partial E}(z, t)| dzt = \int_0^r P(I_s(\partial E); \mathbb{H}^n) ds.$$

We refer the reader to [54] and [40] for a discussion on coarea formulas. Now formula (2.15) follows proving that

$$\lim_{r \to 0} \frac{1}{2r} \int_0^r P(I_s(\partial E); \mathbb{H}^n) ds = P(E; \mathbb{H}^n).$$

The regularity of ∂E is used at this final step: the Riemannian approximation of the distance function from ∂E is of class C^2, if ∂E is of class C^2.

2.2.3 Integral differential quotients H-perimeter can be also expressed as the limit of certain integral differential quotients.

Let $k_n > 0$ be the following geometric constant

$$k_n = \fint_{B_1} |\langle \nu, z\rangle| dz dt,$$

where $B_1 \subset \mathbb{H}^n$ is the unitary Carnot-Carathéodory ball centered at the origin. By the rotational symmetry of B_1, the definition of k_n is independent of the unit vector $\nu \in \mathbb{R}^{2n}$, $|\nu| = 1$. The following theorem is proved in [62].

Theorem 2.14. *A Borel set $E \subset \mathbb{H}^n$ with finite measure has finite H-perimeter in $\mathbb{H}^n$ if and only if*

$$\liminf_{r\downarrow 0} \frac{1}{r} \int_{\mathbb{H}^n} \fint_{B_r(q)} |\chi_E(p) - \chi_E(q)| \, dp \, dq < \infty.$$

Moreover, if E has also finite Euclidean perimeter then

$$\lim_{r\downarrow 0} \frac{1}{r} \int_{\mathbb{H}^n} \fint_{B_r(q)} |\chi_E(p) - \chi_E(q)| \, dp \, dq = k_n P(E; \mathbb{H}^n). \qquad (2.16)$$

For the proof, we refer to [62], where the result is proved in the setting of BV_H functions. It is an open question whether the identity (2.16) holds dropping the assumption "if E has also finite Euclidean perimeter".

The characterization of H-perimeter in Theorem 2.14 is useful in the theory of rearrangements in the Heisenberg group proposed in [49].

2.3 Rectifiability of the reduced boundary

The reduced boundary of sets with finite H-perimeter needs not be rectifiable in the standard sense. However, it is rectifiable in an intrinsic sense that we are going to explain. The main reference is the paper [27]. A systematic treatment of these topics in the setting of stratified groups can be found in [39].

We need first the notion of C^1_H-regular function.

Definition 2.15 (C^1_H-function). Let $A \subset \mathbb{H}^n$ be an open set. A function $f : A \to \mathbb{R}$ is of class $C^1_H(A)$ if:

1) $f \in C(A)$;
2) the derivatives $X_1 f, \dots, X_n f, Y_1 f, \dots, Y_n f$ in the sense of distributions are (represented by) continuous functions in A.

The horizontal gradient of a function $f \in C^1_H(A)$ is the vector valued mapping $\nabla_H f \in C(A; \mathbb{R}^{2n})$, $\nabla_H f = \big(X_1 f, \dots, X_n f, Y_1 f, \dots, Y_n f\big)$.

For C_H^1-regular functions there is an implicit function theorem (Theorem 6.5 in [27]) that can be used to represent the zero set $\{f = 0\}$ as an "intrinsic Lipschitz graph" (see Section 3.1.4).

Definition 2.16 (H-regular hypersurface). A set $S \subset \mathbb{H}^n$ is an H-regular hypersurface if for all $p \in S$ there exists $r > 0$ and a function $f \in C_H^1(B_r(p))$ such that:

1) $S \cap B_r(p) = \{q \in B_r(p) : f(q) = 0\}$;
2) $|\nabla_H f(p)| \neq 0$.

If $S \subset \mathbb{H}^n$ is a hypersurface of class C^1 in the standard sense, then for any $p \in S$ there exist $r > 0$ and a function $f \in C^1(B_r(p))$ such that $S \cap B_r(p) = \{q \in B_r(p) : f(q) = 0\}$ and $|\nabla f(p)| \neq 0$. However, the set S needs not be an H-regular hypersurface because it may happen that $|\nabla_H f(p)| = 0$ at some (many) points $p \in S$. On the other hand, the following theorem, proved in [35] Theorem 3.1, shows that, in general, H-regular hypersurfaces are not rectifiable.

Theorem 2.17 (Kirchheim-Serra Cassano). *There exists an H-regular surface $S \subset \mathbb{H}^1$ such that*

$$\mathcal{H}^{(5-\epsilon)/2}(S) > 0 \quad \text{for all } \epsilon \in (0,1).$$

In particular, the set S is not 2-rectifiable.

Above, $\mathcal{H}^s$ is the standard s-dimensional Hausdorff measure in $\mathbb{R}^3$. The set S constructed in [35] has Euclidean Hausdorff dimension $5/2$. Any H-regular surface $S \subset \mathbb{H}^1$ can be locally parameterized by a $1/2$-Hölder continuous map $\Phi : \mathbb{R}^2 \to \Phi(\mathbb{R}^2) = S \subset \mathbb{H}^1$, i.e., $d(\Phi(u), \Phi(v)) \leq C|u - v|^{1/2}$ for $u, v \in \mathbb{R}^2$, where $C > 0$ is a constant and d is the Carnot-Carathéodory distance, see Theorem 4.1 in [35].

Definition 2.18. A set $\Gamma \subset \mathbb{H}^n$ is $\mathscr{S}_\varrho^{Q-1}$-rectifiable if there exists a sequence of H-regular hypersurfaces $(S_j)_{j \in \mathbb{N}}$ in $\mathbb{H}^n$ such that

$$\mathscr{S}_\varrho^{Q-1}\left(\Gamma \setminus \bigcup_{j \in \mathbb{N}} S_j\right) = 0.$$

This definition is generalized in [43], where the authors study the notion of a s-rectifiable set in $\mathbb{H}^n$ for any integer $1 \leq s \leq Q - 1$. The definition of s-rectifiability has a different nature according to whether $s \leq n$ or $s \geq n + 1$. Definition 2.18 is relevant because the reduced boundary of sets with finite H-perimeter is rectifiable precisely in this sense. The following theorem is the main result of [27].

Theorem 2.19. *Let $E \subset \mathbb{H}^n$ be a set with locally finite H-perimeter. Then the reduced boundary $\partial^* E$ is $\mathscr{S}_{\varrho}^{Q-1}$-rectifiable.*

The proof of Theorem 2.19 goes as follows, for details see Theorem 7.1 in [27]. By Lusin's theorem there are compact sets $K_j \subset \partial^* E$, $j \in \mathbb{N}$, and a set $N \subset \partial^* E$ such that:

i) $\mu_E(N) = 0$;
ii) $\nu_E : K_j \to \mathbb{R}^{2n}$ is continuous, for each $j \in \mathbb{N}$;
iii) $\partial^* E = N \cup \bigcup_{j \in \mathbb{N}} K_j$.

By a Whitney extension theorem (Theorem 6.8 in [27]), it is possible to construct functions $f_j \in C_H^1(\mathbb{H}^n)$ such that $\nabla_H f_j = \nu_E$ and $f_j = 0$ on K_j. Then the sets $S_j = \{f_j = 0\}$ are H-regular hypersurfaces near K_j and $K_j \subset S_j$.

3   Area formulas, first variation and H-minimal surfaces

3.1   Area formulas

In this section, we derive some area formulas for H-perimeter of sets with regular boundary. In particular, we study sets with Euclidean Lipschitz boundary and sets with "intrinsic Lipschitz boundary".

3.1.1   Sets with Lipschitz boundary   Let $E \subset \mathbb{H}^n$ be a set with Lipschitz boundary and denote by N the Euclidean outer unit normal to ∂E. This vector is defined at $\mathscr{H}^{2n}$-a.e. point of ∂E. Here and hereafter, $\mathscr{H}^{2n}$ denotes the standard $2n$-dimensional Hausdorff measure of $\mathbb{R}^{2n+1}$. Using the projections of $X_1, \ldots, X_n, Y_1, \ldots, Y_n$ along the normal N, we can define the $2n$-dimensional vector field $N_H : \partial E \to \mathbb{R}^{2n}$

$$N_H = \big(\langle X_1, N \rangle, \ldots, \langle X_n, N \rangle, \langle Y_1, N \rangle \ldots, \langle Y_n, N \rangle \big), \qquad (3.1)$$

where the vector fields X_j, Y_j and N are identified with elements of $\mathbb{R}^{2n+1}$ and $\langle \cdot, \cdot \rangle$ is the standard scalar product.

Proposition 3.1. *Let $E \subset \mathbb{H}^n$ be a set with Lipschitz boundary. Then the H-perimeter of E in an open set $A \subset \mathbb{H}^n$ is*

$$P(E; A) = \int_{\partial E \cap A} |N_H| d\mathscr{H}^{2n}, \qquad (3.2)$$

where N is the Euclidean (outer) unit normal to ∂E and $|N_H|$ is the Euclidean norm of N_H.

Proof. For any $\varphi \in C^1_c(A;\mathbb{R}^{2n})$ let $V = \sum_{j=1}^n \varphi_j X_j + \varphi_{n+j} Y_j$ be the horizontal vector field with coordinates φ. By the standard divergence theorem and the Cauchy-Schwarz inequality, we have

$$\begin{aligned}\int_E \mathrm{div}_H\varphi\, dzdt &= \int_E \mathrm{div} V\, dzdt = \int_{\partial E} \langle V, N\rangle d\mathscr{H}^{2n}\\ &= \int_{\partial E} \sum_{j=1}^n \varphi_j\langle X_j, N\rangle + \varphi_{n+j}\langle Y_j, N\rangle d\mathscr{H}^{2n}\\ &\le \int_{\partial E} \sum_{j=1}^n |\varphi||N_H| d\mathscr{H}^{2n},\end{aligned}$$

and taking the supremum with $\|\varphi\|_\infty \le 1$ it follows that $P(E;A) \le \int_{\partial E\cap A} |N_H| d\mathscr{H}^{2n}$.

The opposite inequality can be obtained by approximation. By Lusin's theorem, for any $\epsilon > 0$ there exists a compact set $K \subset \partial E \cap A$ such that

$$\int_{(\partial E\setminus K)\cap A} |N_H| d\mathscr{H}^{2n} < \epsilon,$$

and $N_H : K \to \mathbb{R}^{2n}$ is continuous and nonzero. By Tietze's theorem, there exists $\psi \in C_c(A;\mathbb{R}^{2n})$ such that $\|\psi\|_\infty \le 1$ and $\psi = N_H/|N_H|$ on K. By mollification there exists $\varphi \in C^1_c(A;\mathbb{R}^{2n})$ such that $\|\varphi\|_\infty \le 1$ and $\|\psi - \varphi\|_\infty < \epsilon$. For such a test function φ we have

$$\int_E \mathrm{div}_H\varphi\, dzdt \ge (1-\epsilon)\int_{\partial E\cap A} |N_H| d\mathscr{H}^{2n} - 2\epsilon.$$

This ends the proof. □

3.1.2 Formulas for the horizontal inner normal Let $E \subset \mathbb{H}^n$ be a set with Lipschitz boundary and let $\varphi \in C^1_c(\mathbb{H}^n;\mathbb{R}^{2n})$. From the Gauss-Green formula (2.6) and from the standard divergence theorem, we have

$$\int_{\partial E} \langle\varphi, N_H\rangle d\mathscr{H}^{2n} = \int_E \mathrm{div}_H\varphi\, dzdt = -\int_{\mathbb{H}^n} \langle\varphi, \nu_E\rangle d\mu_E.$$

It follows that the perimeter measure has the following representation

$$\mu_E = |N_H|\mathscr{H}^{2n} \llcorner \partial E,$$

and the measure theoretic inner normal is

$$\nu_E = -\frac{N_H}{|N_H|} \quad \mu_E\text{-a.e. on } \partial E. \tag{3.3}$$

Next, we express ν_E in terms of a defining function for the boundary. Assume that ∂E is a C^1-surface and $f \in C^1(A)$ is a defining function for ∂E, i.e., $\partial E \cap A = \{p \in A : f(p) = 0\}$ with $|\nabla f| \neq 0$ and $f < 0$ inside E. Then the outer Euclidean normal to ∂E is

$$N = \frac{\nabla f}{|\nabla f|} \quad \text{on } \partial E \cap A,$$

and therefore the vector N_H introduced in (3.1) is

$$N_H = \frac{\nabla_H f}{|\nabla f|} \quad \text{on } \partial E \cap A.$$

From (3.3), we conclude that the horizontal inner normal is given by

$$\nu_E = -\frac{\nabla_H f}{|\nabla_H f|} \quad \text{on } \partial E \cap A, \ |\nabla_H f| \neq 0. \tag{3.4}$$

Let N_E be the horizontal vector with coordinates ν_E in the basis $X_1, \dots, X_n, Y_1, \dots, Y_n$. The vector N_E can be recovered in the following way. Fix on $\mathbb{H}^n$ the Riemannian metric making $X_1, \dots, X_n, Y_1, \dots, Y_n$, T orthonormal. The Riemannian exterior normal to the surface $\{f = 0\}$ is the vector

$$N_R = \frac{\nabla_R f}{|\nabla_R f|_R}.$$

where $\nabla_R f = \sum_{j=1}^n (X_j f) X_j + (Y_j f) Y_j + (Tf)T$ is the gradient of f and $|\nabla_R f|_R$ is its Riemannian length. Let $\pi_R : T_p\mathbb{H}^n \to H_p$ be the orthogonal projection onto the horizontal plane. Then the vector N_E is precisely

$$N_E = \frac{\pi_R(N_R)}{|\pi_R(N_R)|_R}.$$

3.1.3 Area formula for t-graphs We specialize formula (3.2) to the case of t-graphs. Let $D \subset \mathbb{R}^{2n} = \mathbb{C}^n$ be an open set and let $f : D \to \mathbb{R}$ be a function. The set $E_f = \{(z,t) \in \mathbb{H}^n : t > f(z),\ z \in D\}$ is called t-epigraph of f. The set $\mathrm{gr}(f) = \{(z,t) \in \mathbb{H}^n : t = f(z),\ z \in D\}$ is called t-graph of f.

Proposition 3.2 (Area formula for t-graphs). *Let $D \subset \mathbb{R}^{2n}$ be an open set and let $f : D \to \mathbb{R}$ be a Lipschitz function. Then we have*

$$P(E_f; D \times \mathbb{R}) = \int_D |\nabla f(z) + 2z^\perp| dz, \tag{3.5}$$

where $z^\perp = (x,y)^\perp = (-y, x)$.

Proof. The outer normal to $\partial E_f \cap (D \times \mathbb{R}) = \mathrm{gr}(f)$ is $N = (\nabla f, -1)/\sqrt{1+|\nabla f|^2}$, and so, for any $j = 1, \ldots, n$, we have

$$\langle N, X_j \rangle = \frac{\partial_{x_j} f - 2y_j}{\sqrt{1+|\nabla f|^2}}, \qquad \langle N, Y_j \rangle = \frac{\partial_{y_j} f + 2y_j}{\sqrt{1+|\nabla f|^2}},$$

and thus

$$|N_H| = \frac{|\nabla f + 2z^{\perp}|}{\sqrt{1+|\nabla f|^2}}.$$

By formula (3.2) and by the standard area formula for graphs, we obtain

$$P(E_f; D \times \mathbb{R}) = \int_{\mathrm{gr}(f)} |N_H| d\mathscr{H}^{2n} = \int_D |\nabla f(z) + 2z^{\perp}| dz \qquad \square$$

The area formula (3.5) is the starting point of many investigations on H-minimal surfaces. Epigraphs of the form $E_f = \{t > f(z)\}$ are systematically studied in [71]. In particular, in Theorem 3.2 of [71] the authors compute the relaxed functional in $L^1(D)$ of the area functional $\mathscr{A}\colon C^1(D) \to [0, \infty]$

$$\mathscr{A}(f) = \int_D |\nabla f(z) + 2z^{\perp}| dz.$$

They also prove existence of minimizers with a trace constraint when D is a bounded open set with Lipschitz boundary (Theorem 1.4) and they show that minimizers are locally bounded (Theorem 1.5). The Lipschitz regularity of minimizers under the bounded slope condition is proved in [63].

3.1.4 Area formula for intrinsic graphs Let $S \subset \mathbb{H}^n$ be a C^1_H-regular hypersurface. Then we have $S = \{f = 0\}$ with $f \in C^1_H$ satisfying $|\nabla_H f| \neq 0$. Up to a change of coordinates, we can assume that locally we have $X_1 f > 0$. Then each integral line of X_1 meets S in one single point: S is a graph along X_1. These considerations lead to the following definitions.

The line flow of the vector field X_1 starting from the point $(z, t) \in \mathbb{H}^n$ is

$$\exp(sX_1)(z, t) = (z + s\mathrm{e}_1, t + 2y_1 s), \quad s \in \mathbb{R},$$

where $\mathrm{e}_1 = (1, 0, \ldots, 0) \in \mathbb{R}^{2n}$ and $z = (x, y) \in \mathbb{C}^n = \mathbb{R}^{2n}$, with $x = (x_1, \ldots, x_n)$ and $y = (y_1, \ldots, y_n)$.

We fix a domain of initial points. The most natural choice is to consider the vertical hyperplane $W = \{(z, t) \in \mathbb{H}^n : x_1 = 0\}$, that is identified with $\mathbb{R}^{2n}$ with the coordinates $w = (x_2, \ldots, x_n, y_1, \ldots, y_n, t)$.

Definition 3.3 (Intrinsic epigraph and graph). Let $D \subset W$ be a set and let $\varphi : D \to \mathbb{R}$ be a function. The set

$$E_\varphi = \{\exp(sX_1)(w) \in \mathbb{H}^n \ : \ s > \varphi(w),\ w \in D\}$$

is called *intrinsic epigraph* of φ along X_1. The set

$$\mathrm{gr}(\varphi) = \{\exp(\varphi(w)X_1)(w) \in \mathbb{H}^n \ : \ w \in D\} \tag{3.6}$$

is called *intrinsic graph* of φ along X_1.

In Definition 3.8, there is an equivalent point of view on intrinsic graphs.

We are going to introduce a nonlinear gradient for functions $\varphi : D \to \mathbb{R}$. First, let us introduce the Burgers' operator $\mathscr{B} : \mathrm{Lip}_{\mathrm{loc}}(D) \to L^\infty_{\mathrm{loc}}(D)$

$$\mathscr{B}\varphi = \frac{\partial \varphi}{\partial y_1} - 4\varphi \frac{\partial \varphi}{\partial t}. \tag{3.7}$$

Next, notice that the vector fields $X_2, \dots, X_n, Y_2, \dots, Y_n$ can be naturally restricted to W.

Definition 3.4 (Intrinsic gradient). The *intrinsic gradient* of a function $\varphi \in \mathrm{Lip}_{\mathrm{loc}}(D)$ is the vector valued mapping $\nabla^\varphi \varphi \in L^\infty_{\mathrm{loc}}(D; \mathbb{R}^{2n-1})$

$$\nabla^\varphi \varphi = (X_2\varphi, \dots, X_n\varphi, \mathscr{B}\varphi, Y_2\varphi, \dots, Y_n\varphi).$$

When $n = 1$, the definition reduces to $\nabla^\varphi \varphi = \mathscr{B}\varphi$.

With abuse of notation, we define the cylinder over $D \subset W$ along X_1 as the set

$$D \cdot \mathbb{R} = \{\exp(sX_1)(w) \in \mathbb{H}^n \ : \ w \in D \text{ and } s \in \mathbb{R}\}.$$

When $D \subset W$ is open, the cylinder $D \cdot \mathbb{R}$ is an open set in $\mathbb{H}^n$. The general version of the following proposition is presented in Theorem 3.9.

Proposition 3.5. *Let $D \subset W$ be an open set and let $\varphi : D \to \mathbb{R}$ be a Lipschitz function. Then the H-perimeter of the intrinsic epigraph E_φ in the cylinder $D \cdot \mathbb{R}$ is*

$$P(E_\varphi; D \cdot \mathbb{R}) = \int_D \sqrt{1 + |\nabla^\varphi \varphi|^2}\, dw, \tag{3.8}$$

where dw is the Lebesgue measure in $\mathbb{R}^{2n}$.

Proof. We prove the claim in the case $n = 1$. The intrinsic graph mapping $\Phi : D \to \mathbb{H}^1$ is $\Phi(y,t) = \exp(\varphi(y,t)X)(0,y,t) = (\varphi, y, t+2y\varphi)$, and thus

$$\Phi_y \wedge \Phi_t = \begin{vmatrix} \mathrm{e}_1 & \mathrm{e}_2 & \mathrm{e}_3 \\ \varphi_y & 1 & 2\varphi + 2\varphi_y \\ \varphi_t & 0 & 1 + 2y\varphi_t \end{vmatrix} = (1 + 2y\varphi_t)\mathrm{e}_1 + (2\varphi\varphi_t - \varphi_y)\mathrm{e}_2 - \varphi_t \mathrm{e}_3.$$

The Euclidean outer normal to the intrinsic graph $\partial E_\varphi \cap (D \cdot \mathbb{R})$ is the vector $N = -\Phi_y \wedge \Phi_t / |\Phi_y \wedge \Phi_t|$ and thus

$$\langle N, X\rangle = \frac{-1}{|\Phi_y \wedge \Phi_t|} \quad \text{and} \quad \langle N, Y\rangle = \frac{\varphi_y - 4\varphi\varphi_t}{|\Phi_y \wedge \Phi_t|} = \frac{\mathscr{B}\varphi}{|\Phi_y \wedge \Phi_t|}.$$

From formula (3.2) and from the standard area formula for graphs, we obtain

$$\begin{aligned} P(E_\varphi; D \cdot \mathbb{R}) &= \int_{\partial E_\varphi \cap D \cdot \mathbb{R}} |N_H| d\mathscr{H}^2 \\ &= \int_D \sqrt{\frac{1}{|\Phi_y \wedge \Phi_t|^2} + \frac{(\mathscr{B}\varphi)^2}{|\Phi_y \wedge \Phi_t|^2}} |\Phi_y \wedge \Phi_t| dy dt \\ &= \int_D \sqrt{1 + |\nabla^\varphi \varphi|^2} dy dt. \end{aligned}$$

□

The area formula (3.8) was originally proved for boundaries that are C^1_H-regular hypersurfaces (see [27] Theorem 6.5 part (vi) and [4] Proposition 2.22). It was later generalized to *intrinsic Lipschitz* graphs.

Definition 3.6. Let $D \subset W = \mathbb{R}^{2n}$ be an open set and let $\varphi \in C(D)$ be a continuous function.

i) We say that $\mathscr{B}\varphi$ exists in the sense of distributions and is represented by a locally bounded function, $\mathscr{B}\varphi \in L^\infty_{\mathrm{loc}}(D)$, if there exists a function $\psi \in L^\infty_{\mathrm{loc}}(D)$ such that for all $\vartheta \in C^1_c(D)$ there holds

$$\int_D \vartheta\, \psi\, dw = -\int_D \Big\{\varphi \frac{\partial \vartheta}{\partial y_1} - 2\varphi^2 \frac{\partial \vartheta}{\partial t}\Big\} dw.$$

ii) We say that the intrinsic gradient $\nabla^\varphi \varphi \in L^\infty_{\mathrm{loc}}(D; \mathbb{R}^{2n-1})$ exists in the sense of distributions if $X_1\varphi, \ldots, X_n\varphi, \mathscr{B}\varphi, Y_2\varphi, \ldots, Y_n\varphi$ are represented by locally bounded functions in D.

We introduce intrinsic Lipschitz graphs along any direction. Theorem 3.9 below relates such graphs to the boundedness of the intrinsic gradient $\nabla^\varphi\varphi$.

Let $\nu \in \mathbb{R}^{2n}$, $|\nu| = 1$, be a unit vector that is identified with $(\nu, 0) \in \mathbb{H}^n$. For any $p \in \mathbb{H}^n$, we let $\nu(p) = \langle p, \nu\rangle\nu \in \mathbb{H}^n$ and we define $\nu^\perp(p) \in \partial H_\nu \subset \mathbb{H}^n$ as the unique point such that

$$p = \nu^\perp(p) \cdot \nu(p). \tag{3.9}$$

Recall that $\|\cdot\|_\infty$ is the box-norm introduced in (1.11).

Definition 3.7 (Intrinsic cones). i) The (open) cone with vertex $0 \in \mathbb{H}^n$, axis $\nu \in \mathbb{R}^{2n}$, $|\nu| = 1$, and aperture $\alpha \in (0, \infty]$ is the set

$$C(0, \nu, \alpha) = \big\{p \in \mathbb{H}^n : \|\nu^\perp(p)\|_\infty < \alpha\|\nu(p)\|_\infty\big\}. \tag{3.10}$$

ii) The cone with vertex $p \in \mathbb{H}^n$, axis $\nu \in \mathbb{R}^{2n}$, and aperture $\alpha \in (0, \infty]$ is the set $C(p, \nu, \alpha) = p \cdot C(0, \nu, \alpha)$.

Definition 3.8 (Intrinsic Lipschitz graphs). Let $D \subset \partial H_\nu$ be a set and let $\varphi : D \to \mathbb{R}$ be a function.

i) The *intrinsic graph* of φ is the set

$$\mathrm{gr}(\varphi) = \big\{p \cdot \varphi(p)\nu \in \mathbb{H}^n : p \in D\big\}. \tag{3.11}$$

ii) The function φ is *L-intrinsic Lipschitz* if there exists $L \geq 0$ such that for any $p \in \mathrm{gr}(\varphi)$ there holds

$$\mathrm{gr}(\varphi) \cap C(p, \nu, 1/L) = \emptyset. \tag{3.12}$$

When $\nu = \mathrm{e}_1$, the definition in (3.11) reduces to the definition in (3.6). Namely, let $\varphi : D \to \mathbb{R}$ be a function with $D \subset W = \{x_1 = 0\}$. For any $w \in D$, we have the identity

$$\exp(\varphi(w)X_1)(w) = w \cdot (\varphi(w)\mathrm{e}_1),$$

where $\varphi(w)\mathrm{e}_1 = (\varphi(w), 0\ldots, 0) \in \mathbb{H}^n$. Then the intrinsic graph of φ is the set

$$\mathrm{gr}(\varphi) = \big\{w \cdot (\varphi(w)\mathrm{e}_1) \in \mathbb{H}^n \ : \ w \in D\big\}.$$

The notion of intrinsic Lipschitz function of Definition 3.8 is introduced in [30]. The cones (3.10) are relevant in the theory of H-convex sets [5]. The following theorem is the final result of many contributions.

Theorem 3.9. *Let $\nu = e_1$, $D \subset \partial H_\nu$ be an open set, and $\varphi : D \to \mathbb{R}$ be a continuous function. The following statements are equivalent:*

A) *We have $\nabla^\varphi \varphi \in L^\infty_{\mathrm{loc}}(D; \mathbb{R}^{2n-1})$.*

B) *For any $D' \Subset D$, the function $\varphi : D' \to \mathbb{R}$ is intrinsic Lipschitz.*

Moreover, if A) *or* B) *holds then the intrinsic epigraph $E_\varphi \subset \mathbb{H}^n$ has locally finite H-perimeter in the cylinder $D \cdot \mathbb{R}$, the inner horizontal normal to ∂E_φ is*

$$\nu_{E_\varphi}(w \cdot \varphi(w)) = \left(\frac{1}{\sqrt{1 + |\nabla^\varphi \varphi(w)|^2}}, \frac{-\nabla^\varphi \varphi(w)}{\sqrt{1 + |\nabla^\varphi \varphi(w)|^2}} \right), \quad \text{for } \mathscr{L}^{2n}\text{-a.e. } w \text{ on } D, \tag{3.13}$$

and, for any $D' \subset D$, we have

$$P(E_\varphi; D' \cdot \mathbb{R}) = \int_{D'} \sqrt{1 + |\nabla^\varphi \varphi|^2} dw = c_n \mathscr{S}_Q^{Q-1}(\mathrm{gr}(\varphi) \cap D' \cdot \mathbb{R}). \tag{3.14}$$

The equivalence between A) and B) is a deep result that is proved in [7], Theorem 1.1. Formula (3.13) for the normal and the area formula (3.14) are proved in [16] Corollary 4.2 and Corollary 4.3, respectively. A related result can be found in [56], where it is proved that if $E \subset \mathbb{H}^n$ is a set with finite H-perimeter having controlled normal ν_E, say $\langle \nu_E, e_1 \rangle \geq k > 0$ μ_E-a.e., then the reduced boundary $\partial^* E$ is an intrinsic Lipschitz graph along X_1.

3.2 First variation and H-minimal surfaces

In this section, we deduce the minimal surface equation for H-minimal surfaces in the special but important case of t-graphs. We show that H-minimal surfaces in $\mathbb{H}^1$ are ruled surfaces. These facts have been observed by several authors.

In Section 3.2.2, we review some results established in [12] and [14] about the characteristic set of surfaces in $\mathbb{H}^1$ with "controlled curvature", see Theorem 3.15 below.

3.2.1 First variation of the area for t-graphs Let $D \subset \mathbb{R}^{2n}$ be an open set and let $f \in C^2(D)$ be a function. Assume that the t-epigraph of f, the set

$$E = \{(z, t) \in \mathbb{H}^n : t > f(z),\ z \in D\},$$

is H-perimeter minimizing in the cylinder $A = D \times \mathbb{R}$. This means that if $F \subset \mathbb{H}^n$ is a set such that $E \Delta F \Subset A$ then $P(E; A) \leq P(F; A)$. Here and in the following, $E\Delta F = E \setminus F \cup F \setminus E$ denotes the symmetric difference of sets.

Let $\Sigma(f) = \{z \in D : \nabla f(z) + 2z^\perp = 0\}$ be the *characteristic set of* f. At points $p = (z, f(z)) \in \partial E$ with $z \in \Sigma(f)$ we have $T_p\partial E = H_p$, the horizontal plane and the tangent plane to ∂E at p coincide. These points are called *characteristic points* of the surface $S = \partial E$. The set of characteristic points of S is denoted by $\Sigma(S)$.

By the area formula (3.5), we have

$$P(E; A) = \int_D |\nabla f(z) + 2z^\perp| dz = \int_{D\setminus\Sigma(f)} |\nabla f + 2z^\perp| dz.$$

By the minimality of E, for any $\epsilon \in \mathbb{R}$ and $\varphi \in C_c^\infty(D)$ we have

$$\begin{aligned}\int_{D\setminus\Sigma(f)} |\nabla f + 2z^\perp| dz &\leq \int_D |\nabla f + \epsilon\nabla\varphi + 2z^\perp| dz \\ &= \int_{D\setminus\Sigma(f)} |\nabla f + \epsilon\nabla\varphi + 2z^\perp| dz + |\epsilon| \\ &\quad \times \int_{\Sigma(f)} |\nabla\varphi| dz = \psi(\epsilon).\end{aligned}$$

If $f \in C^2$ then $\Sigma(f)$ is (contained in) a C^1 hypersurface of D, see Section 3.2.2, and therefore $|\Sigma(f)| = 0$. If we only have $f \in C^1$, this is no longer true. When $|\Sigma(f)| = 0$, the function ψ is differentiable at $\epsilon = 0$ and the minimality of E implies $\psi'(0) = 0$. We deduce that for any test function φ we have

$$\int_{D\setminus\Sigma(f)} \frac{\langle \nabla f + 2z^\perp, \nabla\varphi\rangle}{|\nabla f + 2z^\perp|} dz = 0.$$

If $\varphi \in C_c^1(D \setminus \Sigma(f))$, we can integrate by parts with no boundary contribution obtaining

$$\int_{D\setminus\Sigma(f)} \operatorname{div}\Big(\frac{\nabla f + 2z^\perp}{|\nabla f + 2z^\perp|}\Big)\varphi dz = 0. \tag{3.15}$$

When the support of φ intersects $\Sigma(f)$, there is a contribution to the first variation due to the characteristic set, see Theorem 3.17. From (3.15), we deduce that the function f satisfies the following partial differential equation

$$\operatorname{div}\Big(\frac{\nabla f + 2z^\perp}{|\nabla f + 2z^\perp|}\Big) = 0 \quad \text{in } D \setminus \Sigma(f). \tag{3.16}$$

This is the *H-minimal surface* equation for f, in the case of t-graphs. It is a degenerate elliptic equation. A solution $f \in C^2(D)$ to (3.16) is calibrated and the epigraph of f is H-perimeter minimizing over the cylinder $D \setminus \Sigma(f) \times \mathbb{R}$

Definition 3.10 (H-curvature and H-minimal graphs). For any $f \in C^2(D)$ and $z \in D \setminus \Sigma(f)$, the number

$$H(z) = \operatorname{div}\Big(\frac{\nabla f(z) + 2z^\perp}{|\nabla f(z) + 2z^\perp|}\Big),$$

is called *H-curvature* of the graph of f at the point $(z, f(z))$. If $H = 0$ we say that $\operatorname{gr}(f)$ is an H-minimal graph (surface).

We specialize the analysis to the dimension $n = 1$, where the minimal surface equation (3.16) has a clear geometric meaning. If $n = 1$, then $\partial E \cap (D \times \mathbb{R}) = \operatorname{gr}(f)$ is a 2-dimensional surface.

At noncharacteristic points $p = (z, f(z)) \in \partial E$ with $z \in D \setminus \Sigma(f)$, we have $\dim(T_p\partial E \cap H_p) = 1$. A section of $T_p\partial E \cap H_p$ is the vector field

$$V = \frac{1}{|\nabla f + 2z^\perp|}\big(- (f_y + 2x)X + (f_x - 2y)Y\big).$$

Let $\gamma : (-\delta, \delta) \to \mathbb{H}^1$, $\delta > 0$, be the curve such that $\gamma(0) = p \in \partial E$ and $\dot\gamma = V(\gamma)$. The curve γ is horizontal because V is horizontal. Moreover, we have $\gamma(t) \in \partial E$ for all $t \in (-\delta, \delta)$ because V is tangent to ∂E.

Consider the vector fields in $D \setminus \Sigma(f)$

$$N_f(z) = \frac{\nabla f + 2z^\perp}{|\nabla f + 2z^\perp|} \quad \text{and} \quad N_f(z)^\perp = \frac{\big(- f_y - 2x, f_x - 2y\big)}{|\nabla f + 2z^\perp|}.$$

The vector field $N_f^\perp$ is the projection of V onto the xy-plane. The horizontal projection of γ, the curve $\kappa = (\gamma_1, \gamma_2)$, satisfies $\kappa(0) = z_0$ and solves the differential equation $\dot\kappa = N_f^\perp(\kappa)$. Then the vector N_f is a normal vector to the curve κ.

Viceversa, let κ be the solution of $\dot\kappa = N_f^\perp(\kappa)$ and $\kappa(0) = z_0$ and let γ be the horizontal lift of κ with $\gamma(0) = p = (z_0, f(z_0)) \in \partial E$. Then γ solves $\dot\gamma = V(\gamma)$ and is contained in ∂E.

We summarize these observations in the following proposition.

Proposition 3.11. *Let $S = \operatorname{gr}(f) \subset \mathbb{H}^1$ be the graph of a function $f \in C^1(D)$. Then:*

1) *The horizontal projection κ of a horizontal curve γ contained in $S \setminus \Sigma(S)$ solves $\dot\kappa = N_f^\perp(\kappa)$.*

2) *The horizontal lift γ of a curve κ solving $\dot\kappa = N_f^\perp(\kappa)$ in $D \setminus \Sigma(f)$ is contained in S, if γ starts from S.*

Now it is straightforward to prove the following result.

Theorem 3.12 (Structure of H-minimal surfaces). *Let $D \subset \mathbb{C}$ be an open set and let $f \in C^2(D)$ be a function such that* $\mathrm{gr}(f)$ *is an H-minimal surface. Then for any $z_0 \in D \setminus \Sigma(f)$ there exists a horizontal line segment contained in* $\mathrm{gr}(f)$ *and passing through* $(z_0, f(z_0))$.

Proof. Let γ be the horizontal curve passing through $p = (z_0, f(z_0))$ and contained in $\mathrm{gr}(f)$. The horizontal projection κ solves $\dot\kappa = N_f(\kappa)^\perp$. The minimal surface equation (3.16) reads

$$\operatorname{div} N_f(z) = 0 \quad \text{in } D \setminus \Sigma(f),$$

where N_f is a unit normal vector field of κ. Thus κ is a curve with curvature 0 and thus it is a line segment. Its horizontal lift is also a line segment. □

Remark 3.13. If $H : D \setminus \Sigma(f) \to \mathbb{R}$ is the H-curvature of the graph of f, then the partial differential equation

$$\operatorname{div}\Big(\frac{\nabla f(z) + 2z^\perp}{|\nabla f(z) + 2z^\perp|}\Big) = H(z), \quad \text{in } D \setminus \Sigma(f) \subset \mathbb{C},$$

implies that an integral curve κ of the vector field $N_f^\perp$ has curvature $H(\kappa)$. When H is a nonzero constant, κ is a circle. This is relevant in the Heisenberg isoperimetric problem.

Equation (3.16) can be given a meaning along integral curves of $N_f^\perp$ without assuming the full C^2 regularity of f, see [13]. See also Section 4.3 for the problem of integrating the H-curvature equation for a convex function f.

3.2.2 Characteristic points Let $D \subset \mathbb{C}^{2n}$ be an open set and let $f \in C^2(D)$. Consider the mapping $\Phi : D \to \mathbb{R}^{2n}$

$$\Phi(z) = \nabla f(z) + 2z^\perp, \quad z \in D.$$

The point $z = x + iy \in \Sigma(f)$ is characteristic if and only if $\Phi(z) = 0$, namely,

$$\begin{cases} \Phi_1(z) = \nabla_x f(z) - 2y = 0 \\ \Phi_2(z) = \nabla_y f(z) + 2x = 0. \end{cases}$$

If $z_0 \in \Sigma(f)$ is a point such that $\det(J\Phi(z_0)) \neq 0$ then Φ is a local C^1 diffeomorphism at z_0 and thus z_0 is an isolated point of $\Sigma(f)$.

In general, for any $z_0 \in \Sigma(f)$ there exists $\epsilon > 0$ such that $\Sigma(f) \cap \{|z - z_0| < \epsilon\}$ is *contained* in the graph of a C^1 function. For instance, in the case $n = 1$ we have

$$|\partial_y \Phi_1(z)| + |\partial_x \Phi_2(z)| = |f_{xy}(z) - 2| + |f_{xy}(z) + 2| \neq 0,$$

and the claim follows from the implicit function theorem. We used the C^2 regularity of f to have equality of mixed derivatives $f_{xy} = f_{yx}$.

When f is less than C^2-regular, the characteristic set $\Sigma(f)$ may be large.

Theorem 3.14 (Balogh). *Let $D = (0,1) \times (0,1) \subset \mathbb{R}^2$ be the square. For any $\epsilon > 0$ there exists a function $f \in \bigcap_{0<\alpha<1} C^{1,\alpha}(D)$ such that $|\Sigma(f)| > 1 - \epsilon$.*

This theorem is proved in [6] by the following construction. Given a continuous mapping $F : D \to \mathbb{R}^2$ one has to find a function such that $\nabla f = F$ on a large subset of D. The construction starts from a Cantor type subset of D with large measure. The function f is defined in a recursive way starting from suitable means of F in the subsquares of D generating the Cantor set.

The following theorem, proved in [14], shows that if H-curvature is suitably bounded near characteristic points then $\Sigma(f)$ consists, for $n = 1$, either of isolated points or, locally, of C^1 graphs over intervals. Generalizations to the case $f \in C^1(D)$, with some further technical assumptions, are given in [15]. For surfaces of class C^2 the curvature H needs not be integrable for the standard area element near the characteristic set, see [22].

Theorem 3.15 (Cheng-Hwang-Malchiodi-Yang). *Let $D \subset \mathbb{C}$ be an open set, $f \in C^2(D)$ and $z_0 \in \Sigma(f)$. Assume that:*

1) $\det(J\Phi(z_0)) = 0$.
2) *For any $z \in D \setminus \Sigma(f)$ we have*

$$\mathrm{div}\Big(\frac{\nabla f(z) + 2z^\perp}{|\nabla f(z) + 2z^\perp|}\Big) = H(z), \tag{3.17}$$

where $H : D \setminus \Sigma(f) \to \mathbb{R}$ is a continuous function such that

$$|H(z)| \leq \frac{C}{|z - z_0|}, \quad z \in D \setminus \Sigma(f) \tag{3.18}$$

for some constant $C > 0$.

Then there exists $\epsilon > 0$ such that $\Sigma(f) \cap \{|z - z_0| < \epsilon\}$ is the graph of a C^1 function defined over an open interval.

Proof. Since $\det(J\Phi(z_0)) = 0$ then the Jacobian matrix $J\Phi(z_0)$ has rank at most 1. On the other hand, the antidiagonal of $J\Phi(z_0)$ never vanishes and thus the rank is precisely 1. Up to the sign, there exists a unique unit vector $w \in \mathbb{R}^2$, $|w| = 1$, that is orthogonal to the range of the transposed Jacobian matrix $J\Phi(z_0)^*$.

For $u \in \mathbb{R}^2$, we define the function $\Phi_u : D \to \mathbb{R}$, $\Phi_u = \langle \Phi, u \rangle = u_1(f_x - 2y) + u_2(f_y + 2x)$. If $u \notin \mathrm{Ker}(J\Phi(z_0)^*)$ then

$$\nabla \Phi_u(z_0) = J\Phi(z_0)^* u \neq 0,$$

and thus the equation $\Phi_u = 0$ defines a C^1 curve $\kappa_u : (-s_0, s_0) \to \mathbb{R}^2$, for some $s_0 > 0$, such that $\kappa_u(0) = z_0$ and $\Phi_u(\kappa_u) = 0$. The image of this curve is a graph over an interval. We can assume that $|\dot{\kappa}_u| = 1$. Differentiating $\Phi_u(\kappa_u) = 0$ we obtain $\langle \nabla \Phi_u(\kappa_u), \dot{\kappa}_u \rangle = 0$, and therefore at $s = 0$ we have

$$\langle J\Phi(z_0)^* u, \dot{\kappa}_u(0) \rangle = 0.$$

Then, up to the sign we have $\dot{\kappa}_u(0) = w$. The derivative $\dot{\kappa}_u(0)$ is independent of $u \notin \mathrm{Ker}(J\Phi(z_0)^*)$.

For some small $\epsilon > 0$, we have $\Sigma(f) \cap \{|z - z_0| < \epsilon\} \subset \{\kappa_u(s) \in \mathbb{R}^2 : |s| < s_0\} \cap \{|z - z_0| < \epsilon\}$. We claim that the inclusion is an identity of sets. By contradiction assume that for any $\delta > 0$ there are $0 \le s_1 < s_2 \le \delta$ such that $\kappa_u(s) \notin \Sigma(f)$ for $s_1 < s < s_2$, and $\kappa_u(s_1), \kappa_u(s_2) \in \Sigma(f)$. Without loss of generality, we assume that $s_1 = 0$ and $s_2 = \delta$, where $\delta > 0$ is as small as we wish.

The defining equation $\langle \Phi(\kappa_u), u \rangle = \Phi_u(\kappa_u) = 0$ implies that, for $0 < s < \delta$, the vector

$$N_f(\kappa_u(s)) = \frac{\Phi(\kappa_u(s))}{|\Phi(\kappa_u(s))|} = \pm u^\perp \tag{3.19}$$

is constant, either $+u^\perp$ or $-u^\perp$, where $u^\perp = (-u_2, u_1)$.

There exists a unit vector $v \in \mathbb{R}^2$ such that $v \notin \mathrm{Ker}(J\Phi(z_0)^*)$,

$$\langle u - v, w \rangle \neq 0 \quad \text{and} \quad \langle u + v, w \rangle \neq 0. \tag{3.20}$$

The equation $\Phi_v = 0$ defines a C^1 curve $\kappa_v : (-\bar{s}_0, \bar{s}_0) \to \mathbb{R}^2$ such that $\kappa_v(0) = z_0$, $\dot{\kappa}_v(0) = w$, $|\dot{\kappa}_v| = 1$ and $\Phi_v(\kappa_v) = 0$. There is a number $\bar{\delta} > 0$ such that $\kappa_v(\bar{\delta}) = \kappa_u(\delta)$ and $\kappa_v(s) \notin \Sigma(f)$ for $0 < s < \bar{\delta}$. As above, the equation $\langle \Phi(\kappa_v), v \rangle = \Phi_v(\kappa_v) = 0$ implies that, for $0 < s < \bar{\delta}$, the vector $N_f(\kappa_v(s)) = \pm v^\perp$ is constant.

Let $A \subset \mathbb{R}^2$ be the region enclosed by the curves κ_u restricted to $[0, \delta]$ and κ_v restricted to $[0, \bar{\delta}]$. Integrating the equation (3.17) over A, using the divergence theorem and (3.18), we obtain

$$\begin{aligned}\int_{\partial A} \langle N_f, N\rangle d\mathcal{H}^1 &= \int_A \operatorname{div} N_f(z)\, dz \\ &= \int_A H(z) dz \le C \int_A \frac{1}{|z - z_0|} dz,\end{aligned} \tag{3.21}$$

where N is the exterior normal to ∂A. Namely, along κ_u we have $N = \dot{\kappa}_u^\perp$ and along κ_v we have $N = -\dot{\kappa}_v^\perp$, or viceversa.

Using (3.19), we can compute the integral

$$\begin{aligned}\int_{\kappa_u([0,\delta])} \langle N_f, N\rangle d\mathcal{H}^1 &= \int_0^\delta \langle N_f(\kappa_u(s)), \dot{\kappa}_u^\perp(s)\rangle ds \\ &= \langle \pm u^\perp, \kappa_u(\delta)^\perp - z_0^\perp\rangle \\ &= \langle \pm u, \kappa_u(\delta) - z_0\rangle,\end{aligned}$$

where $\kappa_u(\delta) - z_0 = \delta w + o(\delta)$ as $\delta \to 0$. Analogously, using $\kappa_v(\bar{\delta}) = \kappa_u(\delta)$ we obtain

$$\int_{\kappa_v([0,\bar{\delta}])} \langle N_f, N\rangle d\mathcal{H}^1 = -\int_0^{\bar{\delta}} \langle N_f(\kappa_v(s)), \dot{\kappa}_v^\perp(s)\rangle ds = -\langle \pm v, \kappa_u(\delta) - z_0\rangle,$$

and, therefore, by (3.20) we have for $\delta > 0$ small

$$\Big|\int_{\partial A} \langle N_f, N\rangle d\mathcal{H}^1\Big| \ge \big|\langle u \pm v, \delta w + o(\delta)\big| \ge \frac{\delta}{2} |\langle u \pm v, w\rangle|. \tag{3.22}$$

Fix a parameter $\epsilon > 0$. For $\delta > 0$ small, we have the inclusion $A \subset \{z_0 + rwe^{i\vartheta} \in \mathbb{C} : 0 \le r \le \delta, |\vartheta| \le \epsilon\}$. Using polar coordinates centered at z_0, we find

$$\int_A \frac{1}{|z - z_0|} dz \le 2\epsilon\delta, \tag{3.23}$$

and, from (3.21)-(3.22)-(3.23), we obtain $\frac{\delta}{2}|\langle u \pm v, w\rangle| \le 2\epsilon\delta C$, that is a contradiction if we choose $\epsilon > 0$ such that $4\epsilon C < |\langle u \pm v, w\rangle|$. □

Let $D \subset \mathbb{C}$ be an open set, $f \in C^2(D)$, and assume that $\Sigma(f)$ is a C^1 curve disconnecting D. Then we have the partition

$$D = D^+ \cup D^- \cup \Sigma(f)$$

where $D^+, D^- \subset D$ are disjoint open sets. In [14], Proposition 3.5, it is shown that the vector N_f extends to $\Sigma(f)$ from D^+ and from D^-, separately.

Theorem 3.16. *In the above setting, for any $z_0 \in \Sigma(f)$ the following limits do exist*

$$N_f(z_0)^+ = \lim_{\substack{z\to z_0\\ z\in D^+}} N_f(z),$$
$$N_f(z_0)^- = \lim_{\substack{z\to z_0\\ z\in D^-}} N_f(z),$$

and moreover $N_f(z_0)^+ = -N_f(z_0)^-$.

Proof. Without loss of generality, we assume that $z_0 = 0$. We have either $f_{xy}(0) - 2 \neq 0$ or $f_{xy}(0) + 2 \neq 0$. Assume that $f_{xy}(0) - 2 > 0$. Then $f_x - 2y = 0$ is a defining equation for $\Sigma(f)$ near 0 and $\Sigma(f) = \{(x, \varphi(x)) \in \mathbb{R}^2 : |x| < \delta\}$, where $\varphi \in C^1(-\delta, \delta)$ is such that $\varphi(0) = 0$, and

$$D^+ = \{(x, y) \in D : y > \varphi(x)\} = \{z \in D : f_x(z) - 2y > 0\},$$
$$D^- = \{(x, y) \in D : y < \varphi(x)\} = \{z \in D : f_x(z) - 2y < 0\}.$$

By Cauchy theorem, for any $x \in (-\delta, \delta)$ and for any $y > \varphi(x)$ there exists $\bar\varphi(x) \in (\varphi(x), y)$ such that

$$\frac{f_y(x, y) + 2x}{f_x(x, y) - 2y} = \frac{f_{yy}(x, \bar\varphi(x))}{f_{xy}(x, \bar\varphi(x)) - 2}.$$

When $x \to 0$ and $y \to 0$ we also have $\varphi(\bar x) \to 0$. Then we have

$$\lim_{\substack{z\to 0\\ z\in D^+}} \frac{f_y(z) + 2x}{f_x(z) - 2y} = \frac{f_{yy}(0)}{f_{xy}(0) - 2} = b.$$

Using $f_x(z) - 2y > 0$ on D^+, it follows that

$$N_f(0)^+ = \lim_{\substack{z\to 0\\ z\in D^+}} N_f(z) = \lim_{\substack{z\to 0\\ z\in D^+}} \frac{\nabla f(z) + 2z^\perp}{|\nabla f(z) + 2z^\perp|} = \frac{(1, b)}{\sqrt{1 + b^2}}.$$

An analogous computation using $f_x(z) - 2y < 0$ on D^- shows that

$$N_f(0)^- = \lim_{\substack{z\to 0\\ z\in D^-}} N_f(z) = \lim_{\substack{z\to 0\\ z\in D^-}} \frac{\nabla f(z) + 2z^\perp}{|\nabla f(z) + 2z^\perp|} = -\frac{(1, b)}{\sqrt{1 + b^2}}. \qquad \square$$

For H-minimal graphs, the vectors N_f^+ and N_f^- are tangent to the C^1 curve $\Sigma(f)$. The following theorem and Theorem 3.16 fail when we have only $f \in C^{1,1}$, see Section 5.2.2.

Theorem 3.17. *In the above setting, assume that the epigraph of $f \in C^2(D)$ is H-perimeter minimizing in the cylinder $D \times \mathbb{R}$. Then we have*

$$\langle N_f^+, N\rangle = \langle N_f^-, N\rangle = 0 \quad on\ \Sigma(f),$$

where N is the normal to the C^1 curve $\Sigma(f)$.

Proof. Let $\varphi \in C_c^1(D)$ be a test function and consider the function

$$\psi(\epsilon) = \int_D |\nabla f + \epsilon\nabla\varphi + 2z^\perp|dz, \quad \epsilon \in (-\epsilon_0, \epsilon_0).$$

If the epigraph of f is H-perimeter minimizing then

$$0 = \psi'(0) = \int_D \frac{\langle \nabla f + 2z^\perp, \nabla\varphi\rangle}{|\nabla f + 2z^\perp|}dz.$$

By $|\Sigma(f)| = 0$ and by (3.16), this is equivalent to

$$\int_{D^+} \operatorname{div}\Big(\varphi\frac{\nabla f + 2z^\perp}{|\nabla f + 2z^\perp|}\Big)dz + \int_{D^-} \operatorname{div}\Big(\varphi\frac{\nabla f + 2z^\perp}{|\nabla f + 2z^\perp|}\Big)dz = 0.$$

Denoting by N the exterior unit normal to D^+ along $\Sigma(f)$ and by N_f^+ and N_f^- the traces of N_f onto $\Sigma(f)$ from D^+ and D^-, the divergence theorem gives

$$\begin{aligned} 0 &= \int_{\Sigma(f)} \varphi\langle N, N_f^+\rangle d\mathscr{H}^1 - \int_{\Sigma(f)} \varphi\langle N, N_f^-\rangle d\mathscr{H}^1 \\ &= 2\int_{\Sigma(f)} \varphi\langle N, N_f^+\rangle d\mathscr{H}^1. \end{aligned}$$

In fact, by Theorem 3.16 we have $N_f^- = -N_f^+$. Since φ is arbitrary, we conclude that $\langle N, N_f^+\rangle = 0$ on $\Sigma(f)$. □

3.2.3 First variation of the area functional for intrinsic graphs By (3.14), the H-perimeter of the intrinsic epigraph E_φ along X_1 of an intrinsic Lipschitz function $\varphi : D \to \mathbb{R}$, $D \subset \mathbb{C}^n$ open set, is

$$\mathscr{A}(\varphi) = P(E_\varphi; D \cdot \mathbb{R}) = \int_D \sqrt{1 + |\nabla^\varphi\varphi|^2}dw, \tag{3.24}$$

where $\nabla^\varphi\varphi$ is a distribution represented by $L^\infty(D; \mathbb{R}^{2n-1})$ functions. It is not clear how to compute the first variation of the area functional $\mathscr{A}$ within the class of intrinsic Lipschitz functions. In fact, this class is not a vector space because the Burgers' operator is nonlinear. Even for a smooth function $\psi \in C^\infty(D)$ we have

$$\mathscr{B}(\varphi+\psi) = \varphi_y+\psi_y-4(\varphi+\psi)(\varphi_t+\psi_t) = \mathscr{B}\varphi+\mathscr{B}\psi-4(\varphi\psi_t+\psi\varphi_t),$$

and the distributional derivative φ_t is not represented by an L^∞ function. So, if φ is only intrinsic Lipschitz it may happen that $P(E_{\varphi+\psi}; D\cdot\mathbb{R}) = \infty$ for any small perturbation $\psi \neq 0$. The reason of this phenomenon is that the variation of the intrinsic graph of φ along X_1 is not a contact deformation. On the other hand, if we had $\varphi_t \in L^\infty_{\mathrm{loc}}$ then the intrinsic graph would have the standard Lipschitz regularity.

Assuming the Lipschitz regularity for φ, the first variation for the area functional $\mathscr{A}$ in (3.24), namely the condition

$$\frac{d}{d\epsilon}\mathscr{A}(\varphi+\epsilon\psi) = 0 \quad \text{for any } \psi \in C_c^\infty(D),$$

leads to the following minimal surface equation for a minimizer φ in D:

$$\begin{aligned}\Big(\frac{\partial}{\partial y} - 4\varphi\frac{\partial}{\partial t}\Big)\frac{\mathscr{B}\varphi}{\sqrt{1+|\nabla^\varphi\varphi|^2}} &+ \sum_{j=2}^n X_j\Big(\frac{X_j\varphi}{\sqrt{1+|\nabla^\varphi\varphi|^2}}\Big)\\ &+ Y_j\Big(\frac{Y_j\varphi}{\sqrt{1+|\nabla^\varphi\varphi|^2}}\Big) = 0.\end{aligned} \tag{3.25}$$

This equation, but in a different system of coordinates, is the starting point of the papers [10] and [9], where the authors study the regularity of vanishing viscosity Lipschitz continuous solutions. When $n \geq 2$, vanishing viscosity solutions are C^∞-smooth. When $n = 1$, their intrinsic graph is foliated by horizontal lines.

3.3 First variation along a contact flow

In this section, we present a formula for computing the first variation of H-perimeter for any set with finite H-perimeter. This result can be extended to $\mathscr{S}_\varrho^{Q-1}$-rectifiable sets in the sense of Definition 2.18 and is a joint result with D. Vittone. We give the proof in the smooth case, the technical details for the general case will appear elsewhere. First and second order variation formulas are discussed also in [45], [18], and [31].

Let $A \subset \mathbb{H}^n$ be an open set. A diffeomorphism $\Psi : A \to \mathbb{H}^n$ is said to be a *contact map* if for any $p \in A$ the differential $\Psi_* : T_p\mathbb{H}^n \to T_{\Psi(p)}\mathbb{H}^n$ maps the horizontal space H_p into $H_{\Psi(p)}$:

$$\Psi_*(H_p) = H_{\Psi(p)}, \quad p \in A. \tag{3.26}$$

A one-parameter flow $(\Psi_s)_{s\in\mathbb{R}}$ of diffeomorphisms in $\mathbb{H}^n$ is a *contact flow* if each Ψ_s is a contact map. Contact flows are generated by contact vector fields.

A contact vector field in $\mathbb{H}^n$ is a vector field of the form

$$V_\psi = -4\psi T + \sum_{j=1}^{n} (Y_j\psi)X_j - (X_j\psi)Y_j, \tag{3.27}$$

where $\psi \in C^\infty(\mathbb{H}^n)$ is the *generating function* of the vector field (see [36]). For any compact set $K \subset \mathbb{H}^n$, there exist $\delta = \delta(\psi, K) > 0$ and a flow $\Psi : [-\delta, \delta] \times K \to \mathbb{H}^n$ defined by $\dot{\Psi}(s, p) = V_\psi(\Psi(s, p))$ and $\Psi(0, p) = p$ for any $s \in [-\delta, \delta]$ and $p \in K$. We call Ψ the flow generated by ψ. We also let $\Psi_s = \Psi(s, \cdot)$.

Related to the function ψ, we have, at any point $p \in \mathbb{H}^n$, the real quadratic form $\mathcal{Q}_\psi : H_p \to \mathbb{R}$

$$\begin{aligned}\mathcal{Q}_\psi\Big(\sum_{j=1}^{n} x_jX_j + y_jY_j\Big) = \sum_{i,j=1}^{n} & x_ix_j\, X_jY_i\psi \\ & + x_jy_i\,(Y_iY_j\psi - X_jX_i\psi) \\ & - y_iy_j\, Y_jX_i\psi,\end{aligned} \tag{3.28}$$

where $x_j, y_j \in \mathbb{R}$, and ψ with its derivatives are evaluated at p. In the sequel, we identify a vector $\nu = \nu(p) \in \mathbb{R}^{2n}$, $p \in \mathbb{H}^n$, with the horizontal vector $\sum_{j=1}^{n} \nu_jX_j(p) + \nu_{n+j}Y_j(p)$. The quadratic form $\mathcal{Q}_\psi(\nu)$ is defined accordingly.

Theorem 3.18. *Let $A \subset \mathbb{H}^n$ be an open set and let $\Psi : [-\delta, \delta] \times A \to \mathbb{H}^n$, $\delta = \delta(\psi, A) > 0$, be the flow generated by $\psi \in C^\infty(\mathbb{H}^n)$. Then there exists a constant $C = C(\psi, A) > 0$ such that for any set $E \subset \mathbb{H}^n$ with finite H-perimeter in A we have*

$$\begin{aligned}&\left| P(\Psi_s(E); \Psi_s(A)) - P(E; A) + s\int_A \{4(n+1)T\psi + \mathcal{Q}_\psi(\nu_E)\}d\mu_E \right| \\ &\leq C\, P(E; A)\, s^2\end{aligned} \tag{3.29}$$

for any $s \in [-\delta, \delta]$.

Proof. We prove the theorem when $\partial E \cap A$ is a C^∞ smooth hypersurface. We deduce formula (3.29) from the Taylor expansion for the standard perimeter. Let $E_s = \Psi_s(E)$ and $A_s = \Psi_s(A)$. Then $\partial E_s \cap A_s = \Psi_s(\partial E \cap A)$ is a C^∞ smooth $2n$-dimensional hypersurface. By the area formula (3.2), we have

$$P(E; A) = \int_{\partial E\cap A} K\, d\mathcal{H}^{2n} \quad \text{and} \quad P(E_s; A_s) = \int_{\partial E_s\cap A_s} K_s\, d\mathcal{H}^{2n},$$

where $\mathscr{H}^{2n}$ is the standard $2n$-dimensional Hausdorff measure of $\mathbb{R}^{2n+1}$,

$$K = \Big(\sum_{j=1}^{n} \langle X_j, N\rangle^2 + \langle Y_j, N\rangle^2\Big)^{1/2},$$

$$K_s = \Big(\sum_{j=1}^{n} \langle X_j, N_s\rangle^2 + \langle Y_j, N_s\rangle^2\Big)^{1/2},$$

and N, N_s are the standard Euclidean unit normals to $\partial E \cap A$ and $\partial E_s \cap A_s$, respectively. We fix a coherent orientation.

By the standard Taylor formula for the area, we have

$$\int_{\partial E_s \cap A_s} K_s \, d\mathscr{H}^{2n} = \int_{\partial E \cap A} K_s \circ \Psi_s \, \mathscr{J}\Psi_s \, d\mathscr{H}^{2n}, \tag{3.30}$$

where $\mathscr{J}\Psi_s : \partial E \cap A \to \mathbb{R}$ is the Jacobian determinant of Ψ_s restricted to ∂E:

$$\mathscr{J}\Psi_s = \sqrt{\det\big[J\Psi_s\big|_{\partial E}^{*} \circ J\Psi_s\big|_{\partial E}\big]}. \tag{3.31}$$

This Jacobian determinant has the following first order Taylor expansion in s

$$\mathscr{J}\Psi_s = 1 + s\big(\operatorname{div} V_\psi - \langle (JV_\psi)N, N\rangle\big) + O(s^2) \quad \text{on } \partial E \cap A, \tag{3.32}$$

where $\operatorname{div} V_\psi$ is the standard divergence of the vector field V_ψ generating the flow and JV_ψ is the Jacobian matrix of V_ψ. Here, the vector field V_ψ is identified with the mapping given by the coefficients of V_ψ in the standard basis. The remainder $O(s^2)$ in (3.32) satisfies $|O(s^2)| \le C_1 s^2$ for some constant $C_1 = C_1(\psi, A) > 0$.

We compute the derivative of the function $s \mapsto K_s \circ \Psi_s$. We start from the derivative of $s \mapsto M(s) = N_s(\Psi_s)$. Let us fix a frame $V_1, \dots, V_{2n}$ of orthonormal vector fields (in the standard scalar product) tangent to $\partial E \cap \Omega$. This frame does always exist locally. As the vector fields $J\Psi_s V_1, \dots, J\Psi_s V_{2n}$ are tangent to $\partial E_s \cap \Omega_s$ we can differentiate the identities $\langle J\Psi_s V_i, M(s)\rangle = 0, i = 1, \dots, 2n$. We obtain

$$\langle JV_\psi(\Psi_s)V_i), M(s)\rangle + \langle J\Psi_s V_i, M'(s)\rangle = 0. \tag{3.33}$$

On the other hand, differentiating the identity $|N_s|^2 = 1$ we deduce that $\langle M'(s), N_s(\Psi_s)\rangle = 0$. Using (3.33), we deduce that at the point $s = 0$ we have

$$\begin{aligned} M'(0) &= \sum_{i=1}^{2n} \langle V_i, M'(0)\rangle V_i = -\sum_{i=1}^{2n} \langle (JV_\psi)V_i, N\rangle V_i \\ &= -\sum_{i=1}^{2n} \langle V_i, (JV_\psi)^* N\rangle V_i \\ &= \langle (JV_\psi)^* N, N\rangle N - (JV_\psi)^* N. \end{aligned} \tag{3.34}$$

Using the property of flows, we can repeat the computation for any s and we find the formula

$$M'(s) = \langle (JV_\psi)^* N_s, N_s\rangle N_s - (JV_\psi)^* N_s, \tag{3.35}$$

where the right-hand side is evaluated at Ψ_s.

Now let X be any smooth vector field in $\mathbb{H}^n$ and consider the function $F_X(s) = \langle X, N_s\rangle(\Psi_s)$. The derivative of F_X is

$$F'_X(s) = \langle (JX)V_\psi(\Psi_s), M(s)\rangle + \langle X(\Psi_s), M'(s)\rangle,$$

where JX is the Jacobian matrix of the mapping given by the coefficients of X. We may also use the notation $(JX)V_\psi = V_\psi X$, where V_ψ acts on the coefficients of X. Using (3.35), we obtain

$$\begin{aligned} F'_X(s) &= \langle (JX)V_\psi, N_s\rangle + \big\langle X, \langle (JV_\psi)^* N_s, N_s\rangle N_s - (JV_\psi)^* N_s\big\rangle \\ &= \langle [V_\psi, X], N_s\rangle + \langle (JV_\psi)N_s, N_s\rangle\langle X, N_s\rangle. \end{aligned} \tag{3.36}$$

The right-hand side is evaluated at Ψ_s.

As V_ψ is of the form (3.27), the commutators $[V_\psi, X_j]$ and $[V_\psi, Y_j]$ are horizontal vector fields, i.e., linear combinations of X_i and Y_i. From (3.36) it follows that F'_{X_j} and F'_{Y_j} are homogeneous functions of degree 1 with respect to $\langle X_i, N_s\rangle$ and $\langle Y_i, N_s\rangle$, $i = 1, \ldots, n$.

As Ψ_s is a contact flow, by (3.26) we have $K(p) = 0$ if and only if $K_s(\Psi_s(p)) = 0$. Assuming that $K(p) \neq 0$, we can thus compute the derivative (in the sequel we omit reference to $p \in \partial E \cap A$)

$$\frac{dK_s \circ \Psi_s}{ds} = \frac{1}{K_s} \sum_{j=1}^{n} \langle X_j, N_s\rangle F'_{X_j}(s) + \langle Y_j, N_s\rangle F'_{Y_j}(s), \tag{3.37}$$

and using (3.36) we obtain the formula

$$\begin{aligned} \frac{dK_s \circ \Psi_s}{ds} &= K_s(\langle (JV_\psi)N_s, N_s\rangle \\ &\quad + \frac{1}{K_s} \sum_{j=1}^{n} \Big\langle \langle X_j, N_s\rangle [V_\psi, X_j] + \langle Y_j, N_s\rangle [V_\psi, Y_j], N_s\Big\rangle. \end{aligned} \tag{3.38}$$

The right hand side is evaluated at Ψ_s and it is bounded by K_s. Namely, there exists a constant $C_2 = C_2(\psi, A)$ such that

$$\left|\frac{dK_s \circ \Psi_s}{ds}\right| \le C_2 K_s. \tag{3.39}$$

Then we can interchange integral and derivative in s in the derivative of $P(E_s; A_s)$:

$$\frac{d}{ds}\int_{\partial E\cap A} K_s \circ \Psi_s\, \mathscr{J}\Psi_s\, d\mathscr{H}^{2n} = \int_{\partial E\cap A} \frac{d}{ds}\Big(K_s \circ \Psi_s\, \mathscr{J}\Psi_s\Big)\, d\mathscr{H}^{2n}.$$

A formula for the second derivative of $s \mapsto K_s \circ \Psi_s$ can be obtained starting from (3.37) and using (3.36). We do not compute this formula, here. It suffices to notice that also the second derivative is bounded by K_s, and namely:

$$\left|\frac{d^2 K_s \circ \Psi_s}{ds^2}\right| \le C_3 K_s \tag{3.40}$$

for some $C_3 = C_3(\psi, A) > 0$. This follows again from the formula (3.36). Thus we can differentiate twice in s inside the integral (3.30) defining $P(E_s; A_s)$.

From (3.32) and (3.38), we get the first order Taylor development

$$\begin{aligned} K_s \circ \Psi_s\, \mathscr{J}\Psi_s = K\Big\{1 + s\Big[\mathrm{div}\, V_\psi + \frac{1}{K^2}\sum_{j=1}^{n}\langle N_{X_j}[V_\psi, X_j] \\ + N_{Y_j}[V_\psi, Y_j], N\rangle\Big] + O(s^2)\Big\}, \end{aligned} \tag{3.41}$$

where we let $N_{X_j} = \langle X_j, N\rangle$ and $N_{Y_j} = \langle Y_j, N\rangle$, and $O(s^2)/s^2$ is bounded uniformly in N by some constant $C_4 = C_4(\psi, A) > 0$. Now, using the structure (3.27) of V_ψ, we get

$$\sum_{j=1}^{n}\langle N_{X_j}[V_\psi, X_j] + N_{Y_j}[V_\psi, Y_j], N\rangle = -\mathscr{Q}_\psi\Big(\sum_{j=1}^{n} N_{X_j}X_j + N_{Y_j}Y_j\Big), \tag{3.42}$$

and

$$\mathrm{div}\, V_\psi = -4T\psi + \sum_{j=1}^{n} X_jY_j\psi - Y_jX_j\psi = -4(n+1)T\psi. \tag{3.43}$$

Formula (3.29) follows from (3.30) along with (3.41)–(3.43). □

Remark 3.19. Let $\Gamma \subset \mathbb{H}^n$ be an $\mathscr{S}_\varrho^{Q-1}$-rectifiable set in $\mathbb{H}^n$ in the sense of Definition 2.18. Using the C_H^1-regular surfaces that cover Γ, a unit horizontal normal ν_Γ can be defined $\mathscr{S}_\varrho^{Q-1}$-a.e. on Γ. When Γ is bounded and with finite measure, formula (3.29) reads as follows:

$$\left| \mathscr{S}^{Q-1}(\Psi_s(\Gamma)) - \mathscr{S}^{Q-1}(\Gamma) + s \int_\Gamma \{4(n+1)T\psi + \mathscr{Q}_\psi(\nu_\Gamma)\}d\mathscr{S}^{Q-1} \right|$$
$$\leq C\mathscr{S}^{Q-1}(\Gamma)\, s^2$$

(3.44)

for any $s \in [-\delta, \delta]$, where $\psi \in C^\infty(\mathbb{H}^n)$ is a generating function and $\delta > 0$. The details of the proof of (3.44) will appear elsewhere.

If Γ is locally measure minimizing in an open set $A \subset \mathbb{H}^n$, from (3.44) we deduce the necessary condition

$$\int_\Gamma \{4(n+1)T\psi + \mathscr{Q}_\psi(\nu_\Gamma)\}d\mathscr{S}^{Q-1} = 0$$

for any function $\psi \in C^\infty(A)$.

4 Isoperimetric problem

4.1 Existence of isoperimetric sets and Pansu's conjecture

For a measurable set $E \subset \mathbb{H}^n$ with positive and finite measure, the *isoperimetric quotient* is defined as

$$I(E) = \frac{P(E; \mathbb{H}^n)}{|E|^{(Q-1)/Q}}.$$

The isoperimetric problem consists in minimizing the isoperimetric quotient among all admissible sets

$$C_{\text{isop}} = \inf \{I(E) : E \subset \mathbb{H}^n \text{ measurable set with } 0 < |E| < \infty\}. \quad (4.1)$$

A measurable set $E \subset \mathbb{H}^n$ with $0 < |E| < \infty$ realizing the infimum is called *isoperimetric set*. Isoperimetric sets are defined up to null sets.

If a set E is isoperimetric, then also the left translates $L_p E = p \cdot E$, $p \in \mathbb{H}^n$, are isoperimetric because perimeter and volume are left invariant. Also the dilated sets $\lambda E = \delta_\lambda E$ are isoperimetric, because the isoperimetric quotient is 0-homogeneous, $I(\lambda E) = I(E)$, for any $\lambda > 0$. It follows that the infimum C_{isop} in (4.1) is the infimum of perimeter for fixed volume

$$C_{\text{isop}} = \inf \{P(E; \mathbb{H}^n) : E \subset \mathbb{H}^n \text{ measurable set with } |E| = 1\}. \quad (4.2)$$

Hence, isoperimetric sets are precisely the sets that have least Heisenberg perimeter for given volume.

The infimum in (4.1) is in fact positive, $C_{\mathrm{isop}} > 0$, and we have the isoperimetric inequality

$$P(E; \mathbb{H}^n) \geq C_{\mathrm{isop}} |E|^{\frac{Q-1}{Q}}, \tag{4.3}$$

holding for any measurable set E with finite measure. The constant C_{isop} is the largest constant making true the above inequality (i.e., the *sharp constant*). Isoperimetric sets are precisely the sets for which the inequality (4.3) is an equality.

Inequality (4.3) with a positive nonsharp constant can be obtained by several methods (see, for example, [58], [59], [26], and [33]). The functional analytic proof casts the isoperimetric inequality as a special case of Sobolev-Poincarè inequalities. Indeed, for any $1 \leq p < Q$ there exists a constant $C_{n,p} > 0$ such that

$$C_{p,n} \Big(\int_{\mathbb{H}^n} |u|^{\frac{pQ}{Q-p}} dzdt \Big)^{\frac{Q-p}{pQ}} \leq \Big(\int_{\mathbb{H}^n} |\nabla_H u|^p dzdt \Big)^{1/p} \tag{4.4}$$

for any $u \in C^1_c(\mathbb{H}^n)$. The inequality extends to appropriate Sobolev or BV spaces. The case $p = 1$ is the *geometric case* and reduces to the Heisenberg isoperimetric inequality (4.3). In fact, for the characteristic function of a set $u = \chi_E$ we have

$$\begin{aligned} \int_{\mathbb{H}^n} |\nabla_H u| &= \sup \Big\{ \int_{\mathbb{H}^n} \chi_E \mathrm{div}_H \varphi \, dzdt \; : \; \varphi \in C^1_c(A; \mathbb{R}^{2n}), \; \|\varphi\|_\infty \leq 1 \Big\} \\ &= P(E; \mathbb{H}^n). \end{aligned}$$

Inequality (4.4) can be obtained starting from the potential estimate

$$\begin{aligned} |u(z,t)| &\leq C_n \int_{\mathbb{H}^n} \frac{|\nabla_H u(\zeta, \tau)|}{d((z,t),(\zeta,\tau))^{Q-1}} d\zeta d\tau \\ &= C_n I_{Q-1}(|\nabla_H u|)(z,t), \quad u \in C^1_c(\mathbb{H}^n), \end{aligned}$$

and using the fact that the singular integral operator $I_{Q-1} : L^p(\mathbb{H}^n) \to L^q(\mathbb{H}^n)$ is bounded for $q = pQ/(Q-p)$ and $1 \leq p < Q$.

The existence of isoperimetric sets is established in [38] and follows from a concentration-compactness argument. See also [32] for a proof of existence that avoids to use the concavity of the isoperimetric profile function.

Theorem 4.1 (Leonardi-Rigot). *Let $n \geq 1$. There exists a measurable set $E \subset \mathbb{H}^n$ with $|E| = 1$ realizing the minimum in* (4.2).

Proof. We give a sketch of the proof. Let $(E_j)_{j\in\mathbb{N}}$ be a minimizing sequence of sets for (4.2):

1) $|E_j| = 1$ for all $j \in \mathbb{N}$;
2) $\lim_{j\to\infty} P(E_j; \mathbb{H}^n) = C_{\text{isop}}$.

The key step of the proof is a concentration argument. We claim that there exists an $R > 0$ such that (after a left translation, truncation, and dilation of each E_j) the sequence $(E_j)_{j\in\mathbb{N}}$ can be also assumed to lie in a bounded region. Namely, there exists $R > 0$ such that:

3) $E_j \subset Q_R = \{(z,t) \in \mathbb{H}^n : |x_i|, |y_i|, |t|^2 < R,\ i = 1, \ldots, n\}$ for all $j \in \mathbb{N}$.

Then, by the compactness theorem for $BV_H(Q_R)$ functions (see [33]), there exists a subsequence, still denoted by $(E_j)_{j\in\mathbb{N}}$, that converges in $L^1(\mathbb{H}^n)$ to a set $E \subset \mathbb{H}^n$ such that:

i) $|E| = \lim_{j\to\infty} |E_j| = 1$, by the $L^1(\mathbb{H}^n)$ convergence;
ii) $P(E; \mathbb{H}^n) \le \liminf_{j\to\infty} P(E_j; \mathbb{H}^n) = C_{\text{isop}}$, by the lower semicontinuity of perimeter.

So we have $P(E; \mathbb{H}^n) = C_{\text{isop}}$ with $|E| = 1$, and E is therefore an isoperimetric set. This ends the proof, provided that we show 3). □

Claim 3) follows from the following lemma.

Lemma 4.2. *Let $n \ge 1$. There exist constants $\epsilon_0 > 0, C > 0$, and $R > 0$ such that for each $0 < \epsilon < \epsilon_0$ and for all sets $E \subset \mathbb{H}^n$ such that $|E| = 1$ and $P(E; \mathbb{H}^n) \le (1+\epsilon)C_{\text{isop}}$ there exists a set $F \subset \mathbb{H}^n$ such that:*

i) $|F| = 1$;
ii) $F \subset Q_R = \{(z,t) \in \mathbb{H}^n : |x_i|, |y_i|, |t|^2 < R,\ i = 1, \ldots, n\}$;
iii) $P(F; \mathbb{H}^n) \le \left(1 - C\epsilon^{\frac{Q}{Q-1}}\right)^{-(Q-1)/Q} P(E; \mathbb{H}^n)$.

Proof. For $s \in \mathbb{R}$, let us define the following sets:

$$\Pi_s^- = \{(z,t) \in \mathbb{H}^n : x_1 < s\} \quad \text{and} \quad \Pi_s^+ = \{(z,t) \in \mathbb{H}^n : x_1 > s\}.$$

We also let $\Pi_s = \{(z,t) \in \mathbb{H}^n : x_1 = s\}$. Let $E \subset \mathbb{H}^n$ be a set with $|E| = 1$ and finite H-perimeter. We define the sets

$$E_s^- = E \cap \Pi_s^- \quad \text{and} \quad E_s^+ = E \cap \Pi_s^+.$$

By the Heisenberg isoperimetric inequality (4.3), we have

$$P(E_s^-; \mathbb{H}^n) \ge C_{\text{isop}} |E_s^-|^{\frac{Q-1}{Q}}, \quad P(E_s^+; \mathbb{H}^n) \ge C_{\text{isop}} |E_s^+|^{\frac{Q-1}{Q}}, \tag{4.5}$$

where

$$\begin{aligned} P(E_s^-;\mathbb{H}^n) &= P(E;H_s^-) + P(E_s^-;\Pi_s),\\ P(E_s^+;\mathbb{H}^n) &= P(E;H_s^+) + P(E_s^+;\Pi_s). \end{aligned} \tag{4.6}$$

The number $P(E_s^-;\Pi_s)$ is the standard $2n$-dimension measure of the trace of E_s^- onto Π_s. Analogously, the number $P(E_s^+;\Pi_s)$ is the standard $2n$-dimension measure of the trace of E_s^+ onto Π_s. The function $v(s) = |E_s^-|$ is continuous and increasing. Therefore it is differentiable almost everywhere. Hence, at differentiability points $s \in \mathbb{R}$ of v we have

$$v'(s) = P(E_s^-;\Pi_s) = P(E_s^+;\Pi_s).$$

We do not prove these claims, here. From (4.6) and (4.5), we obtain

$$\begin{aligned} P(E;\mathbb{H}^n) + 2v'(s) &\geq P(E;\Pi_s^-) + P(E;\Pi_s^+) + 2v'(s)\\ &= P(E;\Pi_s^-) + P(E;\Pi_s^+) + P(E_s^-;\Pi_s)\\ &\quad + P(E_s^+;\Pi_s)\\ &= P(E_s^-;\mathbb{H}^n) + P(E_s^+;\mathbb{H}^n)\\ &\geq C_{\text{isop}}\Big\{|E_s^-|^{\frac{Q-1}{Q}} + |E_s^+|^{\frac{Q-1}{Q}}\Big\}. \end{aligned}$$

Using $P(E;\mathbb{H}^n) \leq C_{\text{isop}}(1+\epsilon)$ and $|E| = 1$, the inequality above implies

$$C_{\text{isop}}(1+\epsilon) + 2v'(s) \geq C_{\text{isop}}\big\{v(s)^{\frac{Q-1}{Q}} + (1-v(s))^{\frac{Q-1}{Q}}\big\},$$

and letting $\psi(v) = v^{\frac{Q-1}{Q}} + (1-v)^{\frac{Q-1}{Q}} - 1$ for $v \in [0,1]$, we finally obtain

$$C_{\text{isop}}\epsilon + 2v'(s) \geq C_{\text{isop}}\psi(v(s)). \tag{4.7}$$

The function ψ is strictly concave with $\psi(0) = \psi(1) = 0$. Then there exist $0 < v_- < v_+ < 1$ such that $\psi(v_-) = \psi(v_+) = 2\epsilon$. By concavity, we have

$$\psi(v) \geq 2\epsilon \quad \text{for all } v_- \leq v \leq v_+.$$

There exist numbers $s_- < s_+$ such that $v(s_-) = v_-$ and $v(s_+) = v_+$. Thus, from (4.7) we get

$$\begin{aligned} s_+ - s_- &\leq \int_{s_-}^{s_+} \frac{C_{\text{isop}}\epsilon + 2v'(s)}{C_{\text{isop}}\psi(v(s))}ds\\ &\leq \frac{1}{2}(s_+ - s_-) + \int_{s_-}^{s_+} \frac{2v'(s)}{C_{\text{isop}}\psi(v(s))}ds\\ &\leq \frac{1}{2}(s_+ - s_-) + \int_0^1 \frac{2}{C_{\text{isop}}\psi(v)}dv. \end{aligned} \tag{4.8}$$

We obtain the bound

$$\frac{s_+ - s_-}{2} \le \widehat{R} = \frac{2}{C_{\text{isop}}} \int_0^1 \frac{1}{\psi(v)} dv < \infty.$$

The set $\widehat{E} = E \cap \{(z,t) \in \mathbb{H}^n : s_- < x_1 < s_+\}$ has volume

$$|\widehat{E}| = |E^-_{s_+}| - |E^-_{s_-}| = 1 - 2v_-.$$

We used the identity $v_+ = 1 - v_-$. The number $0 < v_- < 1/2$ satisfies $\psi(v_-) = 2\epsilon$. There are constants $\epsilon_0 > 0$ and $C > 0$ such that if $0 < \epsilon < \epsilon_0$ we have $v_- \le C\epsilon^{\frac{Q}{Q-1}}$. Let $\lambda > 0$ be such that $|\lambda\widehat{E}| = 1$. Then we have $1 = \lambda^Q |\widehat{E}| \ge \lambda^Q (1 - 2C\epsilon^{\frac{Q}{Q-1}})$, and thus

$$\lambda \le \left(\frac{1}{1 - 2C\epsilon^{\frac{Q}{Q-1}}} \right)^{1/Q}.$$

A calibration argument shows that $P(\widehat{E}; \mathbb{H}^n) \le P(E; \mathbb{H}^n)$. We do not prove this claim, here. So we get

$$P(\lambda\widehat{E}; \mathbb{H}^n) = \lambda^{Q-1} P(\widehat{E}; \mathbb{H}^n) \le \left(\frac{1}{1 - 2C\epsilon^{\frac{Q}{Q-1}}} \right)^{(Q-1)/Q} P(E; \mathbb{H}^n).$$

After a left translation, we may assume that

$$\lambda\widehat{E} \subset \{(z,t) \in \mathbb{H}^n : |x_1| < R\},$$

where we let $R = \lambda\widehat{R}$. Repeating the argument for each coordinate axis, we obtain the claim of the lemma. The argument in the t coordinate requires easy adaptations. □

In 1983, Pansu conjectured a possible solution to the Heisenberg isoperimetric problem, see [59]. The conjecture can be formulated in the following way. Up to a null set, a left translation, and a dilation, the isoperimetric set in $\mathbb{H}^1$ is precisely the set

$$E_{\text{isop}} = \left\{ (z,t) \in \mathbb{H}^1 \ : \ |t| < \arccos|z| + |z|\sqrt{1 - |z|^2}, \ |z| < 1 \right\}. \quad (4.9)$$

Pansu did not give the formula for the conjectured isoperimetric set but he described how to construct it. Let us consider a geodesic $\gamma : [0,\pi] \to \mathbb{H}^1$ joining the point $\gamma(0) = 0$ to the point $\gamma(\pi) = (0,\pi) \in \mathbb{H}^1$. Using the formula (1.14) with $\vartheta = 0$ and $\varphi = 2$, we have the following formula for γ

$$\gamma(s) = \left(\frac{e^{2is} - 1}{2}, s - \sin s \cos \right).$$

The horizontal projection of γ, namely the curve $\kappa(s) = \frac{e^{2is}-1}{2}$, is a circle with diameter 1. Letting $|z| = |\kappa(s)|$ we find $|z|^2 = 1 - \cos^2 s$, and when $s \in [0, \pi/2]$ we get

$$s = \arccos\sqrt{1 - |z|^2}.$$

We can thus define the profile function $\varphi : [0, 1] \to \mathbb{R}$ by letting

$$\begin{aligned}\varphi(|z|) &= s - \sin s \cos s - \frac{\pi}{2}\\ &= \arccos\sqrt{1 - |z|^2} - |z|\sqrt{1 - |z|^2} - \frac{\pi}{2}\\ &= -\arccos|z| - |z|\sqrt{1 - |z|^2}.\end{aligned}$$

The profile φ gives the radial value of the function whose graph is the bottom part of the boundary of the set E_{isop} in (4.9).

Pansu's conjecture is in $\mathbb{H}^1$. Of course, the formula defining E_{isop} in (4.9) makes sense in $\mathbb{H}^n$ for $n \geq 2$ and the conjecture can be naturally extended to any dimension.

Proposition 4.3. *The set $E_{\text{isop}} \subset \mathbb{H}^1$ has the following properties:*

1) *The boundary ∂E_{isop} is of class C^2 but not of class C^3.*
2) *The set E_{isop} is convex.*
3) *The set E_{isop} is axially symmetric.*

Proof. 1) The boundary ∂E_{isop} is of class C^∞ away from the center of the group $Z = \{(0, t) \in \mathbb{H}^1 : t \in \mathbb{R}\}$. We claim that the function $\varphi : [0, 1] \to \mathbb{R}$,

$$\varphi(r) = \arccos r + r\sqrt{1 - r^2},$$

satisfies $\varphi'(0) = \varphi''(0) = 0$ but $\varphi'''(0) \neq 0$. This implies that ∂E_{isop} is of class C^2 but not of class C^3. In fact, we have

$$\varphi'(r) = \frac{-2r^2}{\sqrt{1 - r^2}}, \qquad \varphi''(r) = -2r\frac{2 - r^2}{(1 - r^2)^{3/2}},$$

and thus $\varphi'''(0) = -4 \neq 0$.

2) The set E_{isop} is convex because the function φ satisfies $\varphi'' \leq 0$ on $[0, 1]$ and $\varphi'(0) = 0$.

3) The set E_{isop} is axially symmetric:

$$(z, t) \in E_{\text{isop}} \quad \Rightarrow \quad (\zeta, t) \in E_{\text{isop}} \quad \text{for all } |\zeta| = |z|.$$

In fact, the profile function depends on $|z|$. □

Pansu's conjecture is known to hold assuming some regularity, symmetry, or structure for the isoperimetric set. In the next sections, we describe the following recent results:

1) If $E \subset \mathbb{H}^1$ is isoperimetric and ∂E is of class C^2 then $E = E_{\text{isop}}$, up to dilation and left translation. This result is not known when $n \geq 2$.
2) If $E \subset \mathbb{H}^1$ is isoperimetric and convex then $E = E_{\text{isop}}$, up to dilation and left translation. This result is not known when $n \geq 2$.
3) Let $n \geq 1$. If $E \subset \mathbb{H}^n$ is isoperimetric and axially symmetric then $E = E_{\text{isop}}$, up to a vertical translation and a dilation.
4) Let $n \geq 1$. If $E \subset \mathbb{H}^n$ is contained in a vertical cylinder and has a circular horizontal section, then $E = E_{\text{isop}}$, up to dilation and left translation.

In general, Pansu's conjecture is still open.

4.2 Isoperimetric sets of class C^2

In this section, we show that isoperimetric sets in $\mathbb{H}^1$ of class C^2 are of the form (4.9). This result is due to [69] (Theorems 6.10 and 7.2) and relies upon two facts: the structure of the characteristic set of surfaces of class C^2; the geometric interpretation of the equation for surfaces with constant H-curvature. Both results are limited to $\mathbb{H}^1$.

Theorem 4.4 (Ritoré-Rosales). *Let $E \subset \mathbb{H}^1$ be a bounded isoperimetric set with boundary ∂E of class C^2. Then we have $E = E_{\text{isop}}$, up to dilation and left translation.*

Proof. Let $D \subset \mathbb{C}$ be an open set and let $f \in C^2(D)$ be a function such that

$$\mathrm{gr}(f) = \big\{(z, f(z)) \in \mathbb{H}^1 : z \in D\big\} \subset \partial E.$$

We denote by $\Sigma(f) = \big\{z \in D : \nabla f(z) + 2z^\perp = 0\big\}$ the characteristic set of f. It may be $\Sigma(f) = \emptyset$. We always have $|\Sigma(f)| = 0$.

For $\varphi \in C_c^\infty(D \setminus \Sigma(f))$ and $\epsilon \in \mathbb{R}$ small, consider the set $E_\epsilon \subset \mathbb{H}^1$ that is obtained from E perturbing the piece of boundary of E given by the graph of f, through the function $f + \epsilon\varphi$. Then, for small ϵ we have

$$\frac{P(E;\mathbb{H}^1)}{|E|^{3/4}} = I(E) \leq I(E_\epsilon) = \frac{P(E_\epsilon;\mathbb{H}^1)}{|E_\epsilon|^{3/4}} = \frac{p(\epsilon)}{v(\epsilon)^{3/4}} = \psi(\epsilon), \quad (4.10)$$

where $p(\epsilon) = P(E_\epsilon;\mathbb{H}^1)$ and $p(\epsilon) = |E_\epsilon|$. Using the area formula for H-perimeter (3.5) we find

$$p'(0) = \int_D \frac{\langle \nabla f + 2z^\perp, \nabla\varphi\rangle}{|\nabla f + 2z^\perp|} dz, \quad v'(0) = -\int_D \varphi(z)\, dz.$$

Here, we are assuming that the set E lies above the graph of f. Moreover, we have $\psi' = p'v^{-3/4} - \frac{3}{4}pv^{-7/4}v'$. From (4.10) we deduce that $\psi'(0) = 0$ and thus

$$
\begin{aligned}
0 &= \frac{1}{|E|^{3/4}} \int_D \frac{\langle \nabla f + 2z^\perp, \nabla\varphi\rangle}{|\nabla f + 2z^\perp|} dz + \frac{3}{4}\frac{P(E;\mathbb{H}^1)}{|E|^{7/4}} \int_D \varphi\, dz \\
&= -\frac{1}{|E|^{3/4}} \int_D \varphi \operatorname{div}\Big(\frac{\nabla f + 2z^\perp}{|\nabla f + 2z^\perp|}\Big) dz + \frac{3}{4}\frac{P(E;\mathbb{H}^1)}{|E|^{7/4}} \int_D \varphi\, dz.
\end{aligned}
$$

Since $\varphi \in C_c^\infty(D \setminus \Sigma(f))$ is arbitrary, we deduce that the function f satisfies the partial differential equation

$$
\operatorname{div}\Big(\frac{\nabla f(z) + 2z^\perp}{|\nabla f(z) + 2z^\perp|}\Big) = \frac{3}{4}\frac{P(E;\mathbb{H}^1)}{|E|} =: H, \quad z \in D \setminus \Sigma(f). \tag{4.11}
$$

We conclude that for any $z \in D\setminus\Sigma(f)$ there exists an arc of circle κ_z with curvature H passing through z and such that $\gamma_z = \mathrm{Lift}(\kappa_z)$ is contained in $\mathrm{gr}(f) \subset \partial E$. See Remark 3.13.

Let $\Sigma(\partial E)$ be the characteristic set of ∂E. The above argument shows that for any $p \in \partial E \setminus \Sigma(\partial E)$ there exists a geodesic γ_p contained in $\partial E \setminus \Sigma(\partial E)$ and passing through p. There exists a maximal interval (a, b) such that we have $\gamma_p : (a, b) \to \partial E \setminus \Sigma(\partial E)$. Since E is bounded, γ_p can be extended to a and b with $\gamma(a), \gamma(b) \in \Sigma(\partial E)$.

In a neighborhood of the point $(z_0, t_0) = \gamma(a) \in \Sigma(f)$, the surface ∂E is a graph of the form $t = f(z)$ for some $f \in C^2(D)$ and $D \subset \mathbb{C}$ open set with $z_0 \in D$. This is because the tangent space to ∂E at this point coincides with the horizontal plane. Let (D, f) be the maximal pair such that $\mathrm{gr}(f) \subset \partial E$ with D open set containing z_0 and $f \in C^2(D)$.

By Theorem 3.15, there are two cases:

i) z_0 is an isolated point of $\Sigma(f)$;
ii) Near z_0, $\Sigma(f)$ is a C^1 curve κ_{z_0} passing through z_0.

In the case ii), let κ_{z_0} be the maximal C^1 curve contained in $\Sigma(f)$ and passing through z_0. The curve κ_{z_0} cannot reach the boundary ∂D because this would contradict the maximality of D. The curve κ_{z_0} cannot have limit points inside D that are singular, because of Theorem 3.15. Then κ_{z_0} must be a simple closed curve inside D. But this is not possible because the horizontal lift of κ_{z_0} grows in the t coordinate by an amount that equals 4 times the area of the region enclosed by the simple closed curve.

So we are left with the case $\Sigma(f) = \{z_0\}$ for some $z_0 \in D$. Through any point $z \in D \setminus \{z_0\}$ passes a circle with curvature H starting from z_0.

Now the boundary of E is determined in a neighborhood of $(z_0, f(z_0)) \in \partial E$. The regularity of ∂E forces D to be a circle centered at z_0 and E to be a left translation and dilatation of E_{isop}. □

4.3 Convex isoperimetric sets

We say that a set $E \subset \mathbb{H}^1$ is convex if it is convex for the standard linear structure of $\mathbb{H}^1 = \mathbb{R}^3$. Left translations and dilations preserve convexity. In [53], Pansu's conjecture is proved assuming the convexity of isoperimetric sets. Recall the $E_{\mathrm{isop}} \subset \mathbb{H}^1$ is the set in (4.9).

Theorem 4.5 (Monti-Rickly). *Let $E \subset \mathbb{H}^1$ be a convex (open) isoperimetric set. Then, up to a left translation and a dilation we have $E = E_{\mathrm{isop}}$.*

Using the concentration argument of Theorem 4.1, it is possible to prove the existence of isoperimetric sets within the class of convex sets. However, it is not clear how to compute the first variation remaining inside this class of sets. Theorem 4.5 is not known when $n \geq 2$. It would be also interesting to prove the theorem assuming for isoperimetric sets only H-convexity (convexity along horizontal lines, see [5]) rather than standard convexity.

Here, we describe the technical steps of the proof of Theorem 4.5. For details, we refer the reader to [53]. Let $E \subset \mathbb{H}^1$ be a convex isoperimetric set. Then we have

$$E = \{(z, t) \in \mathbb{H}^1 : z \in D,\ f(z) < t < g(z), \}, \tag{4.12}$$

where $D \subset \mathbb{C} = \mathbb{R}^2$ is a bounded convex open set in the plane, and $-g, f : D \to \mathbb{R}$ are convex functions. In particular, f and g are locally Lipschitz continuous and their first derivatives are locally of bounded variation. The function f satisfies the following partial differential equation

$$\operatorname{div}\left(\frac{\nabla f + 2z^\perp}{|\nabla f + 2z^\perp|}\right) = \frac{3P(E; \mathbb{H}^1)}{4|E|} = H \quad \text{in } D. \tag{4.13}$$

Equation (4.13) can be deduced in the same way as in (4.11), with the difference that the equation is now verified only in the weak sense. As a matter of fact, the vector field

$$N_f(z) = \frac{\nabla f(z) + 2z^\perp}{|\nabla f(z) + 2z^\perp|} \qquad z \in D,$$

is only in $L^\infty(D)$. However, we have $\nabla f(z) + 2z^\perp \in BV_{\mathrm{loc}}(D)$.

The goal is to prove that integral curves of $N_f^\perp$ are circles with curvature H. The vector N_f will be the "normal vector" to the curve.

The first step of the proof of Theorem 4.5 is an improved regularity for solutions of (4.13): the candidate "normal vector" satisfies $N_f \in W^{1,1}_{\mathrm{loc}}(D;\mathbb{R}^2)$, see [53].

The second step of the proof consists in the analysis of the flow of the vector field $v(z) = 2z - \nabla f^\perp(z)$. This vector field is orthogonal to N_f. Since f is convex, we have $v \in BV_{\mathrm{loc}}(\mathrm{int}(D);\mathbb{R}^2)$. Moreover, the distributional divergence of v is in L^∞, in fact $\mathrm{div}\, v = 4$ in $\mathrm{int}(D)$. Thus, by Ambrosio's theory on the Cauchy Problem for vector fields of bounded variation [2], for any compact set $K \subset D$ there exist $r > 0$ and a (unique regular) Lagrangian flow $\Phi : K \times [-r, r] \to D$. In particular, for any $z \in K$, the curve $\gamma_z(s) = \Phi(z, s)$ is an integral curve of v passing through z at time $s = 0$.

The third step of the proof uses the fact that $v/|v|$ is in $W^{1,1}_{\mathrm{loc}}(D;\mathbb{R}^2)$ to show that (a suitable reparameterization of) the integral curve γ_z is twice differentiable in a weak sense. With this regularity, the distributional equation (4.13) can be given a formal meaning along the integral curve γ_z: it says that the curvature of γ_z is the constant H.

Theorem 4.6. *Let $E \subset \mathbb{H}^1$ be a convex isoperimetric set with curvature $H > 0$ (the constant in (4.13)) and let $\Phi : K \times [-r, r] \to D$ be the flow introduced above. Then for a.e. $z \in K$ the curve $s \mapsto \Phi(z, s)$ is an arc of circle with radius $1/H$.*

The shape of a convex isoperimetric set E can now be reconstructed starting from the structure of the characteristic set of ∂E. A point $(z, t) \in \partial E$ is characteristic if the horizontal plane at (z, t) is a supporting plane for E at (z, t). For convex sets, the characteristic set is the disjoint union of at most four compact disjoint horizontal segments, possibly points, see [53]. This property and Theorem 4.6 yield Theorem 4.5 as explained in the final part of the proof of Theorem 4.4.

4.4 Axially symmetric solutions

We denote by $\mathscr{S}$ the family of all measurable subsets $E \subset \mathbb{H}^n$ with $0 < |E| < \infty$ that are axially symmetric:

$$(z,t) \in E \quad \Rightarrow \quad (\zeta, t) \in E \quad \text{for all } |\zeta| = |z|.$$

The isoperimetric problem in the family $\mathscr{S}$ consists in proving existence and classifying all minimizers of the infimum problem

$$C^{\mathscr{S}}_{\mathrm{isop}} = \inf\big\{ I(E) : E \in \mathscr{S} \big\}. \tag{4.14}$$

A set $E \in \mathscr{S}$ for which the infimum in (4.14) is attained is called an *axially symmetric isoperimetric set*. Clearly, we have $C^{\mathscr{S}}_{\text{isop}} \geq C_{\text{isop}}$. Even though we believe that $C^{\mathscr{S}}_{\text{isop}} = C_{\text{isop}}$, we are not able to prove this.

In the axially symmetric setting, Pansu's conjecture amounts to show that the solution to Problem (4.14) is the set

$$E_{\text{isop}} = \left\{ (z,t) \in \mathbb{H}^n : |t| < \arccos|z| + |z|\sqrt{1-|z|^2},\ |z| < 1 \right\}. \quad (4.15)$$

for any dimension $n \geq 1$. This result is proved in [48] and, in this section, we present the scheme of the proof.

Theorem 4.7 (Monti). *The infimum $C^{\mathscr{S}}_{\text{isop}} > 0$ is attained and any axially symmetric isoperimetric set coincides with the set E_{isop} in (4.15), up to a dilation, a vertical translation, and a Lebesgue negligible set.*

By a rearrangement argument, Theorem 4.7 can be reduced to a one dimensional problem. The first step is the reduction to an isoperimetric problem in the half plane $\mathbb{R}^2_+ = \mathbb{R}^+ \times \mathbb{R}$.

Using spherical coordinates in $\mathbb{C}^n$, a measurable axially symmetric set $E \subset \mathbb{H}^n$ is generated by a measurable set $F \subset \mathbb{R}^2_+$ (and viceversa), and we have the following formula

$$P(E;\mathbb{H}^n) = \omega_{2n-1}\mathscr{Q}(F;\mathbb{R}^2_+), \quad (4.16)$$

where $\mathscr{Q}(\cdot;\mathbb{R}^2_+)$ is a weighted perimeter functional in the half-plane

$$\mathscr{Q}(F;\mathbb{R}^2_+) = \sup\left\{ \int_F \{\partial_r(r^{2n-1}\psi_1) + \partial_t(2r^{2n}\psi_2)\}\,drdt : \psi \in C^1_c(\mathbb{R}^2_+;\mathbb{R}^2),\ \|\psi\|_\infty \leq 1 \right\}. \quad (4.17)$$

Above, $\omega_{2n-1} = \mathscr{H}^{2n-1}(\mathbb{S}^{2n-1})$ is the standard surface measure of the $(2n-1)$-dimensional unit sphere. For any axially symmetric set $E \subset \mathbb{H}^n$, the volume transforms according to the following rule

$$|E| = \omega_{2n-1}\int_F r^{2n-1}drdt = \omega_{2n-1}V(F), \quad (4.18)$$

where $V(\cdot)$ is a volume functional in the half-plane. From (4.17) and (4.18), the axially symmetric isoperimetric problem (4.14) transforms into the weighted isoperimetric problem in the half plane

$$C^{\mathscr{S}}_{\text{isop}} = \omega^{1/Q}_{2n-1} \inf\left\{ \frac{\mathscr{Q}(F;\mathbb{R}^2_+)}{V(F)^{\frac{Q-1}{Q}}} : F \subset \mathbb{R}^2_+ \text{ such that } 0 < V(F) < \infty \right\}. \quad (4.19)$$

The observation made in [48] is that the isoperimetric quotient for sets $F \subset \mathbb{R}^2_+$ is improved by a certain rearrangement of F in the variable r for fixed t that is tailored to the perimeter $\mathscr{Q}(\cdot;\mathbb{R}^2_+)$. We measure the t-sections of F, the sets $F_t = \{r > 0 : (r,t) \in F\}$, using the line density $\tau(r) = 2r^{2n}$. The function τ is the weight appearing in the definition of the functional $\mathscr{Q}(\cdot;\mathbb{R}^2_+)$ in (4.17). We let

$$\Theta(r) = \int_0^r \tau(s)\,ds = \frac{2}{2n+1} r^{2n+1}, \tag{4.20}$$

and we say that a measurable set $F \subset \mathbb{R}^2_+$ is τ-rearrangeable if the function $f : \mathbb{R} \to [0,+\infty]$

$$f(t) = \int_{F_t} \tau(r)\,dr \tag{4.21}$$

is in $L^1_{\mathrm{loc}}(\mathbb{R})$. In this case, we call the set

$$F^\sharp = \{(r,t) \in \mathbb{R}^2_+ : \Theta(r) < f(t)\} \tag{4.22}$$

the τ-rearrangement of F. The t-sections of $F^\sharp$ are intervals $(0, \Theta^{-1}(f(t)))$ with the same τ-measure as the t-sections F_t.

The following intermediate result is proved in [48].

Theorem 4.8. *Let $F \subset \mathbb{R}^2_+$ be a τ-rearrangeable set. Then:*

i) *We have $\mathscr{Q}(F^\sharp;\mathbb{R}^2_+) \le \mathscr{Q}(F;\mathbb{R}^2_+)$, and in case of equality there holds $F = F^\sharp$, up to a negligible set.*
ii) *We have $V(F^\sharp) \ge V(F)$.*

Using Theorem 4.8, it is easy to find a compact minimizing sequence, thus getting the existence of axially symmetric isoperimetric sets. Moreover, a set F minimizing (4.19) satisfies:

i) $F = F^\sharp$, up to a negligible set;
ii) the sections $F_r = \{t \in \mathbb{R} : (r,t) \in F\}$ are equivalent to intervals, for $\mathscr{L}^1$-a.e. $r \in \mathbb{R}^+$.

Now the boundary of ∂F inside $\mathbb{R}^2_+$ is a Lipschitz curve that can be computed by a standard variational argument. This curve is the profile of the isoperimetric set conjectured by Pansu and, as a matter of fact, it does not depend on the dimension n.

4.5 Calibration argument

In [67], Ritoré proved Pansu's conjecture within a special class of sets by a calibration argument. The sets have one circular horizontal section and are contained in a vertical cylinder, see also [19]. The argument works in any dimension. We let

$$B = \{(z, 0) \in \mathbb{H}^n : |z| < 1\} \quad \text{and} \quad C = \{(z, t) \in \mathbb{H}^n : |z| < 1,\ t \in \mathbb{R}\}.$$

We identify $B = \{|z| < 1\} \subset \mathbb{C}^n$.

Theorem 4.9 (Ritoré). *Let $E \subset \mathbb{H}^n$, $n \geq 1$, be a bounded open set with finite H-perimeter such that:*

i) $B \subset E \subset C$;
ii) $|E| = |E_{\text{isop}}|$, *where E_{isop} is the set in* (4.15).

Then, we have $P(E_{\text{isop}}; \mathbb{H}^n) \leq P(E; \mathbb{H}^n)$.

Proof. Let $\varphi : \bar{B} \to \mathbb{R}$ be the profile function of E_{isop},

$$\varphi(z) = \arccos|z| + |z|\sqrt{1 - |z|^2}, \qquad |z| \leq 1.$$

The function $f : \bar{C} \to \mathbb{R}$, $f(z, t) = |t| - \varphi(z)$, is a defining function for ∂E_{isop}. Let us define the vector field $\psi : \bar{C} \setminus Z \to \mathbb{R}^{2n}$

$$\psi(z, t) = \frac{\nabla_H f(z, t)}{|\nabla_H f(z, t)|}, \qquad 0 < |z| < 1, \quad t \neq 0.$$

The vector field ψ is not defined when $z = 0$ or $t = 0$; it can be extended to $|z| = 1$; it jumps at $t = 0$. In the set $\{0 < |z| < 1,\ t \neq 0\}$, ψ is of class C^∞ and there is a constant $H \neq 0$ such that

$$\operatorname{div}_H \psi(z, t) = H, \qquad 0 < |z| < 1, \quad t \neq 0. \tag{4.23}$$

We consider the following sets:

$$E^+ = E \cap \{t > 0\}, \qquad E^+_{\text{isop}} = E_{\text{isop}} \cap \{t > 0\}$$
$$E^- = E \cap \{t < 0\}, \qquad E^-_{\text{isop}} = E_{\text{isop}} \cap \{t < 0\}.$$

By i), we have $E^+ \Delta E^+_{\text{isop}} \subset C$ and moreover the boundary of $E^+ \Delta E^+_{\text{isop}}$ does not intersect the base B of the cylinder. Let $F^+ = E^+_{\text{isop}} \setminus E^+$ and $G^+ = E^+ \setminus E^+_{\text{isop}}$. Then we have $F^+, G^+ \subset B \times \mathbb{R}^+$ and $E^+ \Delta E^+_{\text{isop}} = F^+ \cup G^+$. Moreover, denoting by $N_H^{F^+}$ and $N_H^{G^+}$ the horizontal outer normals to ∂F^+ and ∂G^+, respectively:

a) $N_H^{F^+} = N_H^{E_{\text{isop}}}$ a.e. on $\partial F^+ \cap \partial E_{\text{isop}}$ and $N_H^{F^+} = -N_H^E$ a.e. on $\partial F^+ \cap \partial E$;

b) $N_H^{G^+} = -N_H^{E_{\text{isop}}}$ a.e. on $\partial G^+ \cap \partial E_{\text{isop}}$ and $N_H^{G^+} = N_H^E$ a.e. on $\partial G^+ \cap \partial E$.

Integrating (4.23) on F^+ we find

$$\begin{aligned} H|F^+| &= \int_{F^+} \operatorname{div}_H \psi(z,t)\,dzdt = \int_{\partial F^+} \langle N_H^{F^+}, \psi\rangle d\mu_{F^+} \\ &= \int_{\partial F^+} \langle N_H^{E_{\text{isop}}}, \psi\rangle d\mu_{E_{\text{isop}}} - \int_{\partial F^+} \langle N_H^E, \psi\rangle d\mu_E \\ &\geq P(E_{\text{isop}}; \partial F^+) - P(E; \partial F^+), \end{aligned} \tag{4.24}$$

because $\langle N_H^{E_{\text{isop}}}, \psi\rangle = 1$ on $\partial F^+ \cap \partial E_{\text{isop}}$ and $\langle N_H^E, \psi\rangle \leq 1$ on $\partial F^+ \cap \partial E$.

In the same way, we find the inequality

$$\begin{aligned} H|G^+| &= \int_{G^+} \operatorname{div}_H \psi(z,t)\,dzdt = \int_{\partial G^+} \langle N_H^{G^+}, \psi\rangle d\mu_{G^+} \\ &= -\int_{\partial G^+} \langle N_H^{E_{\text{isop}}}, \psi\rangle d\mu_{E_{\text{isop}}} + \int_{\partial G^+} \langle N_H^E, \psi\rangle d\mu_E \\ &\leq -P(E_{\text{isop}}; \partial G^+) + P(E; \partial G^+). \end{aligned} \tag{4.25}$$

From (4.24) and (4.25), we obtain

$$\begin{aligned} H(|F^+| - |G^+|) &\geq P(E_{\text{isop}}; \partial F^+) - P(E; \partial F^+) \\ &\quad + P(E_{\text{isop}}; \partial G^+) - P(E; \partial G^+) \\ &= P(E_{\text{isop}}; \{t > 0\}) - P(E; \{t > 0\}). \end{aligned} \tag{4.26}$$

Let $F^- = E^-_{\text{isop}} \setminus E^-$ and $G^- = E^- \setminus E^-_{\text{isop}}$. Then we have $F^-, G^- \subset B \times \mathbb{R}^-$ and $E^- \Delta E^-_{\text{isop}} = F^- \cup G^-$. Computations analogous to the ones above show that

$$H(|F^-| - |G^-|) \geq P(E_{\text{isop}}; \{t < 0\}) - P(E; \{t < 0\}). \tag{4.27}$$

Since $|E| = |E_{\text{isop}}|$ we have $|F^+| + |F^-| = |G^+| + |G^-|$. Adding (4.26) and (4.27), we obtain

$$\begin{aligned} 0 &= H(|F^+| + |F^-| - |G^+| - |G^-|) \\ &\geq P(E_{\text{isop}}; \{t \neq 0\}) - P(E; \{t \neq 0\}) \\ &= P(E_{\text{isop}}; \mathbb{H}^n) - P(E; \mathbb{H}^n). \end{aligned}$$

This concludes the proof. □

Remark 4.10. In [67], Ritoré also discusses the equality case. Namely, in the setting of Theorem 4.9 and assuming that $\partial E \setminus Z$ is a C^1_H-regular surface, he shows that the equality $P(E; \mathbb{H}^n) = P(E_{\text{isop}}; \mathbb{H}^n)$ implies $E = E_{\text{isop}}$.

5 Regularity problem for H-perimeter minimizing sets

The regularity of H-perimeter minimizing boundaries is a challenging open problem. We list the main steps and the main technical difficulties.

1) *Lipschitz approximation.* The first step in the regularity theory of perimeter minimizing sets in $\mathbb{R}^n$ is a good approximation of minimizers. In De Giorgi's original approach, the approximation is made by convolution and the estimates are based on the monotonicity formula. In the Heisenberg group, the validity of a monotonicity formula is not clear, see [21]. A more flexible approach is the approximation of minimizing boundaries by Lipschitz graphs. This scheme works also in the Heisenberg group. An H-minimizing boundary is approximated in measure by an *intrinsic Lipschitz* graph. The estimate involves the notion of horizontal excess, see Theorem 5.9 and [50].

2) *Harmonic approximation.* The minimal set can be blown-up at a point of the reduced boundary by a quantity depending on excess. It can be shown that the corresponding approximating intrinsic Lipschitz functions converge to a limit function. This holds when $n \geq 2$ thanks to a Poincaré inequality valid on vertical hyperplanes, see [17]. We do not present the details, here. It is an open problem to prove that this limit function is harmonic for the natural (linear) sub-Laplacian of the vertical hyperplane.

3) *Decay estimate for excess.* Known estimates for sub-elliptic harmonic functions should give the decay estimate for excess

$$\mathrm{Exc}(E, B_{\alpha r}) \leq C\alpha^2 \mathrm{Exc}(E, B_r), \quad r > 0,$$

for some $0 < \alpha < 1$ and $C > 0$. By standard facts, this implies the Hölder continuity of the horizontal normal on the reduced boundary. In turn, the continuity of the normal implies that the reduced boundary is a C^1_H-regular surface in the sense of Definition 2.16, see [56], and thus it is locally the intrinsic graph of a continuous function φ having Hölder continuous *distributional* intrinsic gradient $\nabla^\varphi \varphi$, see Definition 3.4.

4) *Schauder-type regularity.* The function φ is a local minimizer of the area functional (see (3.14))

$$\mathscr{A}(\varphi) = \int_D \sqrt{1 + |\nabla^\varphi \varphi|^2} dw.$$

It is an open problem to deduce further regularity for φ, beyond the Hölder continuity of the distributional gradient ∇^φ. It is not even clear how to prove that φ solves the minimal surface equation (3.25).

This is the state of the art on the regularity of H-perimeter minimizing boundaries. In Section 5.3, we present the Lipschitz approximation of

H-perimeter minimizing sets, Theorem 5.9, and also the so-called height estimate, giving a certain flatness of the boundary in the regime of small excess, see Theorem 5.10. The proofs are rather technical and are omitted.

In Section 5.2, we also study some examples of nonsmooth minimizers in $\mathbb{H}^1$, including sets with constant horizontal normal. No similar examples of nonsmooth minimizers are known in $\mathbb{H}^n$ with $n \geq 2$.

5.1 Existence and density estimates

We start from the definition of a local minimizer of H-perimeter.

Definition 5.1. A set $E \subset \mathbb{H}^n$ with locally finite H-perimeter in an open set $A \subset \mathbb{H}^n$ is *H-perimeter minimizing in A* if for all $p \in \mathbb{H}^n$ and $r > 0$ and for any $F \subset \mathbb{H}^n$ such that $E \Delta F \Subset B_r(p) \Subset A$ we have

$$P(E; B_r(p)) \leq P(F; B_r(p)). \tag{5.1}$$

The existence of local minimizers with some boundary condition easily follows by a compactness argument. Let $A \subset \mathbb{H}^n$ be a bounded open set and let $B \subset \mathbb{H}^n$ be a set such that $P(B; \mathbb{H}^n) < \infty$. Define the family of sets:

$$\mathscr{F}(A, B) = \big\{ F \subset \mathbb{H}^n : F \text{ has finite } H\text{-perimeter in } \mathbb{H}^n \text{ and } F \Delta B \subset \bar{A} \big\}.$$

Clearly, $\mathscr{F}(A, B) \neq \emptyset$ because $B \in \mathscr{F}(A, B)$. The set B determines a natural boundary condition.

Proposition 5.2. *Let A and B be as above. Then there exists a set $E \in \mathscr{F}(A, B)$ such that*

$$P(E; \mathbb{H}^n) \leq P(F; \mathbb{H}^n) \quad \textit{for all } F \in \mathscr{F}(A, B).$$

Proof. Define the infimum

$$m = \inf \big\{ P(F; \mathbb{H}^n) : F \in \mathscr{F}(A, B) \big\} \geq 0,$$

and let $(E_j)_{j \in \mathbb{N}}$ be a minimizing sequence of sets $E_j \in \mathscr{F}(A, B)$:

$$\lim_{j \to \infty} P(E_j; \mathbb{H}^n) = m.$$

Let $\Omega \subset \mathbb{H}^n$ be a bounded open set such that $\bar{A} \subset \Omega$ and supporting the compact embedding $BV_H(\Omega) \Subset L^1(\Omega)$. The C^2 regularity of the boundary $\partial\Omega$ is a sufficient condition for compactness (see [33] and [52]).

Then we have:

i) $P(E_j; \mathbb{H}^n) \le m + 1$ for all $j \in \mathbb{N}$ large enough;
ii) $|E_j \cap \Omega| \le |\Omega| < \infty$ for all $j \in \mathbb{N}$.

By compactness, there exists a subsequence, still denoted by $(E_j)_{j\in\mathbb{N}}$, and a measurable set $E \subset \mathbb{H}^n$ such that $\chi_{E_j} \to \chi_E$ in $L^1(\Omega)$. Since $\chi_{E_j} = \chi_B$ in $\mathbb{H}^n \setminus \bar{A}$, we can also assume that $\chi_E = \chi_B$ in $\mathbb{H}^n \setminus \bar{A}$, that is $E \in \mathscr{F}(A, B)$. In particular, we have $\chi_{E_j} \to \chi_E$ in $L^1(\mathbb{H}^n)$.

By the lower semicontinuity of perimeter for the L^1 convergence of sets, we obtain

$$P(E; \mathbb{H}^n) \le \liminf_{j\to\infty} P(E_j; \mathbb{H}^n) = m,$$

If now F is a set such that $E\Delta F \Subset B_r(p) \Subset A$, then $F \in \mathscr{F}(A, B)$ and $P(E; \mathbb{H}^n \setminus \bar{B}_r(p)) = P(F; \mathbb{H}^n \setminus \bar{B}_r(p))$. Therefore, we have

$$\begin{aligned} P(E; B_r(p)) &= P(E; \mathbb{H}^n) - P(E; \mathbb{H}^n \setminus \bar{B}_r(p)) \\ &\le P(F; \mathbb{H}^n) - P(F; \mathbb{H}^n \setminus \bar{B}_r(p)) = P(F; B_r(p)). \quad \square \end{aligned}$$

As for the standard perimeter, sets that are H-perimeter minimizing admit lower and upper density estimates with geometric constants.

Lemma 5.3. *If $E \subset \mathbb{H}^n$ is an H-perimeter minimizing set in a ball B_ϱ for some $\varrho > 0$, then we have*

$$P(E; B_\varrho) \le c_1 \varrho^{Q-1}, \tag{5.2}$$

where $c_1 = P(B_1; \mathbb{H}^n)$.

Proof. Let $0 < s < r < \varrho$. Since the sets E and $E \setminus B_s$ agree inside $B_r \setminus \bar{B}_s$, we have

$$P(E; B_r \setminus \bar{B}_s) = P(E \setminus B_s; B_r \setminus \bar{B}_s) = P(E \setminus B_s; B_r) - P(E \setminus B_s; \bar{B}_s).$$

On the other hand, using $P(E \setminus B_s; B_s) = 0$ and (2.13) we obtain

$$\begin{aligned} P(E \setminus B_s; \bar{B}_s) &= P(E \setminus B_s; \partial B_s) = c_n \mathscr{S}_\varrho^{Q-1}(\partial^*(E \setminus B_s) \cap \partial B_s) \\ &\le c_n \mathscr{S}_\varrho^{Q-1}(\partial B_s) = P(B_s; \mathbb{H}^n) = c_1 s^{Q-1}. \end{aligned}$$

The formula $P(B_s; \mathbb{H}^n) = s^{Q-1} P(B_1; \mathbb{H}^n)$ follows by an elementary homogeneity argument. Then we obtain the inequality $P(E \setminus B_s; B_r) \le$

$P(E; B_r \setminus \bar{B}_s) + c_1 s^{Q-1}$. Since E is H-perimeter minimizing in B_ϱ, by (5.1) we get

$$P(E; B_r) \leq P(E \setminus B_s; B_r) \leq P(E; B_r \setminus \bar{B}_s) + c_1 s^{Q-1}.$$

Letting $s \uparrow r$ and using $P(E; B_r) < \infty$, we obtain $P(E; B_r) \leq c_1 r^{Q-1}$. Letting $r \uparrow \varrho$, we obtain (5.2). □

The density estimates from below are proved in [71], Proposition 2.14 (see also Theorem 2.4 therein).

Lemma 5.4. *There exist constants $c_2, c_3 > 0$ depending on $n \geq 1$ such that for any set $E \subset \mathbb{H}^n$ that is H-perimeter minimizing in $B_{2\varrho}$, $\varrho > 0$, we have, for all $p \in \partial E \cap B_\varrho$ and for all $0 < r < \varrho$,*

$$\min\left\{|E \cap B_r(p)|, |B_r(p) \setminus E|\right\} \geq c_2 r^Q, \tag{5.3}$$

and

$$P(E; B_r(p)) \geq c_3 r^{Q-1}. \tag{5.4}$$

For any set, the reduced boundary is a subset of the measure theoretic boundary, $\partial^* E \subset \partial E$, and moreover $\mu_E(\partial E \setminus \partial^* E) = 0$, see Proposition 2.8. For local minimizers the difference $\partial E \setminus \partial^* E$ is also small in terms of Hausdorff measures.

Lemma 5.5. *For any set $E \subset \mathbb{H}^n$ that is H-perimeter minimizing in $\mathbb{H}^n$, we have*

$$\mathscr{S}_\varrho^{Q-1}(\partial E \setminus \partial^* E) = 0. \tag{5.5}$$

Proof. Let $K = \partial E \setminus \partial^* E$, let A be an open set containing K, and fix $\delta > 0$. For any $p \in K$ there is an $0 < r_p < \delta/10$ such that $B_{5r_p}(p) \subset A$. Then $\{B_{r_p}(p) : p \in K\}$ is a covering of K and by the 5-covering lemma, there exists a sequence $p_i \in K, i \in \mathbb{N}$, such that the balls $B_i = B_{r_i}(p_i)$, with $r_i = r_{p_i}$, are pairwise disjoint and

$$K \subset \bigcup_{i \in \mathbb{N}} B_{5r_i}(p_i).$$

It follows that

$$\begin{aligned}\mathscr{S}_\varrho^{Q-1,\delta}(K \cap A) &\leq \sum_{i \in \mathbb{N}} \operatorname{diam}(B_{5r_i}(p_i))^{Q-1} = 10^{Q-1} \sum_{i \in \mathbb{N}} r_i^{Q-1} \\ &\leq 10^{Q-1} c_3^{-1} \sum_{i \in \mathbb{N}} P(E; B_{r_i}(p_i)) \leq 10^{Q-1} c_3^{-1} P(E; A).\end{aligned}$$

Since $\delta > 0$ is arbitrary, we deduce that $\mathscr{S}_\varrho^{Q-1}(K) \leq 10^{Q-1} c_3^{-1} P(E; A)$. As A is arbitrary and, by (2.13), $P(E; K) = 0$, we conclude that $\mathscr{S}_\varrho^{Q-1}(K) = 0$. □

5.2 Examples of nonsmooth H-minimal surfaces

The existence of nonsmooth H-minimal surfaces in $\mathbb{H}^1$ was already observed in [61]. Then this phenomenon was noticed by several authors, see [14,55,66,71]. In the next examples, we prove perimeter minimality of certain H-minimal surfaces by a calibration argument, see [8,55].

5.2.1 A Lipschitz H-minimal surface In this example, we study a local minimizer of H-perimeter with boundary ∂E that is only Lipschitz-regular. The surface ∂E is, however, C^1_H-regular: whereas the standard normal jumps, the horizontal normal is continuous.

In the open half-space $A = \{(z,t) \in \mathbb{H}^1 : y = \mathrm{Im}(z) > 0\}$, consider the set

$$E = \{(z,t) \in A : x = \mathrm{Re}(z) < 0 \text{ and } t < 0\}.$$

The set E has locally finite H-perimeter in A and its boundary $S = \partial E \cap A$ is a Lipschitz surface consisting of two pieces of plane meeting at the singular line $L = \{(z,t) \in A : x = 0 \text{ and } t = 0\}$. The horizontal inner normal $\nu_E : S \to \mathbb{R}^2$ is the restriction to S of the mapping $\varphi : A \to \mathbb{R}^2$

$$\varphi(z,t) = \begin{cases} \dfrac{(-y,x)}{\sqrt{x^2+y^2}} & \text{if } x \le 0, \\ (-1,0) & \text{if } x \ge 0. \end{cases}$$

The function φ is continuous in A and thus S is an H-regular surface. In fact, φ is locally Lipschitz continuous in A.

We claim that E is a local minimizer of H-perimeter in A. Namely, we prove that for any bounded open set $\Omega \subset\subset A$ and for any $F \subset A$ such that $E \Delta F \Subset \Omega$ we have

$$P(E;\Omega) \le P(F;\Omega). \tag{5.6}$$

The proof is a calibration argument and the calibration is provided by the vector field V in A defined by

$$V(z,t) = \varphi_1(z,t)X + \varphi_2(z,t)Y,$$

where $\varphi = (\varphi_1, \varphi_2)$. Then, at points $(z,t) \in A$ where $x \le 0$ we have

$$\begin{aligned} \operatorname{div} V = \operatorname{div}_H \varphi &= X\Big(\frac{-y}{\sqrt{x^2+y^2}}\Big) + Y\Big(\frac{x}{\sqrt{x^2+y^2}}\Big) \\ &= \frac{\partial}{\partial x}\Big(\frac{-y}{\sqrt{x^2+y^2}}\Big) + \frac{\partial}{\partial y}\Big(\frac{x}{\sqrt{x^2+y^2}}\Big) = 0. \end{aligned}$$

Trivially, we have $\operatorname{div} V = 0$ where $x \ge 0$.

Without loss of generality, we assume that F is closed, that $\partial F \cap A$ is a smooth (say, Lipschitz) surface, and that $F \setminus E = \emptyset$ in such a way that $E \Delta F = E \setminus F = E \cap F'$, where $F' = \mathbb{H}^n \setminus F$.

Let N^E, N^F, and $N^{E\cap F'}$ denote the Euclidean outer unit normals to the boundary of ∂E, ∂F, and $\partial(E \cap F')$, respectively. By the divergence theorem, we have

$$0 = \int_{E\cap F'} \operatorname{div} V \, dzdt = \int_{\partial(E\cap F')} \langle V, N^{E\cap F'} \rangle d\mathcal{H}^2 \\ = \int_{\partial E\cap F'} \langle V, N^E \rangle d\mathcal{H}^2 \\ - \int_{\partial F\cap E} \langle V, N^F \rangle d\mathcal{H}^2. \tag{5.7}$$

On ∂E, we have

$$\varphi = \frac{\big(\langle X, N^E \rangle, \langle Y, N^E \rangle\big)}{\sqrt{\langle X, N^E \rangle^2 + \langle Y, N^E \rangle^2}},$$

and thus

$$\langle V, N^E \rangle = \varphi_1 \langle X, N^E \rangle + \varphi_2 \langle Y, N^E \rangle = \sqrt{\langle X, N^E \rangle^2 + \langle Y, N^E \rangle^2}.$$

By the area formula (3.2), it follows that

$$\int_{\partial E\cap F'} \langle V, N^E \rangle d\mathcal{H}^2 = \int_{\partial E\cap F'} \sqrt{\langle X, N^E \rangle^2 + \langle Y, N^E \rangle^2} d\mathcal{H}^2 \\ = P(E; F').$$

On the other hand, on ∂F we have $|\varphi| = 1$ and by the Cauchy-Schwarz inequality we obtain

$$\langle V, N^F \rangle = \varphi_1 \langle X, N^F \rangle + \varphi_2 \langle Y, N^F \rangle \leq \sqrt{\langle X, N^F \rangle^2 + \langle Y, N^F \rangle^2}.$$

So we deduce that

$$\int_{\partial F\cap E} \langle V, N^F \rangle d\mathcal{H}^2 \leq \int_{\partial F\cap E} \sqrt{\langle X, N^F \rangle^2 + \langle Y, N^F \rangle^2} d\mathcal{H}^2 = P(F; E).$$

So (5.7) implies $P(E; F') \leq P(F; E)$, and this is equivalent to (5.6).

5.2.2 An H-minimal intrinsic graph with discontinuous normal In this example, we study an H-minimal intrinsic Lipschitz graph with discontinuous horizontal normal. This surface is a t-graph with standard $C^{1,1}$-regularity.

Let $\varphi : \mathbb{R}^2 \to \mathbb{R}$ be the function $\varphi(y,t) = \mathrm{sgn}(t)\sqrt{|t|}$. The intrinsic epigraph of φ in the sense of Definition 3.3 is the set

$$E = \{(s, y, t + 2ys) \in \mathbb{H}^1 : (y,t) \in \mathbb{R}^2,\ s > \varphi(y,t)\}.$$

The boundary of E is the intrinsic graph of φ:

$$\partial E = \{(\varphi(y,t), y, t + 2y\varphi(y,t)) \in \mathbb{H}^1 : (y,t) \in \mathbb{R}^2\}.$$

The intrinsic gradient of φ in the sense of Definition 3.4 reduces to the Burgers' component

$$\nabla^\varphi \varphi = \mathscr{B}\varphi = \varphi_y - 4\varphi\varphi_t = -2\mathrm{sgn}(t), \quad t \neq 0.$$

Then $\nabla^\varphi \varphi \in L^\infty(\mathbb{R}^2)$ and $\mathrm{gr}(\varphi)$ is an intrinsic Lipschitz graph, see Theorem 3.9. Moreover, by formula (3.13) the horizontal normal to ∂E is

$$\nu_E = \frac{(1, -\nabla^\varphi \varphi)}{\sqrt{1 + |\nabla^\varphi \varphi|^2}} = \frac{1}{\sqrt{5}}(1, 2\mathrm{sgn}(t)).$$

The normal can be extended in a constant way to $\mathbb{H}^1 \setminus \{x \neq 0\}$, when $x > 0$ and $x < 0$, separately.

Letting $x = \mathrm{sgn}(t)\sqrt{|t|}$, we realize that ∂E is the t-graph of the function $f : \mathbb{R}^2 \to \mathbb{R}$, $f(x,y) = x|x| + 2xy$:

$$\partial E = \{(x, y, f(x,y)) \in \mathbb{H}^1 : (x,y) \in \mathbb{R}^2\}.$$

Clearly, we have $f \in C^{1,1}(\mathbb{R}^2)$.

We claim that E is a local minimizer for H-perimeter in $\mathbb{H}^1$. Namely, we prove that for any bounded open set $A \subset \mathbb{H}^1$ and for any measurable set $F \subset \mathbb{H}^1$ with locally finite H-perimeter and such that $E \Delta F \Subset A$ there holds

$$P(E; A) \leq P(F; A). \tag{5.8}$$

Without loss of generality, we assume that $\partial F \cap A$ is a smooth surface. Let $G = E \Delta F$ and consider the subsets of G:

$$G^- = (E \Delta F) \cap \{x < 0\} \quad \text{and} \quad G^+ = (E \Delta F) \cap \{x > 0\}.$$

Let N^E, N^F, N^G be the Euclidean outer normals to $\partial E, \partial F$, and ∂G, respectively. To fix ideas, we assume that $F \setminus E = \emptyset$, so that we have

$$
\begin{aligned}
N^G &= N^E \quad \text{a.e. on } \partial E \cap \bar{G}, \\
N^G &= -N^F \quad \text{a.e. on } \partial F \cap \bar{G}.
\end{aligned}
$$

Define the horizontal vector field V in $\mathbb{H}^1$ by $V = \sqrt{5}(\nu_E^1 X + \nu_E^2 Y)$, where $\nu_E = (\nu_E^1, \nu_E^2)$ is the extended horizontal normal. Namely, we let

$$
V = \begin{cases} X - 2Y & x < 0 \\ X + 2Y & x > 0. \end{cases}
$$

The vector field V is not defined on the plane $x = 0$. When $x \neq 0$ we have $\operatorname{div} V = \operatorname{div}_H \nu_E = 0$. By the divergence theorem applied to G^- and G^+, we obtain

$$
\begin{aligned}
0 &= \int_G \operatorname{div} V\, dzdt = \int_{G^-} \operatorname{div} V\, dzdt + \int_{G^+} \operatorname{div} V\, dzdt \\
&= \int_{\partial G^-} \langle V, N_{G^-} \rangle d\mathcal{H}^2 + \int_{\partial G^+} \langle V, N_{G^+} \rangle d\mathcal{H}^2.
\end{aligned}
$$

We denote by V^- and V^+ the traces of V onto $\{x = 0\}$, from the left and from the right. The integral on ∂G^- is

$$
\begin{aligned}
\int_{\partial G^-} \langle V, N_{G^-} \rangle d\mathcal{H}^2 &= \int_{\partial G \cap \{x<0\}} \langle V, N^G \rangle d\mathcal{H}^2 + \int_{G \cap \{x=0\}} \langle V^-, N^G \rangle d\mathcal{H}^2 \\
&= \int_{\partial E \cap \{x<0\}} \langle V, N^E \rangle d\mathcal{H}^2 - \int_{\partial F \cap \{x<0\}} \langle V, N^F \rangle d\mathcal{H}^2 \\
&\quad + \int_{G \cap \{x=0\}} \langle V^-, \mathrm{e}_1 \rangle d\mathcal{H}^2.
\end{aligned}
$$

Using the identities and inequalities

$$
\begin{aligned}
\langle V, N^E \rangle &= -|N_H^E| = -\sqrt{\langle N^E, X \rangle^2 + \langle N^E, Y \rangle^2} \quad \text{on } \partial E, \\
\langle V, N^F \rangle &\geq -|N_H^F| = -\sqrt{\langle N^F, X \rangle^2 + \langle N^F, Y \rangle^2} \quad \text{on } \partial F,
\end{aligned}
$$

and $\langle V^-, \mathrm{e}_1 \rangle = 1$, we conclude that

$$
\begin{aligned}
\int_{\partial G^-} \langle V, N_{G^-} \rangle d\mathcal{H}^2 \leq & -\int_{\partial E \cap \{x<0\}} |N_H^E| d\mathcal{H}^2 + \int_{\partial F \cap \{x<0\}} |N_H^F| d\mathcal{H}^2 \\
& + \mathcal{H}^2(G \cap \{x = 0\}).
\end{aligned}
$$

A similar computation for G^+ yields

$$\int_{\partial G^+} \langle V, N_{G^+}\rangle d\mathscr{H}^2 \leq -\int_{\partial E\cap\{x>0\}} |N_H^E| d\mathscr{H}^2 + \int_{\partial F\cap\{x>0\}} |N_H^F| d\mathscr{H}^2 - \mathscr{H}^2(G\cap\{x=0\}).$$

Adding the last two inequalities, the contribution from $G \cap \{x = 0\}$ cancels, and using the area formula (3.2), we finally obtain

$$P(E; \{x \neq 0\} \cap A) \leq P(F; \{x \neq 0\} \cap A) \leq P(F; A).$$

Since $P(E; \{x = 0\}) = 0$, this proves the claim (5.8).

5.2.3 Sets with constant horizontal normal In the previous two examples, the calibration is provided by a suitable extension of the horizontal normal ν_E, extension that is divergence free. A special but interesting case of this situation is when the normal is in fact constant. In this section, we describe sets in $\mathbb{H}^1$ that have, locally, constant horizontal normal (see [50]).

For $r > 0$ and $p \in \mathbb{H}^1$, we let

$$\begin{aligned} Q_r &= \{(x, y, t) \in \mathbb{H}^1 : |x| < r,\ |y| < r,\ |t| < r^2\}, \\ Q_r(p) &= p \cdot Q_r. \end{aligned} \tag{5.9}$$

For $r > 0$ and $(y_0, t_0) \in \mathbb{R}^2$, we also define

$$D_r(y_0, t_0) = \{(y, t) \in \mathbb{R}^2 : |y - y_0| < r,\ |t - t_0| < r^2\}, \tag{5.10}$$

and we let $D_r = D_r(0)$

Theorem 5.6. *Let $E \subset \mathbb{H}^1$ be a set with finite $\mathbb{H}$-perimeter in Q_{4r}, $r > 0$, with $0 \in \partial E$. Assume that $\nu_E(p) = (1, 0) \in \mathbb{S}^1$ for μ_E-a.e. $p \in Q_{4r}$. Then there exists a function $g : D_r \to (-r/4, r/4)$ such that:*

i) *We have, up to a negligible set,*

$$E \cap Q_r = \{(x, y, t) \in Q_r : x > g(y, t)\}.$$

ii) *$g(0) = 0$ and for all $(y, t), (y', t') \in D_r$*

$$|g(y, t) - g(y', t')| \leq |y - y'| + \frac{1}{2r}|t - t'|. \tag{5.11}$$

iii) *The graph of g consists of integral lines of the vector field Y.*

Proof. For the sake of simplicity, we assume that E is open. For any $\alpha, \beta \in \mathbb{R}$ with $\alpha \geq 0$, let $Z = \alpha X + \beta Y$. Then, for any $\varphi \in C_c^1(Q_{4r})$ with $\varphi \geq 0$, by the Gauss-Green formula (2.6) we have

$$\int_E Z\varphi \, dz dt = -\alpha \int_{Q_{4r}} \varphi \, d\mu_E \leq 0,$$

that is $Z\chi_E \geq 0$ in the sense of distribution. It follows that

$$p \in E \cap Q_{4r} \quad \Rightarrow \quad \exp(sZ)(p) \in E, \tag{5.12}$$

for all $s > 0$ such that $\exp(sZ)(p) \in Q_{4r}$.

For any point $q \in E \cap Q_{2r}$ consider the set $E_q = q^{-1} \cdot E$. The set E_q has constant measure theoretic normal $(1, 0) \in \mathbb{S}^1$ in Q_{2r}. We can apply (5.12) to the set E_q starting first from the point $0 \in E_q$ and then from a generic point $p = (0, y, 0) \in E_q$ with $|y| < 2r$. We deduce that

$$\{(x, y, t) \in Q_{2r} : x > 0, \ |t| < 4rx\} \subset E_q.$$

In other words, we have

$$q \in E \cap Q_{2r} \quad \Rightarrow \quad q \cdot \{(x, y, t) \in Q_{2r} : x > 0, \ |t| < 4rx\} \subset E. \tag{5.13}$$

From (5.13), it follows that $E \cap Q_{2r} \cap \{y = 0\}$ is a planar set with the cone property, the cones having all axis parallel to the x-axis and aperture $4r$. We deduce that there exists a Lipschitz function $h : (-r^2, r^2) \to \mathbb{R}$ such that:

(a) $\{(x, t) \in \mathbb{R}^2 : (x, 0, t) \in E\} = \{(x, t) \in D_{2r} : x > h(t)\}$;

(b) $|h(t) - h(t')| \leq \dfrac{1}{4r}|t - t'|$ for all $t, t' \in (-r^2, r^2)$.

Since $0 \in \partial E$, we infer that $h(0) = 0$. From (5.13), we also deduce that ∂E consists of integral lines of Y in Q_{2r}. Then we have

$$\partial E \cap Q_{2r} = \{(h(\tau), \sigma, \tau - 2\sigma h(\tau)) \in \mathbb{H}^1 : (\sigma, \tau) \in D_{2r}\}. \tag{5.14}$$

For any $(y, t) \in D_r$, the system of equations

$$\sigma = y, \quad \tau - 2\sigma h(\tau) = t$$

has a unique solution $(\sigma, \tau) \in D_{2r}$. This is an easy consequence of the Banach fixed point theorem. We claim that the solution $\tau = \tau(y, t)$ of the equation $\tau - 2yh(\tau) = t$ is Lipschitz continuous. Namely, by (b), we have for $(y,t), (y',t') \in D_r$

$$\begin{aligned}|\tau(y,t) - \tau(y',t')| &= |t - 2yh(\tau(y,t)) - t' + 2y'h(\tau(y',t'))|\\ &\leq |t-t'| + 2|y||h(\tau(y,t)) - h(\tau(y',t'))|\\ &\quad + 2|h(\tau(y',t'))||y-y'|\\ &\leq |t-t'| + \frac{1}{2}|\tau(y,t) - \tau(y',t')|\\ &\quad + \frac{1}{2r}|\tau(y',t')||y-y'|,\end{aligned}$$

and this implies

$$|\tau(y,t) - \tau(y',t')| \leq 4r|y-y'| + 2|t-t'|. \tag{5.15}$$

The function $g = h \circ \tau$ satisfies i), ii), and iii). In particular, (5.11) follows from (5.15), and $|g(y,t)| < r/4$ follows from (b). □

There are H-perimeter minimizing surfaces in $\mathbb{H}^1$ with a diffuse Lipschitz regularity. In fact, if $g : D_r \to (-r/4, r/4)$ is a function satisfying ii) and iii) of Theorem 5.6, then its x-graph is a Lipschitz surface that has, $\mathscr{H}^2$-a.e., constant horizontal normal. This vector can be used to show that the x-graph of g is locally minimizing H-perimeter.

Remark 5.7. If, in Theorem 5.6, the radius r can be taken arbitrarily large, then from (5.11) we deduce that the function g does not depend on t. Then from statement iii), we deduce that g does not depend on y, either. Thus E is a vertical half-space. This fact is used in Theorem 2.10.

When $n \geq 2$, the situation is different and easier because if the horizontal normal ν_E is constant in a small convex set then, inside this set, E is a vertical hyperplane orthogonal to the normal (see [27]).

5.3 Lipschitz approximation and height estimate

The notion of horizontal excess is natural:

Definition 5.8 (Horizontal excess). Let $E \subset \mathbb{H}^n$ be a set with locally finite H-perimeter. The *horizontal excess* of E in a ball $B_r(p)$, where $p \in \mathbb{H}^n$ and $r > 0$, is

$$\mathrm{Exc}(E, B_r(p)) = \min_{\substack{\nu\in\mathbb{R}^{2n}\\ |\nu|=1}} \frac{1}{r^{Q-1}} \int_{B_r(p)} |\nu_E - \nu|^2 d\mu_E.$$

Intrinsic Lipschitz graphs are introduced in Definition 3.3, the notion of L-intrinsic Lipschitz function is introduced in Definition 3.8. The following theorem is proved in [50].

Theorem 5.9 (Monti). *Let $n \geq 1$ and let $L > 0$ be a constant that is suitably large when $n = 1$. There are constants $k > 1$ and $c(L, n) > 0$ with the following property. For any set $E \subset \mathbb{H}^n$ that is H-perimeter minimizing in B_{kr} with $0 \in \partial E$ and $r > 0$, there exist $\nu \in \mathbb{R}^{2n}$ with $|\nu| = 1$ and an L-intrinsic Lipschitz function $\varphi : H_\nu \to \mathbb{R}$ such that*

$$\mathscr{S}_Q^{Q-1}\big((\mathrm{gr}(\varphi)\Delta\partial E) \cap B_r\big) \leq c(L, n)(kr)^{Q-1}\mathrm{Exc}(E, B_{kr}). \tag{5.16}$$

The following extension of the so-called "height estimate" to H-perimeter minimizing sets will be proved in the forthcoming paper [57].

Let $\nu = (1, 0 \dots, 0) \in \mathbb{R}^{2n}$ and let $W = \partial H_\nu \subset \mathbb{H}^n$ be the vertical hyperplane orthogonal to ν, i.e., $W = \{x_1 = 0\}$. For any $r > 0$ we let

$$D_r = \big\{w \in W : \|w\|_\infty < r\big\},$$

and we define the truncated cylinder over D_r

$$C_r = D_r \cdot (-r, r) = \big\{w \cdot (s\nu) \in \mathbb{H}^n : |s| < r\big\}.$$

The ν-directional excess of E inside the cylinder D_r is

$$\mathrm{Exc}(E, C_r, \nu) = \frac{1}{r^{Q-1}} \int_{C_r} |\nu_E - \nu|^2 d\mu_E.$$

Theorem 5.10 (Monti-Vittone). *Let $n \geq 2$. There exist constants $\epsilon_0 > 0$, $c_0 > 0$, and $k > 0$ such that if $E \subset \mathbb{H}^n$ is an H-perimeter minimizing set in C_{kr} with*

$$\mathrm{Exc}(E, C_{kr}, \nu) \leq \epsilon_0,$$

then we have

$$\begin{aligned} &\sup\Big\{|x_1| = |\mathrm{Re}(z_1)| \in \mathbb{R} : (z, t) \in \partial E \cap C_r\Big\} \\ &\leq c_0 r \,\mathrm{Exc}(E, C_{kr}, \nu)^{\frac{1}{2(Q-1)}}. \end{aligned} \tag{5.17}$$

The proof follows the scheme of [70]. It relies on a nontrivial slicing technique and on a lower dimensional isoperimetric inequality. The estimate (5.17) does not hold when $n = 1$ because of the examples of Section 5.2.3, for which $\mathrm{Exc}(E, B_r) = 0$ but ∂E is not flat.

ACKNOWLEDGEMENTS. We wish to thank Luca Capogna, Valentina Franceschi, Gian Paolo Leonardi, Valentino Magnani, Andrea Malchiodi, Francesco Serra Cassano, Manuel Ritoré, and Davide Vittone for their comments on a preliminary version of the notes.

References

[1] L. AMBROSIO, *Some fine properties of sets of finite perimeter in Ahlfors regular metric measure spaces*, Adv. Math. **159** (2001), no. 1, 51–67.

[2] L. AMBROSIO, *Transport equation and Cauchy problem for BV vector fields*, Invent. Math. **158** (2004), no. 2, 227–260.

[3] L. AMBROSIO, B. KLEINER and E. LE DONNE, *Rectifiability of sets of finite perimeter in Carnot groups: existence of a tangent hyperplane*, J. Geom. Anal. **19** (2009), no. 3, 509–540.

[4] L. AMBROSIO, F. SERRA CASSANO and D. VITTONE, *Intrinsic regular hypersurfaces in Heisenberg groups*, J. Geom. Anal. **16** (2006), no. 2, 187–232.

[5] G. ARENA, A. O. CARUSO and R. MONTI, *Regularity properties of H-convex sets*, J. Geom. Anal. **22** (2012), no. 2, 583–602.

[6] Z. BALOGH, *Size of characteristic sets and functions with prescribed gradients*, J. Reine Angew. Math. **564** (2003), 63–83.

[7] F. BIGOLIN, L. CARAVENNA and F. SERRA CASSANO, *Intrinsic Lipschitz graphs in Heisenberg groups and continuous solutions of a balance equation*, Ann. I. H. Poincaré - AN (2014), http://dx.doi..org/10.1016/j.anihpc.2014.05.001.

[8] V. BARONE ADESI, F. SERRA CASSANO, D. VITTONE, The Bernstein problem for intrinsic graphs in Heisenberg groups and calibrations. Calc. Var. Partial Differential Equations 30 (2007), no. 1, 17–49.

[9] L. CAPOGNA, G. CITTI and M. MANFREDINI, *Regularity of non-characteristic minimal graphs in the Heisenberg group* $\mathbb{H}^1$, Indiana Univ. Math. J. **58** (2009), no. 5, 2115–2160.

[10] L. CAPOGNA, G. CITTI and M. MANFREDINI, *Smoothness of Lipschitz minimal intrinsic graphs in Heisenberg groups* $\mathbb{H}^n$, $n > 1$, J. Reine Angew. Math. **648** (2010), 75–110.

[11] L. CAPOGNA, D. DANIELLI, S. D. PAULS and J. TYSON, "An Introduction to the Heisenberg Group and the Sub-Riemannian Isoperimetric Problem", Progress in Mathematics, Vol. 259. Birkhäuser Verlag, Basel, 2007. xvi+223 pp.

[12] J.-H. CHENG, J.-F. HWANG and P. YANG, *Existence and uniqueness for p-area minimizers in the Heisenberg group*, Math. Ann. **337** (2007), no. 2, 253–293.

[13] J.-H. CHENG, J.-F. HWANG and P. YANG, *Regularity of C^1 smooth surfaces with prescribed p-mean curvature in the Heisenberg group*, Math. Ann. **344** (2009), no. 1, 1–35.

[14] J.-H. CHENG, J.-F. HWANG, A. MALCHIODI and P. YANG, *Minimal surfaces in pseudohermitian geometry*, Ann. Sc. Norm. Super. Pisa Cl. Sci. (5) **4** (2005), 129–177.

[15] J.-H. CHENG, J.-F. HWANG, A. MALCHIODI, P. YANG, *A Codazzi-like equation and the singular set for C^1 smooth surfaces in the Heisenberg group*, J. Reine Angew. Math. **671** (2012), 131–198.

[16] G. CITTI, M. MANFREDINI A. PINAMONTI and F. SERRA CASSANO, *Smooth approximation for intrinsic Lipschitz functions in the Heisenberg group*, Calc. Var. Partial Differential Equations **49** (2014), no. 3-4, 1279–1308.

[17] G. CITTI, M. MANFREDINI A. PINAMONTI and F. SERRA CASSANO, *Poincaré type inequality for intrinsic Lipschitz continuous vector fields in the Heisenberg group*, preprint 2013.

[18] D. DANIELLI, N. GAROFALO and D.-M. NHIEU, *Sub-Riemannian calculus on hypersurfaces in Carnot groups*, Adv. Math. **215** (2007), no. 1, 292–378.

[19] D. DANIELLI, N. GAROFALO and D.-M. NHIEU, *A partial solution of the isoperimetric problem for the Heisenberg group*, Forum Math. **20** (2008), no. 1, 99–143.

[20] D. DANIELLI, N. GAROFALO and D.-M. NHIEU, *A notable family of entire intrinsic minimal graphs in the Heisenberg group which are not perimeter minimizing*, Amer. J. Math. **130** (2008), no. 2, 317–339.

[21] D. DANIELLI, N. GAROFALO and D.-M. NHIEU, *Sub-Riemannian calculus and monotonicity of the perimeter for graphical strips*, Math. Z. **265** (2010), no. 3, 617–637.

[22] D. DANIELLI, N. GAROFALO and D.-M. NHIEU, *Integrability of the sub-Riemannian mean curvature of surfaces in the Heisenberg group* Proc. Amer. Math. Soc. **140** (2012), no. 3, 811–821.

[23] D. DANIELLI, N. GAROFALO, D.-M. NHIEU and S. D. PAULS, *Instability of graphical strips and a positive answer to the Bernstein problem in the Heisenberg group $\mathbb{H}^1$*, J. Differential Geom. **81** (2009), no. 2, 251–295.

[24] D. DANIELLI, N. GAROFALO, D.-M. NHIEU and S. D. PAULS, *The Bernstein problem for embedded surfaces in the Heisenberg group $\mathbb{H}^1$*, Indiana Univ. Math. J. **59** (2010), no. 2, 563–594.

[25] H. FEDERER, "Geometric Measure Theory", Die Grundlehren der mathematischen Wissenschaften, Band 153 Springer-Verlag New York Inc., New York 1969 xiv+676 pp.

[26] B. FRANCHI, S. GALLOT and R. L. WHEEDEN, *Sobolev and isoperimetric inequalities for degenerate metrics*, Math. Ann. **300** (1994), 557–571.

[27] B. Franchi, R. Serapioni and F. Serra Cassano, *Rectifiability and perimeter in the Heisenberg group*, Math. Ann. **321** (2001), 479–531.

[28] B. Franchi, R. Serapioni, F. Serra Cassano, *Intrinsic Lipschitz graphs in Heisenberg groups*, J. Nonlinear Convex Anal. **7** (2006), no. 3, 423–441.

[29] B. Franchi, R. Serapioni and F. Serra Cassano, *Regular submanifolds, graphs and area formula in Heisenberg groups*, Adv. Math. **211** (2007), no. 1, 152–203.

[30] B. Franchi, R. Serapioni and F. Serra Cassano, *Differentiability of intrinsic Lipschitz functions within Heisenberg groups*, J. Geom. Anal. **21** (2011), no. 4, 1044–1084.

[31] M. Galli, *First and second variation formulae for the sub-Riemannian area in three-dimensional pseudo-Hermitian manifolds*, Calc. Var. Partial Differential Equations **47** (2013), no. 1-2, 117–157.

[32] M. Galli and M. Ritoré, *Existence of isoperimetric regions in contact sub-Riemannian manifolds*, J. Math. Anal. Appl. **397** (2013), no. 2, 697–714.

[33] N. Garofalo and D.-M. Nhieu, *Isoperimetric and Sobolev inequalities for Carnot–Carathéodory spaces and the existence of minimal surfaces*, Comm. Pure Appl. Math. **49** (1996), 1081–1144.

[34] R. K. Hladky and S. D. Pauls, *Constant mean curvature surfaces in sub-Riemannian geometry*, J. Differential Geom. **79** (2008), no. 1, 111–139.

[35] B. Kirchheim and F. Serra Cassano, *Rectifiability and parameterization of intrinsic regular surfaces in the Heisenberg group*, Ann. Sc. Norm. Super. Pisa Cl. Sci. (5) **3** (2004), no. 4, 871–896.

[36] A. Korányi and H. M. Reimann, *Foundations for the theory of quasiconformal mappings on the Heisenberg group*, Adv. Math. **111** (1995), no. 1, 1–87.

[37] G. P. Leonardi and S. Masnou, *On the isoperimetric problem in the Heisenberg group* $\mathbb{H}^n$, Ann. Mat. Pura Appl. (4) **184** (2005), 533–553.

[38] G. P. Leonardi and S. Rigot, *Isoperimetric sets on Carnot groups*, Houston J. Math. **29** (2003), 609–637.

[39] V. Magnani, *Characteristic points, rectifiability and perimeter measure on stratified groups*, J. Eur. Math. Soc. **8** (2006), no. 4, 585–609.

[40] V. Magnani, *Area implies coarea*, Indiana Univ. Math. J. **60** (2011), no. 1, 77–100.

[41] V. MAGNANI, *On a measure theoretic area formula, Proceedings of The Royal Society of Edinburgh*, 2014, to appear.

[42] V. MAGNANI, Personal communication.

[43] P. MATTILA, R. SERAPIONI and F. SERRA CASSANO, *Characterizations of intrinsic rectifiability in Heisenberg groups*, Ann. Sc. Norm. Super. Pisa Cl. Sci. (5) **9** (2010), no. 4, 687–723.

[44] F. MONTEFALCONE, *Some relations among volume, intrinsic perimeter and one-dimensional restrictions of BV functions in Carnot groups*, Ann. Sc. Norm. Super. Pisa Cl. Sci. (5) **4** (2005), no. 1, 79–128.

[45] F. MONTEFALCONE, *Hypersurfaces and variational formulas in sub-Riemannian Carnot groups*, J. Math. Pures Appl. (9) **87** (2007), no. 5, 453–494.

[46] R. MONTI, *Some properties of Carnot-Carathéodory balls in the Heisenberg group*, Atti Accad. Naz. Lincei Cl. Sci. Fis. Mat. Natur. Rend. Lincei (9) Mat. Appl. **11** (2001), no. 3, 155–167.

[47] R. MONTI, *Brunn–Minkowski and isoperimetric inequality in the Heisenberg group*, Ann. Acad. Sci. Fenn. Math. **28** (2003), 99–109.

[48] R. MONTI, *Heisenberg isoperimetric problem. The axial case*, Adv. Calc. Var. **1** (2008), no. 1, 93–121.

[49] R. MONTI, *Rearrangements in metric spaces and in the Heisenberg group*, J. Geom. Anal. **24** (2014), n. 4, 1673–1715.

[50] R. MONTI, *Lipschitz approximation of H-perimeter minimizing boundaries*, Calc. Var. Partial Differential Equations **50** (2014), no. 1-2, 171–198.

[51] R. MONTI and D. MORBIDELLI, *Isoperimetric inequality in the Grushin plane*, J. Geom. Anal. **14** (2004), 355–368.

[52] R. MONTI and D. MORBIDELLI, *Regular domains in homogeneous groups*, Trans. Amer. Math. Soc. **357** (2005), no. 8, 2975–3011.

[53] R. MONTI and M. RICKLY, *Convex isoperimetric sets in the Heisenberg group*, Ann. Sc. Norm. Super. Pisa Cl. Sci. (5) **8** (2009), no. 2, 391–415.

[54] R. MONTI and F. SERRA CASSANO, *Surface measures in Carnot–Carathéodory spaces*, Calc. Var. Partial Differential Equations **13** (2001), 339–376.

[55] R. MONTI, F. SERRA CASSANO and D. VITTONE, *A negative answer to the Bernstein problem for intrinsic graphs in the Heisenberg group*, Boll. Unione Mat. Ital. (9) **1** (2008), no. 3, 709–727.

[56] R. MONTI and D. VITTONE, *Sets with finite H-perimeter and controlled normal*, Math. Z. **270** (2012), no. 1-2, 351–367.

[57] R. MONTI and D. VITTONE, *Height estimate and slicing formulas in the Heisenberg group*, preprint 2014.

[58] P. PANSU, *Une inégalité isopérimétrique sur le groupe de Heisenberg*, C. R. Acad. Sci. Paris Sér. I Math. **295** (1982), 127–130.

[59] P. PANSU, *An isoperimetric inequality on the Heisenberg group. Conference on differential geometry on homogeneous spaces (Turin, 1983)*, Rend. Sem. Mat. Univ. Politec. Torino, Special Issue (1983), 159–174.

[60] S. D. PAULS, *Minimal surfaces in the Heisenberg group*, Geom. Dedicata **104** (2004), 201–231.

[61] S. D. PAULS, H-*minimal graphs of low regularity in* $\mathbb{H}^1$, Comment. Math. Helv. **81** (2006), 337–381.

[62] D. PRANDI, "Rearrangements in Metric Spaces", Master Thesis, Chapter 2, available at http://www.math.unipd.it/ monti/tesi/TESI-finale.pdf

[63] A. PINAMONTI, F. SERRA CASSANO, G. TREU and D. VITTONE, *BV Minimizers of the area functional in the Heisenberg group under the bounded slope condition*, Ann. Sc. Norm. Super. Pisa Cl. Sci., 2014, to appear.

[64] I. PLATIS, *Straight ruled surfaces in the Heisenberg group*, J. Geom. **105** (2014), no. 1, 119–138.

[65] S. RIGOT, *Counterexample to the Besicovitch covering property for some Carnot groups equipped with their Carnot-Carathéodory metric*, Math. Z. **248** (2004), no. 4, 827–848.

[66] M. RITORÉ, *Examples of area-minimizing surfaces in the sub-Riemannian Heisenberg group* $\mathbb{H}^1$ *with low regularity*, Calc. Var. Partial Differential Equations **34** (2009), no. 2, 179–192.

[67] M. RITORÉ, *A proof by calibration of an isoperimetric inequality in the Heisenberg group* $\mathbb{H}^n$, Calc. Var. Partial Differential Equations **44** (2012), no. 1-2, 47–60.

[68] M. RITORÉ and C. ROSALES, *Rotationally invariant hypersurfaces with constant mean curvature in the Heisenberg group* $\mathbb{H}^n$, J. Geom. Anal. **16** (2006), no. 4, 703–720.

[69] M. RITORÉ and C. ROSALES, *Area-stationary surfaces in the Heisenberg group* $\mathbb{H}^1$, Adv. Math. **219** (2008), no. 2, 633–671.

[70] R. SCHOEN and L. SIMON, *A new proof of the regularity theorem for rectifiable currents which minimize parametric elliptic functionals*, Indiana Univ. Math. J. **31** (1982), no. 3, 415–434.

[71] F. SERRA CASSANO and D. VITTONE, *Graphs of bounded variation, existence and local boundedness of non-parametric minimal surfaces in Heisenberg groups*, Adv. Calc. Var. **7** (2014), 409–492. DOI: 10.1515/acv-2013-0105.

Regularity of higher codimension area minimizing integral currents

Emanuele Spadaro

Abstract. This lecture notes are an expanded and revised version of the course *Regularity of higher codimension area minimizing integral currents* that I taught at the *ERC-School on Geometric Measure Theory and Real Analysis*, held in Pisa, September 30th - October 30th 2013.

The lectures aim to explain partially without proofs the main steps of a new proof of the partial regularity of area minimizing integer rectifiable currents in higher codimension, due originally to F. Almgren, which is contained in a series of papers in collaboration with C. De Lellis (University of Zürich).

Contents

1 Introduction

The subject of this course is the study of the regularity of *minimal surfaces*, considered in the sense of *area minimizing integer rectifiable currents*. This is a very classical topic and stems from many diverse questions and applications. Among the most known there is perhaps the so called *Plateau problem*, consisting in finding the submanifolds of least possible volume among all those submanifolds with a fixed boundary.

Plateau problem. Let M be a $(m+n)$-dimensional Riemannian manifold and $\Gamma \subset M$ a compact $(m-1)$-dimensional oriented submanifold. Find an m-dimensional oriented submanifold Σ with boundary Γ such that

$$\mathrm{vol}_m(\Sigma) \leq \mathrm{vol}_m(\Sigma'),$$

for all oriented submanifolds $\Sigma' \subset M$ such that $\partial \Sigma' = \Gamma$.

It is a well-known fact that the solution of the Plateau problem does not always exist. For example, consider $M = \mathbb{R}^4$, $n = m = 2$ and Γ the smooth Jordan curve parametrized in the following way:

$$\Gamma = \left\{(\zeta^2, \zeta^3) : \zeta \in \mathbb{C},\ |\zeta| = 1\right\} \subset \mathbb{C}^2 \simeq \mathbb{R}^4,$$

where we use the usual identification between $\mathbb{C}^2$ and $\mathbb{R}^4$, and we choose the orientation of Γ induced by the anti-clockwise orientation of the unit circle $|\zeta| = 1$ in $\mathbb{C}$. It can be shown (and we will come back to this point in the next sections) that there exist no smooth solutions to the Plateau problem for such fixed boundary, and the *(singular) immersed* 2-dimensional disk

$$S = \left\{(z, w)\ :\ z^3 = w^2,\ |z| \leq 1\right\} \subset \mathbb{C}^2 \simeq \mathbb{R}^4,$$

oriented in such a way that $\partial S = \Gamma$, satisfies

$$\mathcal{H}^2(S) < \mathcal{H}^2(\Sigma),$$

for all smooth, oriented 2-dimensional submanifolds $\Sigma \subset \mathbb{R}^4$ with $\partial\Sigma = \Gamma$. Here and in the following we denote by $\mathcal{H}^k$ the k-dimensional Hausdorff measure, which for $k \in \mathbb{N}$ corresponds to the ordinary k-volume on smooth k-dimensional submanifolds.

This fact motivates the introduction of *weak solutions* to the Plateau problem, and the main questions about their existence and regularity.

1.1 Integer rectifiable currents

One of the most successful theories of generalized submanifolds is the one by H. Federer and W. Fleming in [19] on integer rectifiable currents (see also [8,9] for the special case of codimension one generalized submanifolds). From now on, in order to keep the technicalities to a minimum level, we assume that our ambient Riemannian manifold M is Euclidean.

Definition 1.1 (Integer rectifiable currents). An integer rectifiable current T of dimension m in $\mathbb{R}^{m+n}$ is a triple $T = (R, \tau, \theta)$ such that:

(i) R is a *rectifiable set, i.e.* $R = \bigcup_{i\in\mathbb{N}} C_i$ with $\mathcal{H}^m(R_0) = 0$ and $C_i \subset M_i$ for every $i \in \mathbb{N} \setminus \{0\}$, where M_i are m-dimensional oriented C^1 submanifolds of $\mathbb{R}^{m+n}$;

(ii) $\tau : R \to \Lambda_m$ is a measurable map, called *orientation*, taking values in the space of m-vectors such that, for $\mathcal{H}^m$-a.e. $x \in C_i$, $\tau(x) = v_1 \wedge \cdots \wedge v_m$ with $\{v_1, \ldots, v_m\}$ an oriented orthonormal basis of $T_x M_i$;

(iii) $\theta : R \to \mathbb{Z}$ is a measurable function, called *multiplicity*, which is integrable with respect to $\mathcal{H}^m$.

An integer rectifiable current $T = (R, \tau, \theta)$ induces a continuous linear functional (with respect to the natural Fréchet topology) on smooth, compactly supported m-dimensional differential forms ω, denoted by $\mathscr{D}^m$, acting as follows

$$T(\omega) = \int_R \theta \, \langle \omega, \tau \rangle \, d\mathcal{H}^m.$$

Remark 1.2. The continuous linear functionals defined in the Fréchet space $\mathscr{D}^m$ are called *m-dimensional currents.*

Remark 1.3. Note that the submanifold M_i in Definition 1.1 are only C^1 regular. This restriction is not redundant, but it is connected to several aspects of the theory of rectifiable sets.

For an integer rectifiable current T, one can define the analog of the boundary and the volume for smooth submanifolds.

Definition 1.4 (Boundary and mass). Let $T = (R, \tau, \theta)$ be an integer rectifiable current in $\mathbb{R}^{m+n}$ of dimension m. The *boundary* of T is defined as the $(m-1)$-dimensional current acting as follows

$$\partial T(\omega) := T(d\omega) \quad \forall\, \omega \in \mathscr{D}^{m-1}.$$

The *mass* of T is defined as the quantity

$$\mathbf{M}(T) := \int_R |\theta|\, d\mathcal{H}^m.$$

Note that, in the case $T = (\Sigma, \tau_\Sigma, 1)$ is the current induced by an oriented submanifold Σ with boundary $\partial\Sigma$, with τ_Σ a continuous orienting vector for Σ and similarly $\tau_{\partial\Sigma}$ for its boundary, then by Stoke's Theorem $\partial T = (\partial\Sigma, \tau_{\partial\Sigma}, 1)$ and $\mathbf{M}(T) = \mathrm{vol}_m(\Sigma)$.

Finally we recall that the space of currents is usually endowed with the weak* topology (often called in this context *weak* topology).

Definition 1.5 (Weak topology). We say that a sequence of currents $(T_l)_{l\in\mathbb{N}}$ weakly converges to some current T, and we write $T_l \rightharpoonup T$, if

$$T_l(\omega) \to T(\omega) \quad \forall\, \omega \in \mathscr{D}^m.$$

The Plateau problem has now a straightforward generalization in this context of integer rectifiable currents.

Generalized Plateau problem. Let Γ be a compactly supported $(m-1)$-dimensional integer rectifiable current in $\mathbb{R}^{m+n}$ with $\partial\Gamma = 0$. Find an m-dimensional integer rectifiable current T such that $\partial T = \Gamma$ and

$$\mathbf{M}(T) \leq \mathbf{M}(S),$$

for every S integer rectifiable with $\partial S = \Gamma$.

The success of the theory of integer rectifiable currents is linked ultimately to the possibility to solve the generalized Plateau problem, due to the closure theorem by H. Federer and W. Fleming proven in their pioneering paper [19].

Theorem 1.6 (Federer and Fleming [19]). *Let $(T_l)_{l\in\mathbb{N}}$ be a sequence of m-dimensional integer rectifiable currents in $\mathbb{R}^{m+n}$ with*

$$\sup_{l\in\mathbb{N}} \big(\mathbf{M}(T_l) + \mathbf{M}(\partial T_l)\big) < +\infty,$$

and assume that $T_l \rightharpoonup T$. Then, T is an integer rectifiable current.

It is then natural to ask about the regularity properties of the solutions to the generalized Plateau problem, called in the sequel *area minimizing* integer rectifiable currents.

1.2 Partial regularity in higher codimension

The regularity theory for area minimizing integer rectifiable currents depends very much on the dimension of the current and its *codimension* in the ambient space (*i.e.*, using the same letters as above, if T is an m-dimensional current in $\mathbb{R}^{m+n}$, the codimension is n).

In this course we are interested in the general case of currents with higher codimensions $n > 1$. The case $n = 1$ is usually treated separately, because different techniques can be used and more refined results can be proven (see [10,20,28,30,32,33] for the interior regularity and [3,23] for the boundary regularity). In higher codimension the most general result is due to F. Almgren [5] and concerns the interior partial regularity up to a (relatively) closed set of dimension at most $m - 2$.

Theorem 1.7 (Almgren [5]). *Let T be an m-dimensional area minimizing integer rectifiable current in $\mathbb{R}^{m+n}$. Then, there exists a closed set* $\mathrm{Sing(T)}$ *of Hausdorff dimension at most $m - 2$ such that in $\mathbb{R}^{m+n} \setminus$* $(\mathrm{spt}\,(\partial T) \cup \mathrm{Sing(T)})$ *the current T is induced by the integration over a smooth oriented submanifold of $\mathbb{R}^{m+n}$.*

In the next pages I will give an overview of the new proof of Theorem 1.7 given in collaboration with C. De Lellis in a series of papers [13–17]. Although our proof is considerably simpler than the original one, it remains quite involved: this text is, therefore, meant as a survey of the techniques and the various steps of the proof, and can be considered an introduction to the reading of the papers [14,15,17].

Remark 1.8. The interior partial regularity can be proven for integer rectifiable currents in a Riemannian manifold M. In [5] Almgren proves the result for C^5 regular ambient manifolds M, while our papers [14,15,17] extend this result to $C^{3,\alpha}$ regular manifolds.

Further notation and terminology

Given an m-dimensional integer rectifiable current $T = (R, \tau, \theta)$, we shall often use the following standard notation:

$$\|T\| := |\theta|\,\mathcal{H}^m \llcorner R, \quad \vec{T} := \tau \quad \text{and} \quad \mathrm{spt}\,(T) := \mathrm{spt}\,(\|T\|).$$

The regular and the singular part of a current are defined as follows.

$$\mathrm{Reg(T)} := \big\{\mathrm{x} \in \mathrm{spt}\,(\mathrm{T}) \;:\; \mathrm{spt}\,(\mathrm{T}) \cap \mathrm{B_r(x)} \text{ is induced by a smooth submanifold for some } r > 0\big\},$$

$$\mathrm{Sing(T)} := \mathrm{spt}\,(\mathrm{T}) \setminus \big(\mathrm{spt}\,(\partial \mathrm{T}) \cup \mathrm{Reg(T)}\big).$$

2 The blowup argument: a glimpse of the proof

The main idea of the proof of Theorem 1.7 is to detect the singularities of an area minimizing current by a blowup analysis. For any $r > 0$ and $x \in \mathbb{R}^{m+n}$, let $\iota_{x,r}$ denote the map

$$\iota_{x,r} : y \mapsto \frac{y - x}{r},$$

and set $T_{x,r} := (\iota_{x,r})_\sharp T$, where $_\sharp$ is the push-forward operator, namely

$$(\iota_{x,r})_\sharp T(\omega) := T(\iota^*_{x,r}\omega) \quad \forall\, \omega \in \mathscr{D}^m.$$

By the classical monotonicity formula (see, *e.g.*, [2, Section 5]), for every $r_k \downarrow 0$ and $x \in \mathrm{spt}\,(T) \setminus \mathrm{spt}\,(\partial T)$, there exists a subsequence (not relabeled) such that

$$T_{x,r_k} \rightharpoonup S,$$

where S is a cone without boundary (*i.e.* $S_{0,r} = S$ for all $r > 0$ and $\partial S = 0$) which is locally area minimizing in $\mathbb{R}^{m+n}$. Such a cone will be called, as usual, *a tangent cone to T at x*.

The idea of the blowup analysis dates back to De Giorgi's pioneering paper [10] and has been used in the context of codimension one currents to recognize singular points and regular points, because in this case the tangent cones to singular and regular points are in fact different.

2.1 Flat tangent cones do not imply regularity

This is not the case for higher codimension currents. In order to illustrate this point, let us consider the current $T_{\mathscr{V}}$ induced by the complex curve considered above:

$$\mathscr{V} = \left\{(z, w) \;:\; z^3 = w^2,\ |z| \le 1\right\} \subset \mathbb{C}^2 \simeq \mathbb{R}^4.$$

It is simple to show that $T_{\mathscr{V}}$ is an area minimizing integer rectifiable current (cp. [18, 5.4.19]), which is singular in the origin. Nevertheless, the unique tangent cone to $T_{\mathscr{V}}$ at 0 is the current $S = (\mathbb{R}^2 \times \{0\}, e_1 \wedge e_2, 2)$ which is associated to the integration on the horizontal plane $\mathbb{R}^2 \times \{0\} \simeq \{w = 0\}$ with multiplicity two. The tangent cone is actually regular, although the origin is a singular point!

2.2 Non-homogeneous blowup

One of the main ideas by Almgren is then to extend this reasoning to different types of blowups, by rescaling differently the "horizontal directions", namely those of a flat tangent cone at the point, and the "vertical" ones, which are the orthogonal complement to the former. In this way, in place of preserving the geometric properties of the rectifiable current T, one is led to preserve the *energy* of the associated *multiple valued function*.

In order to explain this point, let us consider again the current $T_{\mathscr{V}}$. The support of such current, namely the complex curve $\mathscr{V}$, can be viewed as the graph of a function which associates to any $z \in \mathbb{C}$ with $|z| \leq 1$ *two* points in the w-plane:

$$z \mapsto \{w_1(z), w_2(z)\} \quad \text{with } w_i(z)^2 = z^3 \text{ for } i = 1, 2. \tag{2.1}$$

Then the right rescaling according to Almgren is the one producing in the limit a multiple valued *harmonic function* preserving the *Dirichlet energy* (for the definitions see the next sections). In the case of $\mathscr{V}$, the correct rescaling is the one fixing $\mathscr{V}$. For every $\lambda > 0$, we consider $\Phi_\lambda : \mathbb{C}^2 \to \mathbb{C}^2$ given by

$$\Phi_\lambda(z, w) = (\lambda^2 z, \lambda^3 w),$$

and note that $(\Phi_\lambda)_\sharp T_{\mathscr{V}} = T_{\mathscr{V}}$ for every $\lambda > 0$. Indeed, in the case of $\mathscr{V}$ the functions w_1 and w_2, being the two determinations of the square root of z^3, are already harmonic functions (at least away from the origin).

2.3 Multiple valued functions

Following these arguments, we have then to face the problem of defining harmonic multiple valued functions, and to study their singularities. Abstracting from the above example, we consider the multiple valued functions from a domain in $\mathbb{R}^m$ which take a fixed number $Q \in \mathbb{N} \setminus \{0\}$ of values in $\mathbb{R}^n$. This functions will be called in the sequel *Q-valued functions*.

The definition of harmonic Q-valued functions is a simple issue around any "regular point" $x_0 \in \mathbb{R}^m$, for it is enough to consider just the superposition of classical harmonic functions (possibly with a constant integer multiplicity), *i.e.*

$$\mathbb{R}^m \supset B_r(x_0) \ni x \mapsto \{u_1(x), \ldots, u_Q(x)\} \in (\mathbb{R}^n)^Q, \tag{2.2}$$

with u_i harmonic and either $u_i = u_j$ or $u_i(x) \neq u_i(x)$ for every $x \in B_r(x_0)$.

The issue becomes much more subtle around the singular points. As it is clear from the example (2.1), in a neighborhood of the origin there is no representation of the map $z \mapsto \{w_1(z), w_2(z)\}$ as in (2.2). In this case the two values $w_1(z)$ and $w_2(z)$ cannot be ordered in a consistent way (due to the *branch point* at 0), and hence cannot be distinguished one from the other. We are then led to consider a multiple valued function as a map taking Q values in the quotient space $(\mathbb{R}^n)^Q/\sim$ induced by the symmetric group $\mathbf{S}_Q$ of permutation of Q indices: namely, given points $P_i, S_i \in \mathbb{R}^n$,

$$(P_1, \dots, P_Q) \sim (S_1, \dots, S_Q)$$

if there exists $\sigma \in \mathbf{S}_Q$ such that $P_i = S_{\sigma(i)}$ for every $i = 1, \dots, Q$.

Note that the space $(\mathbb{R}^n)^Q/\sim$ is a *singular metric space* (for a naturally defined metric, see the next section). Therefore, harmonic maps with values in $(\mathbb{R}^n)^Q/\sim$ have to be carefully defined, for instance by using the metric theory of harmonic functions developed in [22,25,26] (cp. also [13,27]).

Remark 2.1. Note that the integer rectifiable current induced by the graph of a Q-valued function (under suitable hypotheses, cp. [16, Proposition 1.4]) belongs to a subclass of currents, sometimes called "positively oriented", *i.e.* such that the tangent planes make at almost every point a positive angle with a fixed plane. Nevertheless, as it will become clear along the proof, it is enough to consider this subclass as model currents in order to conclude Theorem 1.7.

2.4 The need of centering

A major geometric and analytic problem has to be addressed in the blowup procedure sketched above. In order to make it apparent, let us discuss another example. Consider the complex curve $\mathscr{W}$ given by

$$\mathscr{W} = \left\{(z, w) \ : \ (w - z^2)^2 = z^5, \ |z| \leq 1\right\} \subset \mathbb{C}^2.$$

As before, $\mathscr{W}$ can be associated to an area minimizing integer rectifiable current $T_{\mathscr{W}}$ in $\mathbb{R}^4$, which is singular at the origin. It is easy to prove that the unique tangent plane to $T_{\mathscr{W}}$ at 0 is the plane $\{w = 0\}$ taken with multiplicity two. On the other hand, by simple analytical considerations, the only nontrivial inhomogeneous blowup in these vertical and horizontal coordinates is given by

$$\Phi_\lambda(z, w) = (\lambda\, z, \lambda^2\, w),$$

and $(\Phi_\lambda)_\sharp T_{\mathscr{W}}$ converges as $\lambda \to +\infty$ to the current induced by the *smooth* complex curve $\{w = z^2\}$ taken with multiplicity two. In other words, the inhomogeneous blowup did not produce in the limit any singular current and cannot be used to study the singularities of $T_{\mathscr{W}}$.

For this reason it is essential to "renormalize" $T_{\mathscr{W}}$ by averaging out its regular first expansion, on top of which the singular branching behavior happens. In the case we handle, the regular part of $T_{\mathcal{W}}$ is exactly the smooth complex curve $\{w = z^2\}$, while the singular branching is due to the determinations of the square root of z^5. It is then clear why one can look for parametrizations of $\mathcal{W}$ defined in $\{w = z^2\}$, so that the singular map to be considered reduces to

$$z \mapsto \{u_1(z), u_2(z)\} \quad \text{with } u_1(z)^2 = z^5.$$

The regular surface $\{w = z^2\}$ is called *center manifold* by Almgren, because it behaves like (and in this case it is exactly) the average of the sheets of the current in a suitable system of coordinates. In general the determination of the center manifold is not straightforward as in the above example, and actually constitutes the most intricate part of the proof.

2.5 Excluding an infinite order of contact

Having taken care of the geometric problem of the averaging, in order to be able to perform successfully the inhomogeneous blowup, one has to be sure that the first singular expansion of the current around its regular part does not occur with an infinite order of contact, because in that case the blowup would be by necessity zero.

This issue involves one of the most interesting and original ideas of F. Almgren, namely a new monotonicity formula for the so called *frequency function* (which is a suitable ratio between the energy and a zero degree norm of the function parametrizing the current). This is in fact the right monotone quantity for the inhomogeneous blowups introduced before, and it allows to show that the first singular term in the "expansion" of the current does not occur with infinite order of contact and actually leads to a nontrivial limiting current.

2.6 The persistence of singularities

Finally, in order to conclude the proof we need to assure that the singularities of the current do transfer to singularities of the limiting multiple valued function, which can be studied with more elementary techniques. This is in general not true in a pointwise sense, but it becomes true in a measure theoretic sense as soon as the singular set is supposed to have positive $\mathcal{H}^{m-2+\alpha}$ measure, for some $\alpha > 0$.

The contradiction is then reached in the following way: starting from an area minimizing current with a big singular set ($\mathcal{H}^{m-2+\alpha}$ positive measure), one can perform the analysis outlined before and will end up with a multiple valued function having a big set of singularities, thus giving the desired contradiction.

2.7 Sketch of the proof

The rigorous proof of Theorem 1.7 is actually much more involved and complicated than the rough outline given in the previous section, and can be found either in [5] or in the recent series of papers [13–17]. In this lecture notes we give some more details of this recent new proof, and comments on some of the subtleties which were hidden in the general discussion above. Since the proof is very lengthly, we start with a description of the strategy.

The proof is done by contradiction. We will, indeed, always assume the following in the sequel.

Contradiction assumption: there exist numbers $m \geq 2, n \geq 1, \alpha > 0$ and an area minimizing m-dimensional integer rectifiable current T in $\mathbb{R}^{m+n}$ such that

$$\mathcal{H}^{m-2+\alpha}(\mathrm{Sing(T)}) > 0.$$

Note that the hypothesis $m \geq 2$ is justified because, for $m = 1$ an area minimizing current is locally the union of finitely many non-intersecting open segments.

The aim of the proof is now to show that there exist suitable points of Sing(T) where we can perform the blowup analysis outlined in the previous section. This process consists of different steps, which we next list in a way which does not require the introduction of new notation but needs to be further specified later.

(A) Find a point $x_0 \in \mathrm{Sing(T)}$ and a sequence of radii $(r_k)_k$ with $r_k \downarrow 0$ such that:

(A_1) the rescaling currents $T_{x_0,r_k} := (\iota_{x_0,r_k})_\sharp T$ converge to a flat tangent cone;

(A_2) $\mathcal{H}^{m-2+\alpha}(\mathrm{Sing}(\mathrm{T}_{x_0,r_k}) \cap \mathrm{B}_1) > \eta > 0$ for some $\eta > 0$ and for every $k \in \mathbb{N}$.

Note that both conclusions hold for suitable subsequences, which in principle may not coincide. What we need to prove is that we can select a point and a subsequence satisfying both.

(B) Construction of the center manifold $\mathcal{M}$ and of a normal Lipschitz approximation $N : \mathcal{M} \to \mathbb{R}^{m+n}/\sim$.

This is the most technical part of the proof, and most of the conclusions of the next steps will intimately depend on this construction.

(C) The center manifold that one constructs in step (B) can only be used in general for a finite number of radii r_k of step (A). The reason is that in general its degree of approximation of the average of the minimizing currents T is under control only up to a certain distance from the singular point under consideration. This leads us to define the sets where the approximation works, called in the sequel *intervals of flattening*, and to define an entire *sequence of center manifolds* which will be used in the blowup analysis.

(D) Next we will take care of the problem of the infinite order of contact. This is done in two part. For the first one we derive the *almost monotonicity formula* for a variant of Almgren's frequency function, deducing that the order of contact remains finite within each center manifold of the sequence in (C).

(E) Then one needs to compare different center manifolds and to show that the order of contact still remains finite. This is done by exploiting a deep consequence of the construction in (C) which we call *splitting before tilting* after the inspiring paper by T. Rivière [29].

(F) With this analysis at hand, we can pass into the limit our blowup sequence and conclude the convergence to the graph of a harmonic Q-valued function u.

(G) Finally, we discuss the capacitary argument leading to the persistence of the singularities, to show that the function u in (F) needs to have a singular set with positive $\mathcal{H}^{m-2+\alpha}$ measure, thus contradicting the partial regularity estimate for such multiple valued harmonic functions.

In the remaining part of this course we give a more detailed description of the steps above, referring to the original papers [13–17] for the complete proofs.

3 Q-valued functions and rectifiable currents

Since the final contradiction argument relies on the regularity theory of multiple valued functions, we start recalling the main definitions and results concerning them, and the way they can be used to approximate integer rectifiable currents. The reference for this part of the theory is [13,16,17,34].

3.1 Q-valued functions

We start by giving a metric structure to the space $(\mathbb{R}^n)^Q/\sim$ of unordered Q-tuples of points in $\mathbb{R}^n$, where $Q \in \mathbb{N} \setminus \{0\}$ is a fixed number. It is immediate to see that this space can be identified with the subset of positive measures of mass Q which are the sum of integer multiplicity Dirac delta:

$$(\mathbb{R}^n)^Q/\sim \quad \simeq \quad \mathcal{A}_Q(\mathbb{R}^n) := \left\{ \sum_{i=1}^{Q} [\![P_i]\!] \,:\, P_i \in \mathbb{R}^n \right\},$$

where $[\![P_i]\!]$ denotes the Dirac delta at P_i. We can then endow $\mathcal{A}_Q$ with one of the distances defined for (probability) measures, for example the Wasserstein distance of exponent two: for every $T_1 = \sum_i [\![P_i]\!]$ and $T_2 = \sum_i [\![S_i]\!] \in \mathcal{A}_Q(\mathbb{R}^n)$, we set

$$\mathcal{G}(T_1, T_2) := \min_{\sigma \in \mathbf{S}_Q} \sqrt{\sum_{i=1}^{Q} \left|P_i - S_{\sigma(i)}\right|^2},$$

where we recall that $\mathbf{S}_Q$ denotes the symmetric group of Q elements.

A Q-function simply a map $f : \Omega \to \mathcal{A}_Q(\mathbb{R}^n)$, where $\Omega \subset \mathbb{R}^m$ is an open domain. We can then talk about measurable (with respect to the Borel σ-algebra of $\mathcal{A}_Q(\mathbb{R}^n)$), bounded, uniformly-, Hölder- or Lipschitz-continuous Q-valued functions.

More importantly, following the pioneering approach to weakly differentiable functions with values in a metric space by L. Ambrosio [6], we can also define the class of Sobolev Q-valued functions $W^{1,2}$.

Definition 3.1 (Sobolev Q-valued functions). Let $\Omega \subset \mathbb{R}^m$ be a bounded open set. A measurable function $f : \Omega \to \mathcal{A}_Q$ is in the Sobolev class $W^{1,2}$ if there exist m functions $\varphi_j \in L^2(\Omega)$ for $j = 1, \dots, m$, such that

(i) $x \mapsto \mathcal{G}(f(x), T) \in W^{1,2}(\Omega)$ for all $T \in \mathcal{A}_Q$;
(ii) $\left|\partial_j \mathcal{G}(f, T)\right| \leq \varphi_j$ almost everywhere in Ω for all $T \in \mathcal{A}_Q$ and for all $j \in \{1, \dots, m\}$, where $\partial_j \mathcal{G}(f, T)$ denotes the weak partial derivatives of the functions in (i).

By simple reasonings, one can infer the existence of minimal functions $|\partial_j f|$ fulfilling (ii):

$$|\partial_j f| \leq \varphi_j \text{ a.e. for any other } \varphi_j \text{ satisfying (ii),}$$

We set

$$|Df|^2 := \sum_{j=1}^{m} \left|\partial_j f\right|^2, \tag{3.1}$$

and define the Dirichlet energy of a Q-valued function as (cp. also [25–27] for alternative definitions)

$$\mathrm{Dir}(f) := \int_\Omega |Df|^2.$$

A Q-valued function f is said *Dir-minimizing* if

$$\int_\Omega |Df|^2 \le \int_\Omega |Dg|^2 \tag{3.2}$$
$$\text{for all } g \in W^{1,2}(\Omega, \mathcal{A}_Q) \text{ with } \mathcal{G}(f,g)|_{\partial\Omega} = 0,$$

where the last inequality is meant in the sense of traces.

The main result in the theory of Q-valued functions is the following.

Theorem 3.2. *Let $\Omega \subset \mathbb{R}^m$ be a bounded open domain with Lipschitz boundary, and let $g \in W^{1,2}(\Omega, \mathcal{A}_Q(\mathbb{R}^n))$ be fixed. Then, the following holds.*

(i) *There exists a Dir-minimizing function f solving the minimization problem* (3.2).
(ii) *Every such function f belongs to $C^{0,\kappa}_{\mathrm{loc}}(\Omega, \mathcal{A}_Q(\mathbb{R}^n))$ for a dimensional constant $\kappa = \kappa(m, Q) > 0$.*
(iii) *For every such function f, $|Df| \in L^p_{\mathrm{loc}}(\Omega)$ for some dimensional constant $p = p(m, n, Q) > 2$.*
(iv) *There exists a relatively closed set* Sing(u) $\subset \Omega$ *of Hausdorff dimension at most $m-2$ such that the graph of u outside* Sing(u), i.e. *the set*

$$\text{graph}\, u|_{\Omega\setminus\Sigma} = \{(x,y) \;:\; x \in \Omega \setminus \Sigma,\ y \in \text{spt}\,(u(x))\},$$

is a smoothly embedded m-dimensional submanifold of $\mathbb{R}^{m+n}$.

Remark 3.3. We refer to [13,34] for the proofs and more refined results in the case of of two dimensional domains. Moreover, for some results concerning the boundary regularity we refer to [24], and for an improved estimate of the singular set to [21].

We close this section by some considerations on the Q-valued functions. For the reasons explained in the previous section, a Q-valued function has to be considered as an intrinsic map taking values in the non-smooth space of Q-points $\mathcal{A}_Q$, and cannot be reduced to a "superposition" of a number Q of functions. Nevertheless, in many situations it is possible to handle Q-valued functions as a superposition. For example, as shown in [13, Proposition 0.4] every measurable function $f : \mathbb{R}^m \to \mathcal{A}_Q(\mathbb{R}^n)$ can be written (not uniquely!) as

$$f(x) = \sum_{i=1}^{Q} [\![f_i(x)]\!] \quad \text{for } \mathcal{H}^m\text{-a.e. } x, \tag{3.3}$$

with $f_1, \ldots, f_Q : \mathbb{R}^m \to \mathbb{R}^n$ measurable functions.

Similarly, for weakly differentiable functions it is possible to define a notion of pointwise approximate differential (cp. [13, Corollary 2,7])

$$Df = \sum_i [\![Df_i]\!] \in \mathcal{A}_Q(\mathbb{R}^{n\times m}),$$

with the property that at almost every x it holds $Df_i(x) = Df_j(x)$ if $f_i(x) = f_j(x)$. Note, however, that the functions f_i do not need to be weakly differentiable in (3.3), for the Q-valued function f has an approximate differential.

3.2 Graph of Lipschitz Q-valued functions

There is a canonical way to give the structure of integer rectifiable currents to the graph of a Lipschitz Q-valued function.

To this aim, we consider *proper* Q-valued functions, *i.e.* measurable functions $F : M \to \mathcal{A}_Q(\mathbb{R}^{m+n})$ (where M is any m-dimensional submanifold of $\mathbb{R}^{m+n}$) such that there is a measurable selection $F = \sum_i [\![F_i]\!]$ for which

$$\bigcup_i \overline{(F_i)^{-1}(K)}$$

is compact for every compact $K \subset \mathbb{R}^{m+n}$. It is then obvious that if there exists such a selection, then *every* measurable selection shares the same property.

By a simple induction argument (cp. [16, Lemma 1.1]), there are a countable partition of M in bounded measurable subsets M_i ($i \in \mathbb{N}$) and Lipschitz functions $f_i^j : M_i \to \mathbb{R}^{m+n}$ ($j \in \{1, \dots, Q\}$) such that

(a) $F|_{M_i} = \sum_{j=1}^Q \left[\!\left[f_i^j\right]\!\right]$ for every $i \in \mathbb{N}$ and $\mathrm{Lip}(f_i^j) \le \mathrm{Lip}(F)$ $\forall i, j$;
(b) $\forall\, i \in \mathbb{N}$ and $j, j' \in \{1, \dots, Q\}$, either $f_i^j \equiv f_i^{j'}$ or $f_i^j(x) \neq f_i^{j'}(x)$ $\forall x \in M_i$;
(c) $\forall\, i$ we have $DF(x) = \sum_{j=1}^Q \left[\!\left[Df_i^j(x)\right]\!\right]$ for a.e. $x \in M_i$.

We can then give the following definition.

Definition 3.4 (Q-valued push-forward). Let M be an oriented submanifold of $\mathbb{R}^{m+n}$ of dimension m and let $F : M \to \mathcal{A}_Q(\mathbb{R}^{m+n})$ be a proper Lipschitz map. Then, we define the push-forward of M through F as the current

$$\mathbf{T}_F = \sum_{i,j} (f_i^j)_\sharp [\![M_i]\!],$$

where M_i and f_i^j are as above: that is,

$$\mathbf{T}_F(\omega) := \sum_{i\in\mathbb{N}} \sum_{j=1}^{Q} \int_{M_i} \langle \omega(f_i^j(x)), Df_i^j(x)_\sharp \vec{e}(x) \rangle \, d\mathcal{H}^m(x) \quad \forall \omega \in \mathscr{D}^m(\mathbb{R}^n). \tag{3.4}$$

One can prove that the current in Definition 3.4 does not depend on the decomposition chosen for M and f, and moreover is integer rectifiable (cp. [16, Proposition 1.4])

A particular class of push-forwards are given by graphs.

Definition 3.5 (**Q-graphs**). Let $f = \sum_i [\![f_i]\!] : \mathbb{R}^m \to \mathcal{A}_Q(\mathbb{R}^n)$ be Lipschitz and define the map $F : M \to \mathcal{A}_Q(\mathbb{R}^{m+n})$ as $F(x) := \sum_{i=1}^{Q} [\![(x, f_i(x))]\!]$. Then, $\mathbf{T}_F$ is the *current associated to the graph* $\mathrm{Gr}(f)$ and will be denoted by $\mathbf{G}_f$.

The main result concerning the push-forward of a Q-valued function is the following (see [16, Theorem 2.1]).

Theorem 3.6 (Boundary of the push-forward). *Let $M \subset \mathbb{R}^{m+n}$ be an m-dimensional submanifold with boundary, $F : M \to \mathcal{A}_Q(\mathbb{R}^{m+n})$ a proper Lipschitz function and $f = F|_{\partial M}$. Then, $\partial \mathbf{T}_F = \mathbf{T}_f$.*

Moreover, the following Taylor expansion of the mass of a graph holds (cp. [16, Corollary 3.3]).

Proposition 3.7 (Expansion of $\mathbf{M}(\mathbf{G}_f)$). *There exist dimensional constants $\bar{c}, C > 0$ such that, if $\Omega \subset \mathbb{R}^m$ is a bounded open set and $f : \Omega \to \mathcal{A}_Q(\mathbb{R}^n)$ is a Lipschitz map with $\mathrm{Lip}(f) \le \bar{c}$, then*

$$\mathbf{M}(\mathbf{G}_f) = Q|\Omega| + \frac{1}{2}\int_\Omega |Df|^2 + \int_\Omega \sum_i \bar{R}_4(Df_i), \tag{3.5}$$

where $\bar{R}_4 \in C^1(\mathbb{R}^{n\times m})$ satisfies $|\bar{R}_4(D)| = |D|^3 \bar{L}(D)$ for $\bar{L} : \mathbb{R}^{n\times m} \to \mathbb{R}$ Lipschitz with $\mathrm{Lip}(\bar{L}) \le C$ and $\bar{L}(0) = 0$.

3.3 Approximation of area minimizing currents

Finally we recall some results on the approximation of area minimizing currents.

To this aim we need to introduce more notation. We consider cylinders in $\mathbb{R}^{m+n}$ of the form $\bar{C}_s(x) := \bar{B}_s(x) \times \mathbb{R}^n$ with $x \in \mathbb{R}^m$.

Since we are interested in interior regularity, we can assume for the purposes of this section that we are always in the following setting: for some open cylinder $\bar{C}_{4r}(x)$ (with $r \leq 1$) and some positive integer Q, the area minimizing current T has compact support in $\bar{C}_{4r}(x)$ and satisfies

$$\mathbf{p}_\sharp T = Q\,[\![\bar{B}_{4r}(x)]\!] \quad \text{and} \quad \partial T \llcorner \bar{C}_{4r}(x) = 0, \tag{3.6}$$

where $\mathbf{p} : \mathbb{R}^{m+n} \to \pi_0 := \mathbb{R}^m \times \{0\}$ is the orthogonal projection.

We introduce next the main regularity parameter for area minimizing currents, namely the *Excess*.

Definition 3.8 (Excess measure). For a current T as above we define the *cylindrical excess* $\mathbf{E}(T, \bar{C}_r(x))$ as follows:

$$\begin{aligned}\mathbf{E}(T, \bar{C}_r(x)) &:= \frac{\|T\|(\bar{C}_r(x))}{\omega_m r^m} - Q \\ &= \frac{1}{2\,\omega_m r^m} \int_{\|T\|(\bar{C}_r(x))} |\vec{T} - \vec{\pi}_0|^2\, \delta\|T\|,\end{aligned}$$

where ω_m is the measure of the m-dimensional unit ball, and $\vec{\pi}_0$ is the m-vector orienting π_0.

The most general approximation result of area minimizing currents is the one due to Almgren, and reproved in [17] with more refined techniques, which asserts that under suitable smallness condition of the excess, an area minimizing current coincides on a big set with a graph of a Lipschitz Q-valued function.

Theorem 3.9 (Almgren's strong approximation). *There exist constants $C, \gamma_1, \varepsilon_1 > 0$ (depending on m, n, Q) with the following property. Assume that T is area minimizing in the cylinder $\bar{C}_{4r}(x)$ and assume that*

$$E := \mathbf{E}(T, \bar{C}_{4r}(x)) < \varepsilon_1.$$

Then, there exist a map $f : B_r(x) \to \mathcal{A}_Q(\mathbb{R}^n)$ and a closed set $K \subset \bar{B}_r(x)$ such that the following holds:

$$\mathrm{Lip}(f) \leq C E^{\gamma_1}, \tag{3.7}$$

$$\mathbf{G}_f \llcorner (K \times \mathbb{R}^n) = T \llcorner (K \times \mathbb{R}^n) \quad \textit{and} \quad |B_r(x) \setminus K| \leq C\, E^{1+\gamma_1}\, r^m, \tag{3.8}$$

$$\left| \|T\|(\bar{C}_r(x)) - Q\,\omega_m\, r^m - \tfrac{1}{2} \int_{B_r(x)} |Df|^2 \right| \leq C\, E^{1+\gamma_1}\, r^m. \tag{3.9}$$

The most important improvement of the theorem above with respect to the preexisting approximation results is the small power E^{γ_1} in the three estimates (3.7) - (3.9). Indeed, this will play a crucial role in the construction of the center manifold. It is worthy mentioning that, when $Q = 1$

and $n = 1$, this approximation theorem was first proved with different techniques by De Giorgi in [10] (cp. also [12, Appendix]).

As a byproduct of this approximation, we also obtain the analog of the so called *harmonic approximation*, which allows us to compare the Lipschitz approximation above with a Dir-minimizing function.

Theorem 3.10 (Harmonic approximation). *Let γ_1, ε_1 be the constants of Theorem 3.9. Then, for every $\bar\eta > 0$, there is a positive constant $\bar\varepsilon_1 < \varepsilon_1$ with the following property. Assume that T is as in Theorem 3.9 and*

$$E := \mathbf{E}(T, \bar{C}_{4r}(x)) < \bar\varepsilon_1.$$

If f is the map in Theorem 3.9, then there exists a Dir*-minimizing function w such that*

$$r^{-2}\int_{B_r(x)} \mathcal{G}(f, w)^2 + \int_{B_r(x)} (|Df| - |Dw|)^2 \\ + \int_{B_r(x)} |D(\eta \circ f) - D(\eta \circ w)|^2 \le \bar\eta\, E\, r^m, \quad (3.10)$$

where $\eta : \mathcal{A}_Q(\mathbb{R}^n) \to \mathbb{R}^n$ is the average map

$$\eta\left(\sum_i [\![P_i]\!]\right) = \frac{1}{Q}\sum_i P_i.$$

4 Selection of contradiction's sequence

In this section we give the details of the first step (A) in Section 2.7, namely the selection of a common subsequence such that the rescaled currents converge to a flat tangent cone and the measure of the singular set remains uniformly bounded below away from zero. For this purpose, we introduce the following notation. We denote by $\mathbf{B}_r(x)$ the open ball of radius $r > 0$ in $\mathbb{R}^{m+n}$ (we do not write the point x if the origin) and, for $Q \in \mathbb{N}$, we denote by $\mathrm{D}_Q(T)$ the points of density Q of the current T, and set

$$\mathrm{Reg}_Q(T) := \mathrm{Reg}(T) \cap \mathrm{D}_Q(T) \quad \text{and} \quad \mathrm{Sing}_Q(T) := \mathrm{Sing}(T) \cap \mathrm{D}_Q(T).$$

The precise properties of the sequence that will be used in the blowup argument are stated in the following proposition. We recall that the main hypothesis at the base of the proof is the contradiction assumption of Section 2.7, which we restate for reader's convenience.

Contradiction assumption: there exist numbers $m \geq 2, n \in \mathbb{N}, \alpha > 0$ and an area minimizing m-dimensional integer rectifiable current T in $\mathbb{R}^{m+n}$ such that

$$\mathcal{H}^{m-2+\alpha}(\mathrm{Sing(T)}) > 0.$$

We introduce the *spherical excess* defined as follows: for a given m-dimensional plane π,

$$\mathbf{E}(T, \mathbf{B}_r(x), \pi) := \frac{1}{2\,\omega_m\,r^m} \int_{\mathbf{B}_r(x)} |\vec{T} - \vec{\pi}|^2 \, d\|T\|,$$

$$\mathbf{E}(T, \mathbf{B}_r(x)) := \min_{\tau} \mathbf{E}(T, \mathbf{B}_r(x), \tau).$$

Proposition 4.1 (Contradiction's sequence). *Under the contradiction assumption, there exist*

1. *constants $m, n, Q \geq 2$ natural numbers and $\alpha, \eta > 0$ real numbers;*
2. *an m-dimensional area minimizing integer rectifiable current T in $\mathbb{R}^{m+n}$ with $\partial T = 0$;*
3. *a sequence $r_k \downarrow 0$*

such that $0 \in \mathrm{D}_Q(T)$ and the following holds:

$$\lim_{k \to +\infty} \mathbf{E}(T_{0,r_k}, \mathbf{B}_{10}) = 0, \tag{4.1}$$

$$\lim_{k \to +\infty} \mathcal{H}^{m-2+\alpha}_{\infty}(\mathrm{D}_Q(T_{0,r_k}) \cap \mathbf{B}_1) > \eta, \tag{4.2}$$

$$\mathcal{H}^m\big((\mathbf{B}_1 \cap \mathrm{spt}\,(T_{0,r_k})) \setminus \mathrm{D}_Q(T_{0,r_k})\big) > 0 \quad \forall\, k \in \mathbb{N}. \tag{4.3}$$

Here $\mathcal{H}^{m-2+\alpha}_{\infty}$ is the Hausdorff premeasure computed without any restriction on the diameter of the sets in the coverings.

By Almgren's stratification theorem and by general measure theoretic arguments, there exist sequences satisfying either (4.1) or (4.2). The two subsequences might, however, be different: we show the existence of one point and a single subsequence along which *both* conclusions hold. The proof of the proposition is based on the following two results.

Theorem 4.2 (Almgren [5, 2.27]). *Let $\alpha > 0$ and let T be an integer rectifiable area minimizing current in $\mathbb{R}^{m+n}$. Then,*

(1) *for $\mathcal{H}^{m-2+\alpha}$-a.e. point $x \in \mathrm{spt}\,(T) \setminus \mathrm{spt}\,(\partial T)$ there exists a subsequence $s_k \downarrow 0$ such that T_{x,s_k} converges to a flat cone;*
(2) *for $\mathcal{H}^{m-3+\alpha}$-a.e. point $x \in \mathrm{spt}\,(T) \setminus \mathrm{spt}\,(\partial T)$, it holds that $\Theta(T, x) \in \mathbb{Z}$.*

Lemma 4.3. *Let S be an m-dimensional area minimizing integral current, which is a cone in $\mathbb{R}^{m+n}$ with $\partial S = 0$, $Q = \Theta(S, 0) \in \mathbb{N} \setminus \{0\}$, and assume that*

$$\mathcal{H}^m\big(D_Q(S)\big) > 0 \quad \textit{and} \quad \mathcal{H}^{m-1}(\mathrm{Sing_Q(S)}) = 0.$$

Then S is an m-dimensional plane with multiplicity Q.

Proof of Proposition 4.1. Let $m > 1$ be the smallest integer for which Theorem 1.7 fails. In view of Almgren's stratification Theorem 4.2, we can assume that there exist an integer rectifiable area minimizing current R of dimension m and a positive integer Q such that the Hausdorff dimension of $\mathrm{Sing_Q(R)}$ is larger than $m-2$. We fix the smallest Q for which such a current R exists and note that by Allard's regularity theorem (cp. [2]) it must be $Q > 1$.

Let $\alpha > 0$ be such that $\mathcal{H}^{m-2+\alpha}(\mathrm{Sing_Q(R)}) > 0$, and consider a density point x_0 for the measure $\mathcal{H}^{m-2+\alpha}$ (without loss of generality $x_0 = 0$). In particular, there exists $r_k \downarrow 0$ such that

$$\lim_{k\to+\infty} \frac{\mathcal{H}^{m-2+\alpha}_\infty\big(\mathrm{Sing_Q(R)} \cap \mathbf{B}_{\mathrm{r_k}}\big)}{r_k^{m-2+\alpha}} > 0\,.$$

Up to a subsequence (not relabeled) we can assume that $R_{0,r_k} \to S$, with S a tangent cone. If S is a multiplicity Q flat plane, then we set $T := R$ and the proposition is proven (indeed, (4.3) is satisfied because $0 \in \mathrm{Sing(R)}$ and $\|R\| \geq \mathcal{H}^m \llcorner \mathrm{spt}\,(R)$).

If S is *not* flat, taking into account the convergence properties of area minimizing currents [31, Theorem 34.5] and the upper semicontinuity of $\mathcal{H}^{m-2+\alpha}_\infty$ under the Hausdorff convergence of compact sets, we deduce

$$\mathcal{H}^{m-2+\alpha}_\infty\big(\mathrm{D}_Q(S) \cap \bar{\mathbf{B}}_1\big) \geq \liminf_{k\to+\infty} \mathcal{H}^{m-2+\alpha}_\infty\big(\mathrm{D}_Q(R_{0,r_k}) \cap \bar{\mathbf{B}}_1\big) > 0. \tag{4.4}$$

We claim that (4.4) implies

$$\mathcal{H}^{m-2+\alpha}_\infty(\mathrm{Sing_Q(S)}) > 0. \tag{4.5}$$

Indeed, if all points of $\mathrm{D}_Q(S)$ are singular, then (4.5) follows from (4.4) directly. Otherwise, $\mathrm{Reg_Q(S)}$ is not empty, thus implying $\mathcal{H}^m(D_Q(S) \cap \mathbf{B}_1) > 0$: we can then apply Lemma 4.3 and infer that, since S is not regular, then $\mathcal{H}^{m-1}(\mathrm{Sing_Q(S)}) > 0$ and (4.5) holds.

We can, hence, find $x \in \mathrm{Sing_Q(S)} \setminus \{0\}$ and $r_k \downarrow 0$ such that

$$\lim_{k\to+\infty} \frac{\mathcal{H}^{m-2+\alpha}_\infty\big(\mathrm{Sing_Q(S)} \cap \mathbf{B}_{\mathrm{r_k}}(\mathrm{x})\big)}{r_k^{m-2+\alpha}} > 0.$$

Up to a subsequence (not relabelled), we can assume that S_{x,r_k} converges to S_1. Since S_1 is a tangent cone to the cone S at $x \neq 0$, S_1 splits off a line, *i.e.* $S_1 = S_2 \times [\![\{t\,e : t \in \mathbb{R}\}]\!]$ for some $e \in \mathbb{S}^{m+n-1}$, for some area minimizing cone S_2 in $\mathbb{R}^{m-1+n}$ and some $v \in \mathbb{R}^{m+n}$ (cp. [31, Lemma 35.5]). Since m is, by assumption, the smallest integer for which Theorem 1.7 fails, $\mathcal{H}^{m-3+\alpha}(\mathrm{Sing(S_2)}) = 0$ and, hence, $\mathcal{H}^{m-2+\alpha}(\mathrm{Sing_Q(S_1)}) = 0$. On the other hand, arguing as for (4.4), we have

$$\mathcal{H}^{m-2+\alpha}_\infty(\mathrm{D}_Q(S_1)\cap\bar{\mathbf{B}}_1) \geq \limsup_{k\to+\infty}\mathcal{H}^{m-2+\alpha}_\infty(\mathrm{D}_Q(S_{x,r_k})\cap\bar{\mathbf{B}}_1) > 0.$$

Thus $\mathrm{Reg_Q(S_1)} \neq \emptyset$ and, hence, $\mathcal{H}^m(D_Q(S_1)) > 0$. We then can apply Lemma 4.3 again and conclude that S_1 is an m-dimensional plane with multiplicity Q. Therefore, the proposition follows taking T a suitable translation of S. □

Proof of Lemma 4.3

We premise the following lemma.

Lemma 4.4. *Let T be an integer rectifiable current of dimension m in $\mathbb{R}^{m+n}$ with locally finite mass and U an open set such that*

$$\mathcal{H}^{m-1}(\partial U\cap\mathrm{spt}\,(T)) = 0 \quad \text{and} \quad (\partial T)\llcorner U = 0.$$

Then $\partial(T\llcorner U) = 0$.

Proof. Consider $V \subset\subset \mathbb{R}^{m+n}$. By the slicing theory

$$S_r := T\llcorner(V\cap U\cap\{\mathrm{dist}\,(x,\partial U) > r\})$$

is a normal current in $\mathbf{N}_m(V)$ for a.e. r. Since

$$\mathbf{M}(T\llcorner(V\cap U) - S_r) \to 0 \quad \text{as} \quad r\downarrow 0,$$

we conclude that $T\llcorner(U\cap V)$ is in the $\mathbf{M}$-closure of $\mathbf{N}_m(V)$. Thus, by [18, 4.1.17], $T\llcorner U$ is a flat chain in $\mathbb{R}^{m+n}$ and by [18, 4.1.12] $\partial(T\llcorner U)$ is a flat chain. Since $\mathrm{spt}\,(\partial(T\llcorner U)) \subset \partial U\cap\mathrm{spt}\,(T)$, we can apply [18, Theorem 4.1.20] to conclude that $\partial(T\llcorner U) = 0$. □

We next prove Lemma 4.3. For each $x \in \mathrm{Reg_Q(S)}$, let r_x be such that $S\llcorner\mathbf{B}_{2r_x}(x) = Q\,[\![\Gamma]\!]$ for some regular submanifold Γ and set

$$U := \bigcup_{x\in\mathrm{Reg_Q(S)}}\mathbf{B}_{r_x}(x).$$

Obviously, $\mathrm{Reg}_Q(S) \subset U$; hence, by assumption, it is not empty. Fix $x \in \mathrm{spt}(S) \cap \partial U$. Let next $(x_k)_{k\in\mathbb{N}} \subset \mathrm{Reg}_Q(S)$ be such that

$$\mathrm{dist}(x, \mathbf{B}_{r_{x_k}}(x_k)) \to 0.$$

We necessarily have that $r_{x_k} \to 0$: otherwise we would have $x \in \mathbf{B}_{2r_{x_k}}(x_k)$ for some k, which would imply $x \in \mathrm{Reg}_Q(S) \subset U$, *i.e.* a contradiction. Therefore, $x_k \to x$ and, by [31, Theorem 35.1],

$$Q = \limsup_{k\to+\infty} \Theta(S, x_k) \leq \Theta(S, x) = \lim_{\lambda\downarrow 0} \Theta(S, \lambda x) \leq \Theta(S, 0) = Q.$$

This implies $x \in \mathrm{D}_Q(S)$. Since $x \in \partial U$, we must then have $x \in \mathrm{Sing}_Q(S)$. Thus, we conclude that $\mathcal{H}^{m-1}(\mathrm{spt}(S) \cap \partial U) = 0$. It follows from Lemma 4.4 that $S' := S \llcorner U$ has 0 boundary in $\mathbb{R}^{m+n}$. Moreover, since S is an area minimizing cone, S' is also an area-minimizing cone. By definition of U we have $\Theta(S', x) = Q$ for $\|S'\|$-a.e. x and, by semi-continuity,

$$Q \leq \Theta(S', 0) \leq \Theta(S, 0) = Q.$$

We apply Allard's theorem [2] and deduce that S' is regular, *i.e.* S' is an m-plane with multiplicity Q. Finally, from $\Theta(S', 0) = \Theta(S, 0)$, we infer $S' = S$. □

5 Center manifold's construction

In this section we describe the procedure for the construction of the center manifold. As mentioned in the introduction, this is the most complicated part of the proof: indeed, the construction of the center manifold comes together with a series of other estimates which will enter significantly in the proof of the main Theorem 1.7. In particular, as an outcome of the procedure we obtain the following several things.

(1) A decomposition of the horizontal plane $\pi_0 = \mathbb{R}^m \times \{0\}$ of "Whitney's type".
(2) A family of interpolating functions defined on the cubes of this decomposition.
(3) A normal approximation taking values in the normal bundle of the center manifold.
(4) A set of criteria (which will in fact determine the Whitney decomposition) which lead to what we call *splitting-before-tilting* estimates.
(5) An family of intervals, called *intervals of flattening*, where the construction will be effective.
(6) A family of pairs cube–ball transforming the estimates on the Whitney decomposition into estimates on balls (thus passing from the cubic lattice of the decomposition to the standard geometry of balls).

5.1 Notation and assumptions

Let us recall the following notation. Given an integer rectifiable current T with compact support, we consider the *spherical* and the *cylindrical* excesses defined as follows, respectively: for given m-planes π, π', we set

$$\mathbf{E}(T, \mathbf{B}_r(x), \pi) := \left(2\omega_m r^m\right)^{-1} \int_{\mathbf{B}_r(x)} |\vec{T} - \vec{\pi}|^2 \, d\|T\|, \tag{5.1}$$

$$\mathbf{E}(T, \bar{C}_r(x, \pi), \pi') := \left(2\omega_m r^m\right)^{-1} \int_{\bar{C}_r(x,\pi)} |\vec{T} - \vec{\pi}'|^2 \, d\|T\|, \tag{5.2}$$

where $\bar{C}_r(x, \pi) = \bar{B}_r(x, \pi) \times \pi^\perp$ is the cylinder over the closed ball $\bar{B}_r(x, \pi)$ or radius r and center x in the m-dimensional plane π. And we consider the *height function* in a set A (we denote by $\mathbf{p}_\pi$ the orthogonal projection on a plane π)

$$\mathbf{h}(T, A, \pi) := \sup_{x,y \in \operatorname{spt}(T) \cap A} |\mathbf{p}_{\pi^\perp}(x) - \mathbf{p}_{\pi^\perp}(y)| .$$

We also set

$$\mathbf{E}(T, \mathbf{B}_r(x)) := \min_\tau \mathbf{E}(T, \mathbf{B}_r(x), \tau) = \mathbf{E}(T, \mathbf{B}_r(x), \pi), \tag{5.3}$$

and we will use $\mathbf{E}(T, \bar{C}_r(x, \pi))$ in place of $\mathbf{E}(T, \bar{C}_r(x, \pi), \pi)$: note that it coincides with the cylindrical excess as defined in Section 3.3 when

$$(\mathbf{p}_\pi)_\sharp T \llcorner \bar{C}_r(x, \pi) = Q \left[\!\left[\bar{B}_r(\mathbf{p}_\pi(x), \pi) \right]\!\right] .$$

In this section we will work with an area minimizing integer rectifiable current T^0 with compact support which satisfies the following assumptions: for some constant $\varepsilon_2 \in (0, 1)$, which we always suppose to be small enough,

$$\Theta(0, T^0) = Q \quad \text{and} \quad \partial T^0 \llcorner \mathbf{B}_{6\sqrt{m}} = 0, \tag{5.4}$$

$$\|T^0\|(\mathbf{B}_{6\sqrt{m}\rho}) \leq \left(\omega_m Q (6\sqrt{m})^m + \varepsilon_2^2\right) \rho^m \quad \forall \rho \leq 1, \tag{5.5}$$

$$E := \mathbf{E}\left(T^0, \mathbf{B}_{6\sqrt{m}}\right) = \mathbf{E}\left(T^0, \mathbf{B}_{6\sqrt{m}}, \pi_0\right) \leq \varepsilon_2^2, \tag{5.6}$$

It follows from standard considerations in geometric measure theory that there are positive constants $C_0(m, n, Q)$ and $c_0(m, n, Q)$ with the following property. If T^0 is as in (5.4) - (5.6), $\varepsilon_2 < c_0$ and $T := T^0 \llcorner \mathbf{B}_{23\sqrt{m}/4}$, then:

$$\partial T \llcorner \bar{C}_{11\sqrt{m}/2}(0, \pi_0) = 0, \tag{5.7}$$

$$(\mathbf{p}_{\pi_0})_\sharp T \llcorner \bar{C}_{11\sqrt{m}/2}(0, \pi_0) = Q \left[\!\left[B_{11\sqrt{m}/2}(0, \pi_0) \right]\!\right], \tag{5.8}$$

$$\mathbf{h}(T, \bar{C}_{5\sqrt{m}}(0, \pi_0)) \leq C_0 \varepsilon_2^{\frac{1}{m}}. \tag{5.9}$$

In particular for each $x \in B_{11\sqrt{m}/2}(0, \pi_0)$ there is a point $p \in \mathrm{spt}(T)$ with $\mathbf{p}_{\pi_0}(p) = x$.

5.2 Whitney decomposition and interpolating functions

The construction of the center manifold is done by following a suitable decomposition of the horizontal plane π_0 into cubes. We denote by $\mathscr{C}^j$, $j \in \mathbb{N}$, the family of dyadic closed cubes L of π_0 with side-length $2^{1-j} =: 2\,\ell(L)$. Next we set $\mathscr{C} := \bigcup_{j \in \mathbb{N}} \mathscr{C}^j$. If H and L are two cubes in $\mathscr{C}$ with $H \subset L$, then we call L an *ancestor* of H and H a *descendant* of L. When in addition $\ell(L) = 2\ell(H)$, H is *a son* of L and L *the father* of H.

Definition 5.1. A Whitney decomposition of $[-4, 4]^m \subset \pi_0$ consists of a closed set $\mathbf{\Gamma} \subset [-4, 4]^m$ and a family $\mathscr{W} \subset \mathscr{C}$ satisfying the following properties:

(w1) $\mathbf{\Gamma} \cup \bigcup_{L \in \mathscr{W}} L = [-4, 4]^m$ and $\mathbf{\Gamma}$ does not intersect any element of $\mathscr{W}$;
(w2) the interiors of any pair of distinct cubes $L_1, L_2 \in \mathscr{W}$ are disjoint;
(w3) if $L_1, L_2 \in \mathscr{W}$ have nonempty intersection, then

$$\frac{1}{2}\ell(L_1) \leq \ell(L_2) \leq 2\,\ell(L_1).$$

Observe that (w1) - (w3) imply

$$\mathrm{dist}\,(\mathbf{\Gamma}, L) := \inf\left\{|x - y| : x \in L, y \in \mathbf{\Gamma}\right\} \geq 2\ell(L) \quad \text{for every } L \in \mathscr{W}.$$

However, we do *not* require any inequality of the form $\mathrm{dist}\,(\mathbf{\Gamma}, L) \leq C\ell(L)$, although this would be customary for what is commonly called Whitney decomposition in the literature.

We denote by $\mathscr{S}^j$ all the dyadic cubes with side-length 2^{1-j} which are not contained in $\mathscr{W}$ and set $\mathscr{S} := \cup_{j \geq N_0} \mathscr{S}^j$ for some big natural number N_0. For each cube $L \in \mathscr{W} \cup \mathscr{S}$, we set $r_L = M_0 \sqrt{m}\ell(L)$, with $M_0 \in \mathbb{N}$ a dimensional constant to be fixed later, and we call its center x_L. We can then find points $p_L \in \mathrm{spt}\,(T)$, with coordinates $p_L = (x_L, y_L) \in \pi_0 \times \pi_0^\perp$, and *interpolating functions*

$$g_L : B_{4r_L}(p_L, \pi_0) \to \pi_0^\perp,$$

such that the following holds: for every $H, L \in \mathscr{W} \cup \mathscr{S}$,

$$\|g_H\|_{C^0} \leq C\, E^{\frac{1}{2m}} \quad \text{and} \quad \|Dg_H\|_{C^{2,\kappa}} \leq C E^{\frac{1}{2}}; \tag{5.10}$$

$$\|g_H - g_L\|_{C^i(B_{r_L}(p_L,\pi_0))} \leq C E^{\frac{1}{2}} \ell(H)^{3+\kappa-i} \tag{5.11}$$
$$\forall\, i \in \{0, \dots, 3\} \text{ if } H \cap L \neq \emptyset;$$

$$|D^3 g_H(x_H) - D^3 g_L(x_L)| \leq C E^{\frac{1}{2}} |x_H - x_L|^\kappa; \tag{5.12}$$

$$\sup_{(x,y)\in \operatorname{spt}(T),\, x\in H} \|g_H - y\|_{C^0} \leq C E^{\frac{1}{2m}} \ell(H), \tag{5.13}$$

for some $\kappa > 0$, and where we used the notation

$$B_r(p_L, \pi_0) := \mathbf{B}_r(p_L) \cap (p_L + \pi_0).$$

It is now very simple to show how to patch all the interpolating functions g_L in order to construct a center manifold. To this aim, we set

$$\mathscr{P}^j := \mathscr{S}^j \cup \left\{L \in \mathscr{W} \,:\, \ell(L) \geq 2^{-j}\right\}.$$

For every $L \in \mathscr{P}^j$ we define

$$\vartheta_L(y) := \vartheta\left(\frac{y - x_L}{\ell(L)}\right),$$

for some fixed $\vartheta \in C_c^\infty\left([-\frac{17}{16}, \frac{17}{16}]^m, [0,1]\right)$ that is identically 1 on $[-1,1]^m$. We can then patch all the interpolating functions using the partition of the unit induced by the ϑ_L, *i.e.*

$$\varphi_j := \frac{\sum_{L\in\mathscr{P}^j} \vartheta_L g_L}{\sum_{L\in\mathscr{P}^j} \vartheta_L}. \tag{5.14}$$

The following theorem is now a very easy consequence of the estimates on the interpolating functions.

Theorem 5.2 (Existence of the center manifold). *Assume to be given a Whitney decomposition $(\Gamma, \mathscr{W})$ and interpolating functions g_H as above. If ε_2 is sufficiently small, then*

(i) *the functions φ_j defined in* (5.14) *satisfy*

$$\|D\varphi_j\|_{C^{2,\kappa}} \leq C E^{\frac{1}{2}} \quad \text{and} \quad \|\varphi_j\|_{C^0} \leq C E^{\frac{1}{2m}},$$

(ii) *φ_j converges to a map $\boldsymbol{\varphi}$ such that $\mathcal{M} := \mathrm{Gr}(\boldsymbol{\varphi}|_{]-4,4[^m})$ is a $C^{3,\kappa}$ submanifold of Σ, called in the sequel* center manifold,

(iii) *for all* $x \in \mathbf{\Gamma}$, *the point* $(x, \boldsymbol{\varphi}(x)) \in \operatorname{spt}(T)$ *and is a multiplicity* Q *point. Setting* $\boldsymbol{\Phi}(y) := (y, \boldsymbol{\varphi}(y))$, *we call* $\boldsymbol{\Phi}(\mathbf{\Gamma})$ *the* contact set.

Proof. Define $\chi_H := \vartheta_H / (\sum_{L \in \mathscr{P}^j} \vartheta_L)$ and observe that

$$\sum \chi_H = 1 \quad \text{and} \quad \|\chi_H\|_{C^i} \leq C_0(i, m, n)\, \ell(H)^{-i} \qquad \forall i \in \mathbb{N}\,. \tag{5.15}$$

Set $\mathscr{P}^j(H) := \{L \in \mathscr{P}^j : L \cap H \neq \emptyset\} \setminus \{H\}$. By construction

$$\frac{1}{2}\ell(L) \leq \ell(H) \leq 2\,\ell(L) \quad \text{for every } L \in \mathscr{P}^j(H),$$

and the cardinality of $\mathscr{P}^j(H)$ is bounded by a geometric constant C_0. The estimate $|\varphi_j| \leq C E^{\frac{1}{2m}}$ follows then easily from (5.10).

For $x \in H$ we write

$$\begin{aligned} \varphi_j(x) &= \Big(g_H \chi_H + \sum_{L \in \mathscr{P}^j(H)} g_L \chi_L\Big)(x) \\ &= g_H(x) + \sum_{L \in \mathscr{P}^j(H)} (g_L - g_H)\chi_L\,(x)\,. \end{aligned} \tag{5.16}$$

Using the Leibniz rule, (5.15), (5.10) and (5.11), for $i \in \{1, 2, 3\}$ we get

$$\begin{aligned} \|D^i \varphi_j\|_{C^0(H)} &\leq \|g_H\|_{C^i} + \sum_{0 \leq l \leq i} \sum_{L \in \mathscr{P}^j(H)} \|g_L - g_H\|_{C^l(H)} \ell(L)^{l-i} \\ &\leq C E^{\frac{1}{2}} \big(1 + \ell(H)^{3+\kappa-i}\big). \end{aligned}$$

Next, using also $[D^3 g_H - D^3 g_L]_\kappa \leq C E^{\frac{1}{2}}$, we obtain

$$\begin{aligned} [D^3 \varphi_j]_{\kappa, H} \leq \sum_{0 \leq l \leq 3} \sum_{L \in \mathscr{P}^j(H)} & \ell(H)^{l-3} \big(\ell(H)^{-\kappa} \|D^l (g_L - g_H)\|_{C^0(H)} \\ &+ [D^l (g_L - g_H)]_{\kappa, H}\big) + [D^3 g_H]_{\kappa, H} \leq C E^{\frac{1}{2}}. \end{aligned}$$

Fix now $x, y \in [-4, 4]^m$, let $H, L \in \mathscr{P}^j$ be such that $x \in H$ and $y \in L$. If $H \cap L \neq \emptyset$, then

$$|D^3 \varphi_j(x) - D^3 \varphi_j(y)| \leq C\big([D^3 \varphi_j]_{\kappa, H} + [D^3 \varphi_j]_{\kappa, L}\big)|x - y|^\kappa. \tag{5.17}$$

If $H \cap L = \emptyset$, we assume without loss of generality $\ell(H) \leq \ell(L)$ and observe that

$$\max\big\{|x - x_H|, |y - x_L|\big\} \leq \ell(L) \leq |x - y|\,.$$

Moreover, by construction φ_j is identically equal to g_H in a neighborhood of its center x_H. Thus, we can estimate

$$\begin{aligned}|D^3\varphi_j(x) - D^3\varphi_j(y)| &\le |D^3\varphi_j(x) - D^3\varphi_j(x_H)| \\ &\quad + |D^3 g_H(x_H) - D^3 g_L(x_L)| \\ &\quad + |D^3\varphi_j(x_L) - D^3\varphi_j(y)| \\ &\le C E^{\frac{1}{2}} \left(|x - x_H|^\kappa + |x_H - x_L|^\kappa + |y - x_L|^\kappa\right) \\ &\le C E^{\frac{1}{2}} |x - y|^\kappa ,\end{aligned}$$

where we used (5.17) and (5.12). The convergence of the sequence φ_j (up to subsequences) and (iii) are now simple consequences of (5.13) (details are left to the reader). □

5.3 Normal approximation

The main feature of the center manifold $\mathcal{M}$ lies actually in the fact that it allows to make a good approximation of the current which turns out to be almost centered by $\mathcal{M}$.

We introduce the following definition.

Definition 5.3 ($\mathcal{M}$-normal approximation). An *$\mathcal{M}$-normal approximation* of T is given by a pair $(\mathcal{K}, F)$ such that

(A1) $F : \mathcal{M} \to \mathcal{A}_Q(\mathbf{U})$ is Lipschitz and takes the special form

$$F(x) = \sum_i [\![x + N_i(x)]\!] ,$$

with $N_i(x) \perp T_x\mathcal{M}$ for every $x \in \mathcal{M}$ and $i = 1, \ldots, Q$.

(A2) $\mathcal{K} \subset \mathcal{M}$ is closed, contains $\mathbf{\Phi}\big(\mathbf{\Gamma} \cap [-\frac{7}{2}, \frac{7}{2}]^m\big)$ and

$$\mathbf{T}_F \llcorner \mathbf{p}^{-1}(\mathcal{K}) = T \llcorner \mathbf{p}^{-1}(\mathcal{K}).$$

The map $N = \sum_i [\![N_i]\!] : \mathcal{M} \to \mathcal{A}_Q(\mathbf{U})$ is called *the normal part* of F.

As proven in [14, Theorem 2.4], the center manifold $\mathcal{M}$ of the previous section allows to construct an $\mathcal{M}$-normal approximation which does approximate the area minimizing current T. In order to state the result, to each $L \in \mathscr{W}$ we associate a *Whitney region* $\mathcal{L}$ on $\mathcal{M}$ as follows:

$$\mathcal{L} := \mathbf{\Phi}\left(H \cap \left[-\frac{7}{2}, \frac{7}{2}\right]^m\right),$$

where H is the cube concentric to L with $\ell(H) = \frac{17}{16}\ell(L)$. We will use $\|N|_{\mathcal{L}}\|_0$ to denote the quantity $\sup_{x\in\mathcal{L}} \mathcal{G}(N(x), Q[\![0]\!])$.

Theorem 5.4. *Let $\gamma_2 := \frac{\gamma_1}{4}$, with γ_1 the constant of Theorem 3.9. Under the hypotheses of Theorem 5.2, if ε_2 is sufficiently small, then there exist constants $\beta_2, \delta_2 > 0$ and an $\mathcal{M}$-normal approximation $(\mathcal{K}, F)$ such that the following estimates hold on every Whitney region $\mathcal{L}$:*

$$\mathrm{Lip}(N|_{\mathcal{L}}) \leq C E^{\gamma_2} \ell(L)^{\gamma_2} \quad \text{and} \quad \|N|_{\mathcal{L}}\|_{C^0} \leq C E^{\frac{1}{2m}} \ell(L)^{1+\beta_2}, \tag{5.18}$$

$$\int_{\mathcal{L}} |DN|^2 \leq C E\, \ell(L)^{m+2-2\delta_2}, \tag{5.19}$$

$$|\mathcal{L} \setminus \mathcal{K}| + \|\mathbf{T}_F - T\|(\mathbf{p}^{-1}(\mathcal{L})) \leq C E^{1+\gamma_2} \ell(L)^{m+2+\gamma_2}. \tag{5.20}$$

Moreover, for any $a > 0$ and any Borel set $\mathcal{V} \subset \mathcal{L}$, we have

$$\int_{\mathcal{V}} |\eta \circ N| \leq C E \left(\ell(L)^{3+\frac{\beta_2}{3}} + a\, \ell(L)^{2+\frac{\gamma_2}{2}}\right) |\mathcal{V}| + \frac{C}{a} \int_{\mathcal{V}} \mathcal{G}(N, Q[\![\eta \circ N]\!])^{2+\gamma_2}. \tag{5.21}$$

Let us briefly explain the conclusions of the theorem. The estimates in (5.18) and (5.19) concern the regularity properties of the normal approximation N, and will play an important role in many of the subsequent arguments. However, the key properties of N are (5.20) and (5.21): the former estimates the error done in the approximation on every Whitney region; while the latter estimates the L^1 norm of the average of N, which is a measure of the centering of the center manifold. Note that both estimates are in some sense "superlinear" with respect to the relevant parameters: indeed, as it will be better understood later on, they involve either a superlinear power of the excess $E^{1+\gamma_2}$ or the $L^{2+\gamma_2}$ norm of N (which is of higher order with respect to the "natural" L^2 norm).

5.4 Construction criteria

The estimates and the results of the previous two subsections depend very much on the way the Whitney decomposition, the interpolating functions and the normal approximation are constructed.

We start recalling the notation $p_L = (x_L, y_L)$ where L is a dyadic cube, x_L its center and $y_L \in \pi_0^\perp$ is chosen in such a way that $p_L \in \mathrm{spt}\,(T)$. Moreover, we set

$$\mathbf{B}_L := \mathbf{B}_{64 r_L}(p_L),$$

where we recall that $r_L := M_0 \sqrt{m}\, \ell(L)$ for some large constant $M_0 \in \mathbb{N}$.

We define the families of cubes of the Whitney decomposition

$$\mathscr{W} = \mathscr{W}_e \cup \mathscr{W}_h \cup \mathscr{W}_n \quad \text{and} \quad \mathscr{S} \subset \mathscr{C}.$$

We use the notation $\mathscr{S}^j = \mathscr{S} \cap \mathscr{C}^j$, $\mathscr{W}^j = \mathscr{W} \cap \mathscr{C}^j$ and so on.

We recall the notation for the excess,

$$\mathbf{E}(T, \mathbf{B}_r(x)) := \min_\tau \mathbf{E}(T, \mathbf{B}_r(x), \tau) = \mathbf{E}(T, \mathbf{B}_r(x), \pi).$$

The m-dimensional planes π realizing the minimum above are called *optimal planes* of T in a ball $\mathbf{B}_r(x)$ if, in addition, π optimizes the height among all planes that optimize the excess:

$$\begin{aligned}\mathbf{h}(T, \mathbf{B}_r(x), \pi) &= \min\left\{\mathbf{h}(T, \mathbf{B}_r(x), \tau) : \tau \text{ satisfies (5.3)}\right\}\\ &=: \mathbf{h}(T, \mathbf{B}_r(x)).\end{aligned} \tag{5.22}$$

An optimal plane in the ball $\mathbf{B}_L$ is denoted by π_L.

We fix a big natural number N_0, and constants $C_e, C_h > 0$, and we define $\mathscr{W}^i = \mathscr{S}^i = \emptyset$ for $i < N_0$. We proceed with $j \geq N_0$ inductively: if the father of $L \in \mathscr{C}^j$ is *not* in $\mathscr{W}^{j-1}$, then

(EX) $L \in \mathscr{W}^j_e$ if $\mathbf{E}(T, \mathbf{B}_L) > C_e\, E\, \ell(L)^{2-2\delta_2}$;
(HT) $L \in \mathscr{W}^j_h$ if $L \notin \mathscr{W}^j_e$ and $\mathbf{h}(T, \mathbf{B}_L) > C_h E^{\frac{1}{2m}} \ell(L)^{1+\beta_2}$;
(NN) $L \in \mathscr{W}^j_n$ if $L \notin \mathscr{W}^j_e \cup \mathscr{W}^j_h$ but it intersects an element of $\mathscr{W}^{j-1}$;

if none of the above occurs, then $L \in \mathscr{S}^j$.

We finally set

$$\mathbf{\Gamma} := [-4, 4]^m \setminus \bigcup_{L \in \mathscr{W}} L = \bigcap_{j \geq N_0} \bigcup_{L \in \mathscr{S}^j} L. \tag{5.23}$$

Observe that, if $j > N_0$ and $L \in \mathscr{S}^j \cup \mathscr{W}^j$, then necessarily its father belongs to $\mathscr{S}^{j-1}$.

For what concerns the interpolating functions g_L, they are obtained as the result of the following procedure.

(1) Let $L \in \mathscr{S} \cup \mathscr{W}$ and π_L be an optimal plane. Then, $T \llcorner \bar{C}_{32 r_L}(p_L, \pi_L)$ fulfills the assumptions of the approximation Theorem 3.9 in the cylinder $\bar{C}_{32 r_L}(p_L, \pi_L)$, and we can then construct a Lipschitz approximation

$$f_L : B_{8 r_L}(p_L, \pi_L) \to \mathcal{A}_Q(\pi_L^\perp).$$

(2) We let $h_L : B_{7 r_L}(p_L, \pi_L) \to \pi_L^\perp$ be a regularization of the average given by

$$h_L := (\boldsymbol{\eta} \circ f_L) * \varrho_{\ell(L)},$$

where $\varrho \in C^\infty_c(B_1)$ is radial, $\int \varrho = 1$ and $\int |x|^2 \varrho(x)\, dx = 0$.

(3) Finally, we find a smooth map $g_L : B_{4r_L}(p_L, \pi_0) \to \pi_0^\perp$ such that

$$\mathbf{G}_{g_L} = \mathbf{G}_{h_L} \llcorner \bar{C}_{4r_L}(p_L, \pi_0),$$

where we recall that $\mathbf{G}_u$ denotes the current induced by the graph of a function u.

The fact that the above procedure can be applied follows from the choice of the stopping criteria for the construction of the Whitney decomposition. We refer to [14] for a detailed proof. Here we only stress the fact that this construction depends strongly on the choice of the constants involved: in particular, C_e, C_h, β_2, δ_2, M_0 are positive real numbers and N_0 a natural number satisfying in particular

$$\beta_2 = 4\,\delta_2 = \min\left\{\frac{1}{2m}, \frac{\gamma_1}{100}\right\}, \tag{5.24}$$

where γ_1 is the constant of Theorem 3.9, and

$$M_0 \geq C_0(m, n, \bar{n}, Q) \geq 4 \quad \text{and} \quad \sqrt{m} M_0 2^{7-N_0} \leq 1\,. \tag{5.25}$$

Note that β_2 and δ_2 are fixed, while the other parameters are not fixed but are subject to further restrictions in the various statements, respecting a very precise "hierarchy" (cp. [14, Assumption 1.9]).

Finally, we add also a few words concerning the construction of the normal approximation N. In every Whitney region $\mathcal{L}$ the map N is a suitable extension of the reparametrization of the Lipschitz approximation f_L. Then the estimates (5.18), (5.19) and (5.20) follow easily from Theorem 3.9. The most intricate proof is the one of (5.21) for which the choice of the regularization h_L deeply plays a role. The main idea is that, on the optimal plane π_L, the average of the sheets of the minimizing current is almost the graph of a harmonic function. Therefore, a good way to regularize it (which actually would keep it unchanged if it were exactly harmonic) is to convolve with a radial symmetric mollifier. This procedure, which we stress is not the only possible one, will indeed preserve the main properties of the average.

5.5 Splitting before tilting

The above criteria are not just important for the construction purposes, but also lead to a couple of important estimates which will be referred to as *splitting-before-tilting* estimates. Indeed, it is not a case that the powers of the side-length in the (EX) and (HT) criteria look like the powers in the familiar decay of the excess and in the height bound. In fact it turns out that, following the arguments for the height bound and for the decay of the excess, one can infer two further consequences of the Whitney decomposition's criteria.

5.5.1 (HT)-cubes If a dyadic cube L has been selected by the Whitney decomposition procedure for the height criterion, then the $\mathcal{M}$-normal approximation above the corresponding Whitney region needs to have a large pointwise separation (see (5.28) below).

Proposition 5.5 ((HT)-estimate). *If ε_2 is sufficiently small, then the following conclusions hold for every $L \in \mathscr{W}_h$:*

$$\Theta(T, p) \leq Q - \frac{1}{2} \quad \forall\, p \in \mathbf{B}_{16r_L}(p_L), \tag{5.26}$$

$$L \cap H = \emptyset \quad \forall\, H \in \mathscr{W}_n \ \textit{with}\ \ell(H) \leq \frac{1}{2}\ell(L); \tag{5.27}$$

$$\mathcal{G}\big(N(x), Q\,[\![\eta \circ N(x)]\!]\big) \geq \frac{1}{4} C_h \mathbf{E}^{\frac{1}{2m}} \ell(L)^{1+\beta_2} \quad \forall\, x \in \mathcal{L}. \tag{5.28}$$

A simple corollary of the previous proposition is the following.

Corollary 5.6. *Given any $H \in \mathscr{W}_n$ there is a chain $L = L_0, L_1, \ldots, L_j = H$ such that:*

(a) *$L_0 \in \mathscr{W}_e$ and $L_i \in \mathscr{W}_n$ for all $i = 1, \ldots, j$;*
(b) *$L_i \cap L_{i-1} \neq \emptyset$ and $\ell(L_i) = \frac{1}{2}\ell(L_{i-1})$ for all $i = 1, \ldots, j$.*

In particular, $H \subset B_{3\sqrt{m}\ell(L)}(x_L, \pi_0)$.

We use this last corollary to partition $\mathscr{W}_n$.

Definition 5.7 (Domains of influence). We first fix an ordering of the cubes in $\mathscr{W}_e$ as $\{J_i\}_{i\in\mathbb{N}}$ so that their side-length decreases. Then $H \in \mathscr{W}_n$ belongs to $\mathscr{W}_n(J_0)$ if there is a chain as in Corollary 5.6 with $L_0 = J_0$. Inductively, $\mathscr{W}_n(J_r)$ is the set of cubes $H \in \mathscr{W}_n \setminus \cup_{i<r} \mathscr{W}_n(J_i)$ for which there is a chain as in Corollary 5.6 with $L_0 = J_r$.

5.5.2 (Ex)-cubes Similarly, if a cube of the Whitney decomposition is selected by the (EX) condition, *i.e.* the excess does not decay at some given scale, then a certain amount of separation between the sheets of the current must also in this case occur.

Proposition 5.8 ((EX)-estimate). *If $L \in \mathscr{W}_e$ and $\Omega = \mathbf{\Phi}(B_{\ell(L)/4}(q, \pi_0))$ for some point $q \in \pi_0$ with* $\mathrm{dist}(L, q) \leq 4\sqrt{m}\,\ell(L)$, *then*

$$C_e \mathbf{E}\ell(L)^{m+2-2\delta_2} \leq \ell(L)^m \mathbf{E}(T, \mathbf{B}_L) \leq C \int_\Omega |DN|^2, \tag{5.29}$$

$$\int_{\mathcal{L}} |DN|^2 \leq C\ell(L)^m \mathbf{E}(T, \mathbf{B}_L) \leq C\ell(L)^{-2} \int_\Omega |N|^2. \tag{5.30}$$

Both propositions above are a typical *splitting-before-tilting* phenomenon in this sense: the key assumption is that the excess has decayed up to a given scale (*i.e.* no "tilting" occurs), while the conclusion is that a certain amount of separation between the sheets of the current ("splitting") holds. We borrowed this terminology from the paper by T. Rivière [29], where a similar phenomenon (but not completely the same) was proved for semi-calibrated two dimensional currents as a consequence of a lower epi-perimetric inequality.

5.6 Intervals of flattening

Here we define the last feature of the construction of the center manifold, namely the so called interval of flattening. A center manifold constitutes a good approximation of the average of the sheets of a current as soon as the errors in Theorem 5.4 are small compared to the distance from the origin. In this case, we are forced to interrupt our blowup analysis and to start a new center manifold. This procedure is explained in details in the following paragraph.

5.6.1 Defining procedure We fix the constant $c_s := \frac{1}{64\sqrt{m}}$ and notice that $2^{-N_0} < c_s$. We set

$$\mathcal{R} := \left\{r \in]0,1] : \mathbf{E}(T, \mathbf{B}_{6\sqrt{m}r}) \leq \varepsilon_3^2\right\}, \tag{5.31}$$

where $\varepsilon_3 > 0$ is a suitably chosen constant, always assumed to be smaller than ε_2. Observe that, if $(s_k) \subset \mathcal{R}$ and $s_k \uparrow s$, then $s \in \mathcal{R}$. We cover $\mathcal{R}$ with a collection $\mathcal{F} = \{I_j\}_j$ of intervals $I_j =]s_j, t_j]$ defined as follows: we start with

$$t_0 := \max\{t : t \in \mathcal{R}\}.$$

Next assume, by induction, to have defined

$$t_0 > s_0 \geq t_1 > s_1 \geq \ldots > s_{j-1} \geq t_j,$$

and consider the following objects:

- $T_j := ((\iota_{0,t_j})_\sharp T) \llcorner \mathbf{B}_{6\sqrt{m}}$, and assume (without loss of generality, up to a rotation) that $\mathbf{E}(T_j, \mathbf{B}_{6\sqrt{m}}, \pi_0) = \mathbf{E}(T_j, \mathbf{B}_{6\sqrt{m}})$;
- let $\mathcal{M}_j$ the corresponding center manifold for T_j, given as the graph of a map $\varphi_j : \pi_0 \supset [-4,4]^m \to \pi_0^\perp$, (for later purposes we set $\mathbf{\Phi}_j(x) := (x, \varphi_j(x))$).

Then, one of the following possibilities occurs:

(Stop) either there is $r\in]0,3]$ and a cube L of the Whitney decomposition $\mathscr{W}^{(j)}$ of $[-4,4]^m\subset\pi_0$ (applied to T_j) such that

$$\ell(L)\geq c_s\, r \qquad \text{and} \qquad L\cap\bar B_r(0,\pi_0)\neq\emptyset; \tag{5.32}$$

(Go) or there exists no radius as in (Stop).

It is possible to show that when (Stop) occurs for some r, such r is smaller than 2^{-5}. This justifies the following:

(1) in case (Go) holds, we set $s_j:=0$, *i.e.* $I_j:=]0,t_j]$, and end the procedure;
(2) in case (Stop) holds we let $s_j:=\bar r\,t_j$, where $\bar r$ is the maximum radius satisfying (Stop). We choose then t_{j+1} as the largest element in $\mathcal{R}\cap]0,s_j]$ and proceed iteratively.

The following are easy consequences of the definition: for all $r\in]\frac{s_j}{t_j},3[$, it holds

$$\mathbf{E}(T_j,\mathbf{B}_r)\leq C\varepsilon_3^2\,r^{2-2\delta_2}, \tag{5.33}$$

$$\sup\{\mathrm{dist}(x,\mathcal{M}_j): x\in\mathrm{spt}\,(T_j)\cap\mathbf{p}_j^{-1}(\mathcal{B}_r(p_j))\}\leq C\,(E^j)^{\frac{1}{2m}}r^{1+\beta_2}, \tag{5.34}$$

where $E^j:=\mathbf{E}(T_j,\mathbf{B}_{6\sqrt{m}})$ and $\mathbf{p}_j$ denotes the nearest point projection on $\mathcal{M}_j$ defined on a neighborhood of the center manifold (for the proof we refer to [15]).

5.7 Families of subregions

Let $\mathcal{M}$ be a center manifold and $\mathbf{\Phi}:\pi_0\to\mathbb{R}^{m+n}$ the paremetrizing map. Set $q:=\mathbf{\Phi}(0)$ and denote by B the projection of the geodesic ball $\mathbf{p}_{\pi_0}(\mathcal{B}_r(q))$, for some $r\in(0,4)$. Since $\|\varphi\|_{C^{3,\kappa}}\leq C\varepsilon_2^{1/m}$ in Theorem 5.2, it is simple to show that B is a C^2 convex set and that the maximal curvature of ∂B is everywhere smaller than $\frac{2}{r}$. Thus, for every $z\in\partial B$ there is a ball $B_{r/2}(y)\subset B$ whose closure touches ∂B at z.

In this section we show how one can partition the cubes of the Whitney decomposition which intersect B into disjoint families which are labeled by pairs $(L,B(L))$ cube–ball enjoying different properties.

Proposition 5.9. *There exists a set* $\mathscr{Z}$ *of pairs* $(L, B(L))$ *with this properties:*

(i) *if* $(L, B(L)) \in \mathscr{Z}$*, then* $L \in \mathscr{W}_e \cup \mathscr{W}_h$*, the radius of* $B(L)$ *is* $\frac{\ell(L)}{4}$*,* $B(L) \subset B$ *and* $\operatorname{dist}(B(L), \partial B) \geq \frac{\ell(L)}{4}$*;*
(ii) *if the pairs* $(L, B(L)), (L', B(L')) \in \mathscr{Z}$ *are distinct, then* L *and* L' *are distinct and* $B(L) \cap B(L') = \emptyset$*;*
(iii) *the cubes* $\mathscr{W}$ *which intersect* B *are partitioned into disjoint families* $\mathscr{W}(L)$ *labeled by* $(L, B(L)) \in \mathscr{Z}$ *such that, if* $H \in \mathscr{W}(L)$*, then* $H \subset B_{30\sqrt{m}\ell(L)}(x_L)$*.*

In this way, every cube of the Whitney decomposition intersecting B can be uniquely associated to a ball $B(L) \subset B$ for some $L \in \mathscr{W}_e \cap \mathscr{W}_h$. This will allow to transfer the estimates form the cubes of the Whitney decomposition to the ball B.

5.7.1 Proof of Proposition 5.9 We start defining appropriate families of cubes and balls.

Definition 5.10 (Family of cubes). We first define a family $\mathcal{T}$ of cubes in the Whitney decomposition $\mathscr{W}$ as follows:

(i) $\mathcal{T}$ includes all $L \in \mathscr{W}_h \cup \mathscr{W}_e$ which intersect B;
(ii) if $L' \in \mathscr{W}_n$ intersects B and belongs to the domain of influence $\mathscr{W}_n(L)$ of the cube $L \in \mathscr{W}_e$ as in Definition 5.7, then $L \in \mathcal{T}$.

It is easy to see that, if r belongs to an interval of flattening, then for every $L \in \mathcal{T}$ it holds that $\ell(L) \leq 3c_s r \leq r$ and $\operatorname{dist}(L, B) \leq 3\sqrt{m}\,\ell(L)$. Therefore, we can also define the following associated balls.

Definition 5.11. For every $L \in \mathcal{T}$, let x_L be the center of L and:

(a) if $x_L \in \overline{B}$, we then set $s(L) := \ell(L)$ and $B^L := B_{s(L)}(x_L, \pi)$;
(b) otherwise we consider the ball $B_{r(L)}(x_L, \pi) \subset \pi$ such that its closure touches $\overline{B}$ at exactly one point $p(L)$, we set $s(L) := r(L) + \ell(L)$ and define $B^L := B_{s(L)}(x_L, \pi)$.

We proceed to select a countable family $\mathscr{T}$ of pairwise disjoint balls $\{B^L\}$. We let $S := \sup_{L \in \mathcal{T}} s(L)$ and start selecting a maximal subcollection $\mathscr{T}_1$ of pairwise disjoint balls with radii larger than $S/2$. Clearly, $\mathscr{T}_1$ is finite. In general, at the stage k, we select a maximal subcollection $\mathscr{T}_k$ of pairwise disjoint balls which do not intersect any of the previously selected balls in $\mathscr{T}_1 \cup \ldots \cup \mathscr{T}_{k-1}$ and which have radii $r \in]2^{-k}S, 2^{1-k}S]$. Finally, we set $\mathscr{T} := \bigcup_k \mathscr{T}_k$.

Definition 5.12 (Family of pairs cube-balls $(L, B(L)) \in \mathscr{L}$). Recalling the convexity properties of B and $\ell(L) \leq r$, it easy to see that there exist balls $B_{\ell(L)/4}(q_L, \pi) \subset B^L \cap B$ which lie at distance at least $\ell(L)/4$ from ∂B. We denote by $B(L)$ one of such balls and by $\mathscr{L}$ the collection of pairs $(L, B(L))$ with $B^L \in \mathscr{T}$.

Next, we partition the cubes of $\mathscr{W}$ which intersect B into disjoint families $\mathscr{W}(L)$ labeled by $(L, B(L)) \in \mathscr{L}$ in the following way. Let $H \in \mathscr{W}$ have nonempty intersection with B. Then, either H is in $\mathcal{T}$ and we set $J := H$, or is in the domain of influence of some $J \in \mathcal{T}$. If $J \neq H$, then the separation between J and H is at most $3\sqrt{m}\ell(J)$ and, hence, $H \subset B_{4\sqrt{m}\ell(J)}(x_J)$. By construction there is a $B^L \in \mathscr{T}$ with $B^J \cap B^L \neq \emptyset$ and radius $s(L) \geq \frac{s(J)}{2}$. We then prescribe $H \in \mathscr{W}(L)$. Observe that

$$s(L) \leq 4\sqrt{m}\,\ell(L) \quad \text{and} \quad s(J) \geq \ell(J).$$

Therefore, it also holds

$$\ell(J) \leq 8\sqrt{m}\,\ell(L) \quad \text{and} \quad |x_J - x_L| \leq 5s(L) \leq 20\sqrt{m}\,\ell(L),$$

thus implying

$$H \subset B_{4\sqrt{m}\,\ell(J)}(x_J) \subset B_{4\sqrt{m}\ell(J)+20\sqrt{m}\,\ell(L)}(x_L) \subset B_{30\sqrt{m}\,\ell(L)}(x_L)\,.$$

6 Order of contact

In this section we discuss the issues in steps (D) and (E) of the sketch of proof in Section 2.7, *i.e.* the order of contact of the normal approximation with the center manifold.

The key word for this part is *frequency function*, which is the monotone quantity discovered by Almgren controlling the vanishing order of a harmonic function. In order to explain this point, we consider first the case of a real valued harmonic function $f : B_1 \subset \mathbb{R}^2 \to \mathbb{R}$ with an expansion in polar coordinates

$$f(r, \theta) = a_0 + \sum_{k=1}^{\infty} r^k \big(a_k \cos(k\theta) + b_k \sin(k\theta)\big).$$

How can one detect the smallest index k such that a_k or b_k is not 0? It is not difficult to show that the quantity

$$I_f(r) := \frac{r \int_{B_r} |\nabla f|^2}{\int_{\partial B_r} |f|^2} \tag{6.1}$$

is monotone increasing in r and its limit as $r \downarrow 0$ gives exactly the smallest non-zero index in the expansion above.

I_f is what Almgren calls the frequency function (and the reason for such terminology is now apparent from the example above), and one of the most striking discoveries of Almgren is that the monotonicity of the frequency remains true for Q-valued functions and in fact allows to obtain a non-trivial blowup limit.

In the next subsections, we see how this discussion generalizes to the case of area minimizing currents, where an *almost monotonicity* formula can be derived for a suitable frequency defined for the $\mathcal{M}$-normal approximation.

6.1 Frequency function's estimate

For every interval of flattening $I_j =]s_j, t_j]$, let N_j be the normal approximation of T_j on $\mathcal{M}_j$. Since the L^2 norm of the trace of N_j may not have any connection to the current itself (remember that N_j misses a set of positive measure of T_j), we need to introduce an averaged version of the frequency function. To this aim, consider the following piecewise linear function $\varphi : [0 + \infty[\to [0, 1]$ given by

$$\varphi(r) := \begin{cases} 1 & \text{for } r \in [0, \frac{1}{2}], \\ 2 - 2r & \text{for } r \in]\frac{1}{2}, 1], \\ 0 & \text{for } r \in]1, +\infty[, \end{cases}$$

and let us define a new frequency function in the following way.

Definition 6.1. For every $r \in]0, 3]$ we define:

$$\mathbf{D}_j(r) := \int_{\mathcal{M}^j} \varphi\left(\frac{d_j(p)}{r}\right) |DN_j|^2(p)\, dp,$$

and

$$\mathbf{H}_j(r) := -\int_{\mathcal{M}^j} \varphi'\left(\frac{d_j(p)}{r}\right) \frac{|N_j|^2(p)}{d(p)}\, dp\,,$$

where $d_j(p)$ is the geodesic distance on $\mathcal{M}_j$ between p and $\mathbf{\Phi}_j(0)$. If we have that $\mathbf{H}_j(r) > 0$, then we define the *frequency function*

$$\mathbf{I}_j(r) := \frac{r\,\mathbf{D}_j(r)}{\mathbf{H}_j(r)}.$$

Note that, by the Coarea formula,

$$\mathbf{H}_j(r) = 2\int_{\mathcal{B}_r\setminus\mathcal{B}_{r/2}(\Phi_j(0))} \frac{|N|^2}{d(p)} = 2\int_{r/2}^{r}\frac{1}{t}\int_{\partial\mathcal{B}_t(\Phi_j(0))}|N_j|^2\,dt, \tag{6.2}$$

whereas, using Fubini,

$$r\,\mathbf{D}_j(r) = \int_{\mathcal{M}_j}|DN_j|^2(x)\int_{\frac{r}{2}}^{r}\mathbf{1}_{]|x|,\infty[}(t)\,dt\,d\mathcal{H}^m(x) = 2\int_{\frac{r}{2}}^{r}\int_{\mathcal{B}_t(\mathbf{\Phi}_j(0))}|DN_j|^2\,dt. \tag{6.3}$$

This explains in which sense $\mathbf{I}_j$ is an average of the quantity introduced by F. Almgren.

The main analytical estimate is then the following.

Theorem 6.2. *If ε_3 in (5.31) is sufficiently small, then there exists a constant $C > 0$ (indepent of j) such that, if $[a, b] \subset [\frac{s}{t}, 3]$ and $\mathbf{H}_j|_{[a,b]} > 0$, then it holds*

$$\mathbf{I}_j(a) \leq C(1 + \mathbf{I}_j(b)). \tag{6.4}$$

To simplify the notation, we drop the index j and omit the measure $\mathcal{H}^m$ in the integrals over regions of $\mathcal{M}$. For the proof of the theorem we need to introduce some auxiliary functions (all absolutely continuous with respect to r). We let $\partial_{\hat{r}}$ denote the derivative along geodesics starting at $\mathbf{\Phi}(0)$. We set

$$\mathbf{E}(r) := -\int_{\mathcal{M}}\varphi'\left(\tfrac{d(p)}{r}\right)\sum_{i=1}^{Q}\langle N_i(p), \partial_{\hat{r}}N_i(p)\rangle\,dp,$$

$$\mathbf{G}(r) := -\int_{\mathcal{M}}\varphi'\left(\tfrac{d(p)}{r}\right)d(p)\,|\partial_{\hat{r}}N(p)|^2\,dp,$$

$$\Sigma(r) := \int_{\mathcal{M}}\varphi\left(\tfrac{d(p)}{r}\right)|N|^2(p)\,dp.$$

The proof of Theorem 6.2 exploits some "integration by parts" formulas, which in our setting are given by the first variations for the minimizing current. We collect these identities in the following proposition, and proceed then with the proof of the theorem.

Proposition 6.3. *There exist dimensional constants $C, \gamma_3 > 0$ such that, if the hypotheses of Theorem 6.2 hold and $\mathbf{I} \geq 1$, then*

$$\left|\mathbf{H}'(r) - \tfrac{m-1}{r}\,\mathbf{H}(r) - \tfrac{2}{r}\,\mathbf{E}(r)\right| \leq C\mathbf{H}(r), \tag{6.5}$$

$$\left|\mathbf{D}(r) - r^{-1}\mathbf{E}(r)\right| \leq C\mathbf{D}(r)^{1+\gamma_3} + C\varepsilon_3^2\,\Sigma(r), \tag{6.6}$$

$$\begin{aligned}\left|\mathbf{D}'(r) - \tfrac{m-2}{r}\,\mathbf{D}(r) - \tfrac{2}{r^2}\,\mathbf{G}(r)\right| &\leq C\mathbf{D}(r) + C\mathbf{D}(r)^{\gamma_3}\mathbf{D}'(r)\\ &\quad + r^{-1}\mathbf{D}(r)^{1+\gamma_3},\end{aligned} \tag{6.7}$$

$$\Sigma(r) + r\,\Sigma'(r) \leq C\,r^2\,\mathbf{D}(r) \leq C r^{2+m}\varepsilon_3^2. \tag{6.8}$$

We assume for the moment the proposition and prove the theorem.

Proof of Theorem 6.2. It enough to consider the case in which $\mathbf{I} > 1$ on $]a, b[$. Set $\Omega(r) := \log \mathbf{I}(r)$. By Proposition 6.3, if ε_3 is sufficiently small, then

$$\frac{\mathbf{D}(r)}{2} \leq \frac{\mathbf{E}(r)}{r} \leq 2\,\mathbf{D}(r), \tag{6.9}$$

from which we conclude that $\mathbf{E} > 0$ over the interval $]a, b'[$. Set for simplicity $\mathbf{F}(r) := \mathbf{D}(r)^{-1} - r\mathbf{E}(r)^{-1}$, and compute

$$-\Omega'(r) = \frac{\mathbf{H}'(r)}{\mathbf{H}(r)} - \frac{\mathbf{D}'(r)}{\mathbf{D}(r)} - \frac{1}{r} \overset{(6.6)}{=} \frac{\mathbf{H}'(r)}{\mathbf{H}(r)} - \frac{r\mathbf{D}'(r)}{\mathbf{E}(r)} - \mathbf{D}'(r)\mathbf{F}(r) - \frac{1}{r}.$$

Again by Proposition 6.3:

$$\frac{\mathbf{H}'(r)}{\mathbf{H}(r)} \overset{(6.5)}{\leq} \frac{m-1}{r} + C + \frac{2}{r}\,\frac{\mathbf{E}(r)}{\mathbf{H}(r)}, \tag{6.10}$$

$$|\mathbf{F}(r)| \overset{(6.6)}{\leq} C\,\frac{r\,(\mathbf{D}(r)^{1+\gamma_3} + \Sigma(r))}{\mathbf{D}(r)\,\mathbf{E}(r)} \overset{(6.9)}{\leq} C\,\mathbf{D}(r)^{\gamma_3 - 1} + C\,\frac{\Sigma(r)}{\mathbf{D}(r)^2}, \tag{6.11}$$

$$\begin{aligned}-\frac{r\mathbf{D}'(r)}{\mathbf{E}(r)} &\overset{(6.7)}{\leq} \left(C - \frac{m-2}{r}\right)\frac{r\mathbf{D}(r)}{\mathbf{E}(r)} - \frac{2}{r}\,\frac{\mathbf{G}(r)}{\mathbf{E}(r)}\\ &\quad + C\,\frac{r\mathbf{D}(r)^{\gamma_3}\mathbf{D}'(r) + \mathbf{D}(r)^{1+\gamma_3}}{\mathbf{E}(r)}\\ &\leq C - \frac{m-2}{r} + \frac{C}{r}\mathbf{D}(r)|\mathbf{F}(r)| - \frac{2}{r}\,\frac{\mathbf{G}(r)}{\mathbf{E}(r)}\\ &\quad + C\mathbf{D}(r)^{\gamma_3 - 1}\mathbf{D}'(r) + C\,\frac{\mathbf{D}(r)^{\gamma_3}}{r}\\ &\overset{(6.8),\,(6.11)}{\leq} C - \frac{m-2}{r} - \frac{2}{r}\,\frac{\mathbf{G}(r)}{\mathbf{E}(r)} + C\mathbf{D}(r)^{\gamma_3 - 1}\mathbf{D}'(r) + C\,r^{\gamma_3 m - 1},\end{aligned} \tag{6.12}$$

where we used the rough estimate $\mathbf{D}(r) \leq C\, r^{m+2-2\delta_2}$ coming from (5.19) of Theorem 5.4 and the condition (Stop).

By Cauchy-Schwartz, we have

$$\frac{\mathbf{E}(r)}{r\mathbf{H}(r)} \leq \frac{\mathbf{G}(r)}{r\mathbf{E}(r)}. \tag{6.13}$$

Thus, by (6.10), (6.12) and (6.13), we conclude

$$\begin{aligned} -\Omega'(r) &\leq C + C\, r^{\gamma_3 m-1} + Cr\mathbf{D}(r)^{\gamma_3-1}\mathbf{D}'(r) - \mathbf{D}'(r)\mathbf{F}(r) \\ &\overset{(6.11)}{\leq} C\, r^{\gamma_3 m-1} + C\mathbf{D}(r)^{\gamma_3-1}\mathbf{D}'(r) + C\frac{\Sigma(r)\mathbf{D}'(r)}{\mathbf{D}(r)^2}. \end{aligned} \tag{6.14}$$

Integrating (6.14) we conclude:

$$\begin{aligned} \Omega(a) - \Omega(b) &\leq C + C\left(\mathbf{D}(b)^{\gamma_3} - \mathbf{D}(a)^{\gamma_3}\right) \\ &\quad + C\left[\frac{\Sigma(a)}{\mathbf{D}(a)} - \frac{\Sigma(b)}{\mathbf{D}(b)} + \int_a^b \frac{\Sigma'(r)}{\mathbf{D}(r)}\, dr\right] \overset{(6.8)}{\leq} C. \end{aligned}$$

□

6.1.1 Proof of Proposition 6.3 The remaining part of this subsection is devoted to give some arguments for the proof of the first variation formulas.

The estimate (6.5) follows from a straightforward computation: using the area formula and setting $y = rz$, we have

$$\mathbf{H}(r) = -r^{m-1} \int_{T_q\mathcal{M}} \frac{\varphi'(|z|)}{|z|} |N|^2(\exp(rz))\, \mathbf{J}\exp(rz)\, dx,$$

and differentiating under the integral sign, we easily get (6.5):

$$\begin{aligned} \mathbf{H}'(r) = &- (m-1)\, r^{m-2} \int_{T_q\mathcal{M}} \frac{\varphi'(|z|)}{|z|} |N|^2(\exp(rz))\, \mathbf{J}\exp(rz)\, dz \\ &- 2\, r^{m-1} \int_{T_q\mathcal{M}} \varphi'(|z|) \sum_i \langle N_i, \partial_{\hat{r}} N_i\rangle\, (\exp(rz))\, \mathbf{J}\exp(rz)\, dz \\ &- r^{m-1} \int_{T_q\mathcal{M}} \frac{\varphi'(|z|)}{|z|} |N|^2(\exp(rz))\, \frac{d}{dr}\mathbf{J}\exp(rz)\, dz \\ = &\; \frac{m-1}{r}\mathbf{H}(r) + \frac{2}{r}\mathbf{E}(r) + O(1)\, \mathbf{H}(r), \end{aligned}$$

where we the following simple fact for the Jacobian of the exponential map $\frac{d}{dr}\mathbf{J}\exp(r\, z) = O(1)$, because $\mathcal{M}$ is a $C^{3,\kappa}$ submanifold and the exponential map exp is a $C^{2,\kappa}$ map.

Similarly, (6.8) follows by simple computation which involve a Poincaré inequality: namely, if $\mathbf{I} \geq 1$, then

$$\int_{\mathcal{B}_r(q)} |N|^2 \leq C r^2 \mathbf{D}(r). \tag{6.15}$$

We refer to [15] for the details of the proof.

Here we try to explain the remaining two estimates, which instead are connected to the first variation $\delta T(X)$ of the area minimizing current T along a vector field X.

The idea is the following: since the first variations of T are zero, we compute them using its approximation N and derive the integral equality in the Proposition 6.3. To understand the meaning of these estimates, consider $u : \mathbb{R}^m \to \mathbb{R}^n$ a harmonic function. Then, computing the variations of the Dirichlet energy of u leads to the following two identities:

$$\int_{B_r} |Du|^2 = \int_{\partial B_r} u \cdot \frac{\partial u}{\partial \nu},$$

$$\int_{\partial B_r} |Du|^2 = \frac{m-2}{r} \int_{B_r} |Du|^2 + 2 \int_{\partial B_r} \left| \frac{\partial u}{\partial \nu} \right|^2,$$

which are the exact analog of (6.6) and (6.7) without any error term. What we need to do is then to replace the Dirichlet energy with the area functional, and to consider the fact that the normal approximation N is only approximately stationary with respect to this functional.

We start fixing a tubular neighborhood $\mathbf{U}$ of $\mathcal{M}$ and the normal projection $\mathbf{p} : \mathbf{U} \to \mathcal{M}$. Observe that $\mathbf{p} \in C^{2,\kappa}$. We will consider:

(1) the *outer variations*, where $X(p) = X_o(p) := \varphi\left(\frac{d(\mathbf{p}(p))}{r}\right)(p - \mathbf{p}(p))$.
(2) the *inner variations*, where $X(p) = X_i(p) := Y(\mathbf{p}(p))$ with

$$Y(p) := \frac{d(p)}{r} \varphi\left(\frac{d(p)}{r}\right) \frac{\partial}{\partial \hat{r}} \quad \forall\, p \in \mathcal{M}.$$

Consider now the map $F(p) := \sum_i [\![p + N_i(p)]\!]$ and the current $\mathbf{T}_F$ associated to its image. Observe that X_i and X_o are supported in $\mathbf{p}^{-1}(\mathcal{B}_r(q))$ but none of them is *compactly* supported. However, it is simple to see that $\delta T(X) = 0$. Then, we have

$$|\delta \mathbf{T}_F(X)| = |\delta \mathbf{T}_F(X) - \delta T(X)|$$
$$\leq \underbrace{\int_{\mathrm{spt}\,(T) \setminus \mathrm{Im}(F)} \left| \mathrm{div}_{\vec{T}} X \right| d\|T\| + \int_{\mathrm{Im}(F) \setminus \mathrm{spt}\,(T)} \left| \mathrm{div}_{\vec{\mathbf{T}}_F} X \right| d\|\mathbf{T}_F\|}_{\mathrm{Err}_4}, \tag{6.16}$$

where $\mathrm{Im}(F)$ is the image of the map $F(x) = \sum_i [\![(x, N_i(x))]\!]$, *i.e.* the support of the current $\mathbf{T}_F$.

Set now for simplicity $\varphi_r(p) := \varphi\left(\frac{d(p)}{r}\right)$. It is not hard to realize that the mass of the current $\mathbf{T}_F$ can be expressed in the following way:

$$\begin{aligned}\mathbf{M}(\mathbf{T}_F) = {} & Q\,\mathcal{H}^m(\mathcal{M}) - Q\int_{\mathcal{M}} \langle H, \eta\circ N\rangle + \frac{1}{2}\int_{\mathcal{M}} |DN|^2 \\ & + \int_{\mathcal{M}} \sum_i \Big(P_2(x, N_i) + P_3(x, N_i, DN_i) + R_4(x, DN_i)\Big),\end{aligned} \tag{6.17}$$

where P_2, P_3 and R_4 are quadratic, cubic and fourth order errors terms (see [16, Theorem 3.2]) One can then compute the first variation of a push-forward current $\mathbf{T}_F$ and obtain (cp. [16, Theorem 4.2])

$$\delta\mathbf{T}_F(X_o) = \int_{\mathcal{M}} \Big(\varphi_r\,|DN|^2 + \sum_{i=1}^{Q} N_i\otimes\nabla\varphi_r : DN_i\Big) + \sum_{j=1}^{3} \mathrm{Err}_j^o, \tag{6.18}$$

where the errors Err_j^o satisfy

$$\mathrm{Err}_1^o = -Q\int_{\mathcal{M}} \varphi_r\langle H_{\mathcal{M}}, \eta\circ N\rangle, \tag{6.19}$$

$$|\mathrm{Err}_2^o| \le C\int_{\mathcal{M}} |\varphi_r||A|^2|N|^2, \tag{6.20}$$

$$|\mathrm{Err}_3^o| \le C\int_{\mathcal{M}} \big(|N||A| + |DN|^2\big)\big(|\varphi_r||DN|^2 + |D\varphi_r||DN||N|\big), \tag{6.21}$$

here $H_{\mathcal{M}}$ is the mean curvature vector of $\mathcal{M}$. Plugging (6.18) into (6.16), we then conclude

$$\big|\mathbf{D}(r) - r^{-1}\mathbf{E}(r)\big| \le \sum_{j=1}^{4} \big|\mathrm{Err}_j^o\big|\,, \tag{6.22}$$

where Err_4^o corresponds to Err_4 of (6.16) when $X = X_o$. Arguing similarly with $X = X_i$ (cp. [16, Theorem 4.3]), we get

$$\begin{aligned}\delta\mathbf{T}_F(X_i) = {} & \frac{1}{2}\int_{\mathcal{M}} \Big(|DN|^2 \mathrm{div}_{\mathcal{M}} Y - 2\sum_{i=1}^{Q} \langle DN_i : (DN_i\cdot D_{\mathcal{M}} Y)\rangle\Big) \\ & + \sum_{j=1}^{3} \mathrm{Err}_j^i\,,\end{aligned} \tag{6.23}$$

where this time the errors Err^i_j satisfy

$$\mathrm{Err}^i_1 = -Q\int_{\mathcal{M}} \left(\langle H_{\mathcal{M}}, \eta\circ N\rangle \operatorname{div}_{\mathcal{M}} Y + \langle D_Y H_{\mathcal{M}}, \eta\circ N\rangle\right), \tag{6.24}$$

$$|\mathrm{Err}^i_2| \le C\int_{\mathcal{M}} |A|^2\left(|DY||N|^2 + |Y||N|\,|DN|\right), \tag{6.25}$$

$$|\mathrm{Err}^i_3| \le C\int_{\mathcal{M}} |Y||A||DN|^2(|N| + |DN|) + |DY|\big(|A|\,|N|^2|DN| + |DN|^4\big)\Big). \tag{6.26}$$

Straightforward computations lead to

$$D_{\mathcal{M}}Y(p) = \varphi'\left(\frac{d(p)}{r}\right)\frac{d(p)}{r^2}\frac{\partial}{\partial\hat{r}}\otimes\frac{\partial}{\partial\hat{r}} + \varphi\left(\frac{d(p)}{r}\right)\left(\frac{\mathrm{Id}}{r} + O(1)\right), \tag{6.27}$$

$$\operatorname{div}_{\mathcal{M}} Y(p) = \varphi'\left(\frac{d(p)}{r}\right)\frac{d(p)}{r^2} + \varphi\left(\frac{d(p)}{r}\right)\left(\frac{m}{r} + O(1)\right). \tag{6.28}$$

Plugging (6.27) and (6.28) into (6.23) and using (6.16) we then conclude

$$\left|\mathbf{D}'(r) - (m-2)r^{-1}\mathbf{D}(r) - 2r^{-2}\mathbf{G}(r)\right| \le C\mathbf{D}(r) + \sum_{j=1}^{4}\left|\mathrm{Err}^i_j\right|. \tag{6.29}$$

Proposition 6.3 is then proved by the estimates of the errors terms done in the next subsection.

6.1.2 Estimates of the errors terms We consider the family of pairs $\mathscr{L} = \{(J_i, B(J_i))\}_{i\in\mathbb{N}}$ introduced in the previous section, and set

$$\mathcal{B}^i := \mathbf{\Phi}(B(J_i)) \quad\text{and}\quad \mathcal{U}_i = \cup_{H\in\mathscr{W}(J_i)}\mathbf{\Phi}(H)\cap\mathcal{B}_r(q).$$

Set $\mathcal{V}_i := \mathcal{U}_i\setminus\mathcal{K}$, where $\mathcal{K}$ is the coincidence set of Theorem 5.4. By a simple application of Theorem 5.4 we derive the following estimates:

$$\int_{\mathcal{U}_i}|\eta\circ N| \le C\,E\,\ell_i^{2+m+\frac{\gamma_2}{2}} + C\int_{\mathcal{U}^i}|N|^{2+\gamma_2}, \tag{6.30}$$

$$\int_{\mathcal{U}_i}|DN|^2 \le C\,E\,\ell_i^{m+2-2\delta_2}, \tag{6.31}$$

$$\|N\|_{C^0(\mathcal{U}_i)} + \sup_{p\in\mathrm{spt}\,(T)\cap\mathbf{p}^{-1}(\mathcal{U}_i)}|p - \mathbf{p}(p)| \le C\,E^{\frac{1}{2m}}\ell_i^{1+\beta_2}, \tag{6.32}$$

$$\mathrm{Lip}(N|_{\mathcal{U}_i}) \le C\,E^{\gamma_2}\ell_i^{\gamma_2}, \tag{6.33}$$

$$\mathbf{M}(T\llcorner\mathbf{p}^{-1}(\mathcal{V}_i)) + \mathbf{M}(\mathbf{T}_F\llcorner\mathbf{p}^{-1}(\mathcal{V}_i)) \le C\,E^{1+\gamma_2}\ell_i^{m+2+\gamma_2}. \tag{6.34}$$

Observe that the separation between $\mathcal{B}^i$ and $\partial\mathcal{B}_r(q)$ is larger than $\ell(J_i)/4$ by Proposition 5.9 (i), and then $\varphi_r(p) = \varphi\big(\frac{d(p)}{r}\big)$ satisfies

$$\inf_{p\in\mathcal{B}^i} \varphi_r(p) \geq (4r)^{-1}\ell_i\,, \tag{6.35}$$

where $\ell_i := \ell(J_i)$. From this and Proposition 5.9 (iii), we also obtain

$$\sup_{p\in\mathcal{U}_i} \varphi_r(p) - \inf_{p\in\mathcal{U}_i} \varphi_r(p) \leq C\mathrm{Lip}(\varphi_r)\ell_i \leq \frac{C}{r}\,\ell_i \overset{(6.35)}{\leq} C \inf_{p\in\mathcal{B}_i} \varphi_r(p)\,,$$

which translates into

$$\sup_{p\in\mathcal{U}_i} \varphi_r(p) \leq C \inf_{p\in\mathcal{B}^i} \varphi_r(p)\,. \tag{6.36}$$

Moreover, by an application of the *splitting-before-tilting* estimates in Proposition 5.5 and Proposition 5.8, we infer that

$$\int_{\mathcal{B}^i} |N|^2 \geq c\, E^{\frac{1}{m}}\ell_i^{m+2+2\beta_2} \quad \text{if } L_i \in \mathscr{W}_h, \tag{6.37}$$

$$\int_{\mathcal{B}^i} |DN|^2 \geq c\, E\, \ell_i^{m+2-2\delta_2} \quad \text{if } L_i \in \mathscr{W}_e\,. \tag{6.38}$$

This easily implies the following estimates under the hypotheses $\mathbf{I} \geq 1$: by applying (6.15), (6.35), (6.37) and (6.38), we get, for suitably chosen $\gamma(t), C(t) > 0$,

$$\begin{aligned}\sup_i E^t\Big[\ell_i^t + \Big(\inf_{\mathcal{B}^i}\varphi_r\Big)^{\frac{t}{2}}\ell_i^{\frac{t}{2}}\Big] &\leq C(t)\sup_i\Big(\int_{\mathcal{B}^i}\varphi_r(|DN|^2+|N|^2)\Big)^{\gamma(t)}\\ &\leq C(t)\mathbf{D}(r)^{\gamma(t)}\,,\end{aligned} \tag{6.39}$$

and similarly

$$\begin{aligned}\sum_i\Big(\inf_{\mathcal{B}^i}\varphi_r\Big)E\,\ell_i^{m+2+\frac{\gamma_2}{4}} &\leq C\sum_i\int_{\mathcal{B}^i}\varphi_r(|DN|^2+|N|^2)\\ &\leq C\mathbf{D}(r)\,,\end{aligned} \tag{6.40}$$

$$\begin{aligned}\sum_i E\,\ell_i^{m+2+\frac{\gamma_2}{4}} &\leq C\int_{\mathcal{B}_r(q)}\big(|DN|^2+|N|^2\big)\\ &\leq C\big(\mathbf{D}(r)+r\mathbf{D}'(r)\big).\end{aligned} \tag{6.41}$$

We can now pass to estimate the errors terms in (6.6) and (6.7) in order to conclude the proof of Proposition 6.3.

Errors of type 1. By Theorem 5.2, the map φ defining the center manifold satisfies $\|D\varphi\|_{C^{2,\kappa}} \leq C\,E^{\frac{1}{2}}$, which in turn implies $\|H_{\mathcal{M}}\|_{L^\infty} + \|DH_{\mathcal{M}}\|_{L^\infty} \leq C\,E^{\frac{1}{2}}$ (recall that $H_{\mathcal{M}}$ denotes the mean curvature of $\mathcal{M}$). Therefore, by (6.36), (6.30), (6.40) and (6.39), we get

$$
\begin{aligned}
\left|\mathrm{Err}_1^o\right| &\leq C \int_{\mathcal{M}} \varphi_r\,|H_{\mathcal{M}}|\,|\eta\circ N| \\
&\leq C\,E^{\frac{1}{2}} \sum_j \Big(\big(\sup_{\mathcal{U}_i} \varphi_r\big)\, E\,\ell_j^{2+m+\gamma_2} + C \int_{\mathcal{U}_j} \varphi_r |N|^{2+\gamma_2} \Big) \\
&\leq C\mathbf{D}(r)^{1+\gamma_3} + C \sum_j E^{\frac{1}{2}}\,\ell_j^{\gamma_2(1+\beta_2)} \int_{\mathcal{U}_j} \varphi_r |N|^2 \leq C\mathbf{D}(r)^{1+\gamma_3}\,,
\end{aligned}
$$

and analogously

$$
\begin{aligned}
\left|\mathrm{Err}_1^i\right| &\leq C\,r^{-1} \int_{\mathcal{M}} (|H_{\mathcal{M}}| + |D_Y H_{\mathcal{M}}|)|\eta\circ N| \\
&\leq C\,r^{-1} E^{\frac{1}{2}} \sum_j \Big(E\,\ell_j^{2+m+\gamma_2} + C \int_{\mathcal{U}_j} |N|^{2+\gamma_2} \Big) \\
&\leq C\,r^{-1}\mathbf{D}(r)^{\gamma} \big(\mathbf{D}(r) + r\,\mathbf{D}'(r)\big).
\end{aligned}
$$

Errors of type 2. From $\|A\|_{C^0} \leq C\|D\varphi\|_{C^2} \leq C E^{\frac{1}{2}} \leq C\varepsilon_3$, it follows that $\mathrm{Err}_2^o \leq C\varepsilon_3^2 \Sigma(r)$. Moreover, since $|DX_i| \leq Cr^{-1}$, (6.15) leads to

$$
\left|\mathrm{Err}_2^i\right| \leq Cr^{-1} \int_{\mathcal{B}_r(p_0)} |N|^2 + C \int \varphi_r |N||DN| \leq C\mathbf{D}(r)\,.
$$

Errors of type 3. Clearly, we have

$$
\begin{aligned}
\left|\mathrm{Err}_3^o\right| \leq\; & \underbrace{\int \varphi_r \left(|DN|^2|N| + |DN|^4\right)}_{I_1} + \underbrace{C\,r^{-1} \int_{\mathcal{B}_r(q)} |DN|^3|N|}_{I_2} \\
& + \underbrace{C\,r^{-1} \int_{\mathcal{B}_r(q)} |DN||N|^2}_{I_3}\,.
\end{aligned}
$$

We estimate separately the three terms (recall that $\gamma_2 > 4\delta_2$):

$$I_1 \le \int_{\mathcal{B}_r(p_0)} \varphi_r (|N|^2|DN| + |DN|^3) \le I_3 + C \sum_j \sup_{\mathcal{U}_j} \varphi_r E^{1+\gamma_2} \ell_j^{m+2+\frac{\gamma_2}{2}}$$
$$\stackrel{(6.40)\,\&\,(6.39)}{\le} I_3 + C\mathbf{D}(r)^{1+\gamma_3},$$

$$I_2 \le C r^{-1} \sum_j E^{1+\frac{1}{2m}+\gamma_2} \ell_j^{m+3+\beta_2+\frac{\gamma_2}{2}}$$
$$\stackrel{(6.36)}{\le} C \sum_j E^{1+\frac{1}{2m}+\gamma_2} \ell_j^{m+2+\beta_2+\frac{\gamma_2}{2}} \inf_{\mathcal{B}^j} \varphi_r \stackrel{(6.40)\,\&\,(6.39)}{\le} C\mathbf{D}(r)^{1+\gamma_3},$$

$$I_3 \le C r^{-1} \sum_j E^{\gamma_2} \ell_j^{\gamma_2} \int_{\mathcal{U}_j} |N|^2 \stackrel{(6.39)}{\le} C r^{-1} \mathbf{D}(r)^{\gamma_3} \int_{\mathcal{B}_r(q)} |N|^2$$
$$\stackrel{(6.15)}{\le} C\mathbf{D}(r)^{1+\gamma_3}$$

For what concerns the inner variations, we have

$$|\mathrm{Err}_3^i| \le C \int_{\mathcal{B}_r(q)} \left(r^{-1}|DN|^3 + r^{-1}|DN|^2|N| + r^{-1}|DN||N|^2 \right).$$

The last integrand corresponds to I_3, while the remaining part can be estimated as follows:

$$\int_{\mathcal{B}_r(q)} r^{-1}(|DN|^3 + |DN|^2|N|) \le C \sum_j r^{-1}(E^{\gamma_2}\ell_j^{\gamma_2} + E^{\frac{1}{2m}} \ell_j^{1+\beta_2}) \int_{\mathcal{U}_j} |DN|^2$$
$$\stackrel{(6.39)}{\le} C r^{-1} \mathbf{D}(r)^{\gamma_3} \int_{\mathcal{B}_r(q)} |DN|^2$$
$$\le C\mathbf{D}(r)^{\gamma_3} \left(\mathbf{D}'(r) + r^{-1}\mathbf{D}(r) \right).$$

Errors of type 4. We compute explicitly

$$|DX_o(p)| \le 2\,|p - \mathbf{p}(p)|\, \frac{|Dd(\mathbf{p}(p), q)|}{r} + \varphi_r(p)\,|D(p - \mathbf{p}(p))|$$
$$\le C \left(\frac{|p - \mathbf{p}(p)|}{r} + \varphi_r(p) \right).$$

It follows readily from (6.16), (6.32) and (6.34) that

$$
\begin{aligned}
|\mathrm{Err}_4^o| &\le \sum_i C\Big(r^{-1} E^{\frac{1}{2m}} \ell_i^{1+\beta_2} + \sup_{\mathcal{U}_i} \varphi_r\Big) E^{1+\gamma_2} \ell_i^{m+2+\gamma_2} \\
&\overset{(6.35)\,\&\,(6.36)}{\le} C \sum_i \Big[E^{\gamma_2} \ell_i^{\frac{\gamma_2}{4}} \Big] \inf_{\mathcal{B}_i} \varphi_r \, E \, \ell_i^{m+2+\frac{\gamma_2}{4}} \\
&\overset{(6.40)\,\&(6.39)}{\le} C\mathbf{D}(r)^{1+\gamma_3}.
\end{aligned}
\tag{6.42}
$$

Similarly, since $|DX_i| \le Cr^{-1}$, we get

$$
\begin{aligned}
\mathrm{Err}_4^i &\le Cr^{-1} \sum_j \Big(E^{\gamma_2} \ell_j^{\frac{\gamma_2}{2}} \Big) E \, \ell_j^{m+2+\frac{\gamma_2}{2}} \\
&\overset{(6.41)\,\&\,(6.39)}{\le} C\mathbf{D}(r)^{\gamma} \left(\mathbf{D}'(r) + r^{-1}\mathbf{D}(r)\right).
\end{aligned}
$$

Remark 6.4. Note that the "superlinear" character of the estimates in Theorem 5.4 has played a fundamental role in the control of the errors.

6.2 Boundness of the frequency

We have proven in the previous subsection that the frequency of the $\mathcal{M}$-normal approximation remains bounded within a center manifold in the corresponding interval of flattening. In order to pass into the limit along the different center manifolds, we need also to show that the frequency remains bounded in passing from one to the other. This is again a consequence of the *splitting-before-tilting* estimates and we provide here some details of the proof, referring to [14] for the complete argument.

To simplify the notation, we set $p_j := \mathbf{\Phi}_j(0)$ and write simply $\mathcal{B}_\rho$ in place of $\mathcal{B}_\rho(p_j)$.

Theorem 6.5 (Boundedness of the frequency functions). *If the intervals of flattening are infinitely many, then there is a number $j_0 \in \mathbb{N}$ such that*

$$
\mathbf{H}_j > 0 \; on \;]\tfrac{s_j}{t_j}, 3[\; for \; all \; j \ge j_0 \quad and \quad \sup_{j \ge j_0} \sup_{r \in]\frac{s_j}{t_j}, 3[} \mathbf{I}_j(r) < \infty . \tag{6.43}
$$

Sketch of the proof. We partition the extrema t_j of the intervals of flattening into two different classes:

(A) those such that $t_j = s_{j-1}$;
(B) those such that $t_j < s_{j-1}$.

If t_j belongs to (A), set $r := \frac{s_{j-1}}{t_{j-1}}$. Let $L \in \mathscr{W}^{(j-1)}$ be a cube of the Whitney decomposition such that $c_s r \leq \ell(L)$ and $L \cap \bar{B}_r(0, \pi) \neq \emptyset$. Since this cube of the Whitney decomposition at step $j-1$ has size comparable with the distance to the origin, and the next center manifold starts at a comparable radius, the splitting property of the normal approximation needs to hold also for the new approximation: namely, one can show that there exists a constant $\bar{c}_s > 0$ such that

$$\int_{\mathbf{B}_2 \cap \mathcal{M}_j} |N_j|^2 \geq \bar{c}_s E^j := \mathbf{E}(T_j, \mathbf{B}_{6\sqrt{m}}),$$

which obviously gives $\mathbf{H}_{N_j}(3) \geq c E^j$, and than $\mathbf{I}_{N_j}(3)$ is smaller than a given constant, independent of j, thus proving the theorem.

In the case t_j belongs to the class (B), then, by construction there is $\eta_j \in]0, 1[$ such that $\mathbf{E}((\iota_{0,t_j})_\sharp T, \mathbf{B}_{6\sqrt{m}(1+\eta_j)}) > \varepsilon_3^2$. Up to extracting a subsequence, we can assume that $(\iota_{0,t_j})_\sharp T$ converges to a cone S: the convergence is strong enough to conclude that the excess of the cone is the limit of the excesses of the sequence. Moreover (since S is a cone), the excess $\mathbf{E}(S, \mathbf{B}_r)$ is independent of r. We then conclude

$$\varepsilon_3^2 \leq \liminf_{j \to \infty, j \in (B)} \mathbf{E}(T_j, \mathbf{B}_3) .$$

Thus, it follows again from the splitting phenomenon (see for details [15, Lemma 5.2]) that $\liminf_{j \to \infty, j \in (B)} \mathbf{H}_{N_j}(3) > 0$. Since $\mathbf{D}_{N_j}(3) \leq C E^j \leq C\varepsilon_3^2$, we achieve that $\limsup_{j \to \infty, j \in (B)} \mathbf{I}_{N_j}(3) > 0$, and conclude as before. □

7 Final blowup argument

We are now ready for the conclusion of the blowup argument, *i.e.* for the discussion of steps (F) and (G) of Section 2.7.

To this aim we recall here the main results obtained so far.

We start with an m-dimensional area minimizing integer rectifiable T in $\mathbb{R}^{m+n}$ with $\partial T = 0$ and $0 \in \mathrm{D}_Q(T)$, such that there exists a sequence of radii $r_k \downarrow 0$ satisfying

$$\lim_{k \to +\infty} \mathbf{E}(T_{0,r_k}, \mathbf{B}_{10}) = 0, \tag{7.1}$$

$$\lim_{k \to +\infty} \mathcal{H}_\infty^{m-2+\alpha}(\mathrm{D}_Q(T_{0,r_k}) \cap \mathbf{B}_1) > \eta > 0, \tag{7.2}$$

$$\mathcal{H}^m\big((\mathbf{B}_1 \cap \mathrm{spt}\,(T_{0,r_k})) \setminus \mathrm{D}_Q(T_{0,r_k})\big) > 0 \quad \forall\, k \in \mathbb{N}, \tag{7.3}$$

for some constant $\alpha, \eta > 0$. In the process of solving the centering problem for such currents we have obtained the following:

1. the intervals of flattening $I_j =]s_j, t_j]$,
2. the center manifolds $\mathcal{M}_j$,
3. the $\mathcal{M}_j$-normal approximations $N_j : \mathcal{M}_j \to \mathcal{A}_Q(\mathbb{R}^{m+n})$,

satisfying the conclusions of Theorem 5.2 and Theorem 5.4. It follows from the very definition of intervals of flattening that each r_k has to belong to one of these intervals. Therefore, in order to fix the ideas and to simplify the notation, we will in the sequel assume that there are infinitely many intervals of flattening and that $r_k \in I_k$: note that this is not a serious restriction, and everything holds true also in the case of finitely many intervals of flattening.

By the analysis of the order of contact and the estimate on the frequency function, see Theorem 6.2 and Theorem 6.5, we have also derived the information

$$\sup_{j\in\mathbb{N}} \sup_{r\in\left]\frac{s_j}{t_j},3\right]} \mathbf{I}_j(r) < +\infty. \tag{7.4}$$

The ultimate consequence of this estimate, thus clarifying the discussion about the non-triviality of the blowup process, is the following proposition.

Proposition 7.1 (Reverse Sobolev). *There exists a constant $C > 0$ with this property: for every $j \in \mathbb{N}$, there exists $\theta_j \in \left]\frac{3r_j}{2t_j}, 3\frac{r_j}{t_j}\right[$ such that*

$$\int_{\mathcal{B}_{\theta_j}(\Phi_j(0))} |DN_j|^2 \leq C\left(\frac{t_j}{r_j}\right)^2 \int_{\mathcal{B}_{\theta_j}(\Phi_j(0))} |N_j|^2 . \tag{7.5}$$

Proof. Set for simplicity $r := \frac{r_j}{t_j}$ and drop the subscript $_j$ in the sequel. Using (6.2), (6.3) and (7.4), there exists $C > 0$ such that

$$\begin{aligned}\int_{\frac{3}{2}r}^{3r} dt \int_{\mathcal{B}_t(\Phi(0))} |DN|^2 &= \frac{3}{2} r\,\mathbf{D}(3r) \leq C\,\mathbf{H}(3r)\\ &= C \int_{\frac{3}{2}r}^{3r} dt \frac{1}{t} \int_{\partial\mathcal{B}_t(\Phi(0))} |N|^2 .\end{aligned}$$

Therefore, there must be $\theta \in [\frac{3}{2}r, 3r]$ satisfying

$$\int_{\mathcal{B}_\theta(\Phi(0))} |DN|^2 \leq \frac{C}{\theta} \int_{\partial\mathcal{B}_\theta(\Phi(0))} |N|^2 . \tag{7.6}$$

This is almost the desired estimate. In oder to replace the boundary integral with a bulk integral in the right hand side of (7.6), we argue by integrating along radii in a similar way to the case of single valued functions.

Fix indeed any $\sigma \in]\theta/2, \theta[$ and any point $x \in \partial\mathcal{B}_\theta(\Phi(0))$. Consider the geodesic line γ passing through x and $\Phi(0)$, and let $\hat{\gamma}$ be the arc on γ having one endpoint $\bar{x}$ in $\partial\mathcal{B}_\sigma(\Phi(0))$ and one endpoint equal to x. Using [13, Proposition 2.1(b)] and the fundamental theorem of calculus, we easily conclude

$$|N(x)| \leq |N(\bar{x})| + \int_{\hat{\gamma}} |DN||N|\,.$$

Integrating this inequality in x and recalling that $\sigma > s/2$ we then easily conclude

$$\int_{\partial\mathcal{B}_\theta(\Phi(0))} |N|^2 \leq C \int_{\partial\mathcal{B}_\sigma(\Phi(0))} |N|^2 + C \int_{\mathcal{B}_\theta(\Phi(0))} |N|\,|DN|\,.$$

We further integrate in σ between $s/2$ and s to achieve

$$\begin{aligned}\theta \int_{\partial\mathcal{B}_\theta(\boldsymbol{\Phi}(0))} |N|^2 &\leq C \int_{\mathcal{B}_\theta(\boldsymbol{\Phi}(0))} \left(|N|^2 + \theta\,|N|\,|DN|\right) \\ &\leq \frac{\theta^2}{2C} \int_{\mathcal{B}_\theta(\boldsymbol{\Phi}(0))} |DN|^2 + C \int_{\mathcal{B}_\theta(\boldsymbol{\Phi}(0))} |N|^2\,. \end{aligned} \tag{7.7}$$

Combining (7.7) with (7.6) we easily conclude (7.5). □

7.1 Convergence to a Dir-minimizer

We can now define the final blowup sequence, because the Reverse Sobolev inequality proven in Proposition 7.1 gives the right radius θ_k for assuring compactness of the corresponding maps. To this aim set $\bar{r}_k := \frac{2}{3}\theta_k\, t_k \in [r_k, 2r_k]$, and rescale the current and the maps accordingly:

$$\bar{T}_k := (\iota_{0,\bar{r}_k})_\sharp T \quad \text{and} \quad \bar{\mathcal{M}}_k := \iota_{0,\bar{r}_k/t_k}\mathcal{M}_k,$$

and $\bar{N}_k : \bar{\mathcal{M}}_k \to \mathbb{R}^{m+n}$ for the rescaled $\bar{\mathcal{M}}_k$-normal approximations given by

$$\bar{N}_k(p) := \frac{t_k}{\bar{r}_k} N_k\left(\frac{\bar{r}_k p}{t_k}\right).$$

Note that the ball $\mathcal{B}_{s_k} \subset \mathcal{M}_k$ is sent into the ball $\mathcal{B}_{\frac{3}{2}} \subset \bar{\mathcal{M}}_k$. Moreover, via some elementary regularity theory of area minimizing currents, one deduces that

1. $\mathbf{E}(\bar{T}_k, \mathbf{B}_{\frac{1}{2}}) \leq C\mathbf{E}(T, \mathbf{B}_{r_k}) \to 0$;
2. $\bar{T}_k$ locally converge (and in the Hausdorff sense for what concerns the supports) to an m-plane with multiplicity Q;

3. $\bar{\mathcal{M}}_k$ locally converge to the flat m-plane (without loss of generality π_0);
4. recalling (7.2),

$$\mathcal{H}^{m-2+\alpha}_{\infty}(\mathrm{D}_Q(\bar{T}_k) \cap \mathbf{B}_1) \geq \eta' > 0\,, \tag{7.8}$$

for some positive constant η'.

We can then consider the following definition for the blow-up maps

$$N_k^b : B_3 \subset \mathbb{R}^m \to \mathcal{A}_Q(\mathbb{R}^{m+n})$$

given by

$$N_k^b(x) := \mathbf{h}_k^{-1} \bar{N}_k(\mathbf{e}_k(x))\,, \quad \text{with } \mathbf{h}_k := \|\bar{N}_k\|_{L^2(\mathcal{B}_{\frac{3}{2}})}, \tag{7.9}$$

where $\mathbf{e}_k : B_3 \subset \mathbb{R}^m \simeq T_{\bar{p}_k}\bar{\mathcal{M}}_k \to \bar{\mathcal{M}}_k$ denotes the exponential map at $\bar{p}_k = t_k\, \mathbf{\Phi}_k(0)/\bar{r}_k$.

Proposition 7.1 implies then that there exists a constant $C > 0$ such that, for every k,

$$\int_{B_{\frac{3}{2}}} |DN_k^b|^2 \leq C. \tag{7.10}$$

Moreover, as a simple consequence of Theorem 5.4 (details left to the readers), we find an exponent $\gamma > 0$ such that

$$\mathrm{Lip}(\bar{N}_k) \leq C\mathbf{h}_k^{\gamma}, \tag{7.11}$$

$$\mathbf{M}((\mathbf{T}_{\bar{F}_k} - \bar{T}_k)\llcorner(\mathbf{p}_k^{-1}(\mathcal{B}_{\frac{3}{2}})) \leq C\mathbf{h}_k^{2+2\gamma}, \tag{7.12}$$

$$\int_{\mathcal{B}_{\frac{3}{2}}} |\eta \circ \bar{N}_k| \leq C\mathbf{h}_k^2\,. \tag{7.13}$$

It then follows from (7.10), $\|N_k^b\|_{L^2(\mathcal{B}_{3/3})} \equiv 1$ and the Sobolev embedding for Q-valued functions (cp. [13, Proposition 2.11]) that up to subsequences (as usual not relabeled) there exists a Sobolev function N_∞^b : $B_{\frac{3}{2}} \to \mathcal{A}_Q(\mathbb{R}^{m+n})$ such that the maps N_k^b converge strongly in $L^2(B_{\frac{3}{2}})$ to N_∞^b. Then from (7.13) we deduce also that

$$\eta \circ N_\infty^b \equiv 0 \quad \text{and} \quad \|N_k^b\|_{L^2(\mathcal{B}_{3/3})} \equiv 1. \tag{7.14}$$

Moreover, since the $\bar{N}_k$ are $\bar{\mathcal{M}}_k$-normal approximations and the $\bar{\mathcal{M}}_k$ converging to the flat m-dimensional plane $\mathbb{R}^m \times \{0\}$, N_∞^b takes values in the space of Q-points of $\{0\} \times \mathbb{R}^n$ (in place of the full $\mathbb{R}^{m+n}$).

To conclude our contradiction argument, we need to prove the N_∞^b is Dir-minimizing.

7.1.1 N^b_∞ is Dir-minimizing Apart from the necessary technicalities, the proof of this claim is very intuitive and relies on the following observation: if the energy of N^b_∞ could be decreased, then one would be able to find a rectifiable current with less mass then $\bar{T}_k$, because the rescaling of N^b_k are done in terms of the L^2 norm $\mathbf{h}_k$ whereas the errors in the normal approximation are superlinear with $\mathbf{h}_k$.

Next we give all the details for this arguments.

We can consider for every $\bar{\mathcal{M}}_k$ an orthonormal frame of $(T\bar{\mathcal{M}}_k)^\perp$,

$$\nu^k_1, \dots, \nu^k_n,$$

with the property (*cf.* [16, Lemma A.1]) that

$$\nu^k_j \to e_{m+j} \quad \text{in } C^{2,\kappa/2}(\bar{\mathcal{M}}_k) \text{ as } k \uparrow \infty \text{ for every } j$$

(here $e_1, \dots, e_{m+n}$ is the standard basis of $\mathbb{R}^{m+n}$).

Given now any Q-valued map $u = \sum_i [\![u_i]\!] : \bar{\mathcal{M}}_k \to \mathcal{A}_Q(\{0\} \times \mathbb{R}^n)$, we can consider the map

$$\mathbf{u}_k : x \mapsto \sum_i [\![(u_i(x))^j \nu^k_j(x)]\!],$$

where we set $(u_i)^j := \langle u_i(x), e_{m+j}\rangle$ and we use Einstein's convention. Then, the differential map $D\mathbf{u}_k := \sum_i [\![D(\mathbf{u}_k)_i]\!]$ is given by

$$D(\mathbf{u}_k)_i = D(u_i)^j \nu^k_j + (u_i)^j D\nu^k_j.$$

Taking into account that $\|D\nu^k_i\|_{C^0} \to 0$ as $k \to +\infty$, we deduce that

$$\left| \int \left(|D\mathbf{u}_k|^2 - |Du|^2\right) \right| \le o(1) \int \left(|u|^2 + |Du|^2\right). \tag{7.15}$$

Note that N^b_k has also the form $\mathbf{u}^b_k$ for some Q-valued function $u^b_k : \bar{\mathcal{M}}_k \to \mathcal{A}_Q(\{0\} \times \mathbb{R}^n)$.

We now show the Dir-minimizing property of N^b_∞. There is nothing to prove if its Dirichlet energy vanishes. We can therefore assume that there exists $c_0 > 0$ such that

$$c_0 \mathbf{h}^2_k \le \int_{B_{\frac{3}{2}}} |D\bar{N}_k|^2. \tag{7.16}$$

We argue by contradiction and assume there is a radius $t \in \left]\frac{5}{4}, \frac{3}{2}\right[$ and a function $f : B_{\frac{3}{2}} \to \mathcal{A}_Q(\{0\} \times \mathbb{R}^n)$ such that

$$f|_{B_{\frac{3}{2}} \setminus B_t} = N^b_\infty|_{B_{\frac{3}{2}} \setminus B_t} \quad \text{and} \quad \mathrm{Dir}(f, B_t) \le \mathrm{Dir}(N^b_\infty, B_t) - 2\,\delta,$$

for some $\delta > 0$.

Using f as a model, we need to find a sequence of functions v_k^b such that they have the same boundary data of N_k^b and less energy. This can be done because of the strong convergence of the traces and the possibility to make an interpolation between two functions with close by traces. This is one of the instances where thinking to multiple valued functions as classical single valued ones may be useful. In any case, the details are given in [17, Proposition 3.5] and lead to competitor functions v_k^b such that, for k large enough,

$$v_k^b|_{\partial B_r} = N_k^b|_{\partial B_r}, \quad \mathrm{Lip}(v_k^b) \leq C\mathbf{h}_k^\gamma,$$
$$\int_{B_{\frac{3}{2}}} |\eta \circ v_k^b| \leq C\mathbf{h}_k^2 \quad \text{and} \quad \int_{B_{\frac{3}{2}}} |Dv_k^b|^2 \leq \int |DN_k^b|^2 - \delta\, \mathbf{h}_k^2,$$

where $C > 0$ is a constant independent of k. Clearly, setting $\tilde{N}_k = v_k^b \mathbf{e}_k^{-1}$ satisfy

$$\tilde{N}_k \equiv \bar{N}_k \quad \text{in } \mathcal{B}_{\frac{3}{2}} \setminus \mathcal{B}_t, \quad \mathrm{Lip}(\tilde{N}_k) \leq C\mathbf{h}_k^\gamma,$$
$$\int_{\mathcal{B}_{\frac{3}{2}}} |\eta \circ \tilde{N}_k| \leq C\mathbf{h}_k^2 \quad \text{and} \quad \int_{\mathcal{B}_{\frac{3}{2}}} |D\tilde{N}_k|^2 \leq \int_{\mathcal{B}_{\frac{3}{2}}} |D\bar{N}_k|^2 - \delta\mathbf{h}_k^2.$$

Consider finally the map $\tilde{F}_k(x) = \sum_i \left[\!\left[x + \tilde{N}_i(x) \right]\!\right]$. The current $\mathbf{T}_{\tilde{F}_k}$ coincides with $\mathbf{T}_{\bar{F}_k}$ on $\mathbf{p}_k^{-1}(\mathcal{B}_{\frac{3}{2}} \setminus \mathcal{B}_t)$. Define the function $\varphi_k(p) = \mathrm{dist}_{\mathcal{M}_k}(0, \mathbf{p}_k(p))$ and consider for each $s \in \left]t, \frac{3}{2}\right[$ the slices $\langle \mathbf{T}_{\tilde{F}_k} - \bar{T}_k, \varphi_k, s\rangle$. By (7.12) we have

$$\int_t^{\frac{3}{2}} \mathbf{M}(\langle \mathbf{T}_{\tilde{F}_k} - \bar{T}_k, \varphi_k, s\rangle) \leq C\mathbf{h}_k^{2+\gamma}.$$

Thus we can find for each k a radius $\sigma_k \in \left]t, \frac{3}{2}\right[$ on which $\mathbf{M}(\langle \mathbf{T}_{\tilde{F}_k} - \bar{T}_k, \varphi_k, \sigma_k\rangle) \leq C\mathbf{h}_k^{2+\gamma}$. By the isoperimetric inequality (see [17, Remark 4.3]) there is a current S_k such that

$$\partial S_k = \langle \mathbf{T}_{\tilde{F}_k} - \bar{T}_k, \varphi_k, \sigma_k\rangle, \quad \mathbf{M}(S_k) \leq C\mathbf{h}_k^{(2+\gamma)m/(m-1)}.$$

Our competitor current is, then, given by

$$Z_k := \bar{T}_k \llcorner (\mathbf{p}_k^{-1}(\mathcal{M}_k \setminus \mathcal{B}_{\sigma_k})) + S_k + \mathbf{T}_{\tilde{F}_k} \llcorner (\mathbf{p}_k^{-1}(\mathcal{B}_{\sigma_k})).$$

Note that Z_k has the same boundary as $\bar{T}_k$. On the other hand, by (7.12) and the bound on $\mathbf{M}(S_k)$, we have

$$\mathbf{M}(\tilde{T}_k) - \mathbf{M}(\bar{T}_k) \leq \mathbf{M}(\mathbf{T}_{\tilde{F}_k}) - \mathbf{M}(\mathbf{T}_{\bar{F}_k}) + C\mathbf{h}_k^{2+2\gamma}. \tag{7.17}$$

Denote by A_k and by H_k respectively the second fundamental forms and mean curvatures of the manifolds $\bar{\mathcal{M}}_k$. Using the Taylor expansion of [16, Theorem 3.2], we achieve

$$\begin{aligned}\mathbf{M}(\tilde{T}_k) - \mathbf{M}(\bar{T}_k) \le \frac{1}{2}\int_{B_\rho}\left(|D\tilde{N}_k|^2 - |D\bar{N}_k|^2\right) \\ + C\|H_k\|_{C^0}\int\left(|\eta\circ\bar{N}_k| + |\eta\circ\tilde{N}_k|\right) \\ + \|A_k\|^2_{C^0}\int\left(|\bar{N}_k|^2 + |\tilde{N}_k|^2\right) + o(\mathbf{h}_k^2) \\ \le -\frac{\delta}{2}\mathbf{h}_k^2 + o(\mathbf{h}_k^2)\,. \qquad (7.18)\end{aligned}$$

Clearly, (7.18) and (7.17) contradict the minimizing property of $\bar{T}_k$ for k large enough and this concludes the proof.

7.2 Persistence of singularities

We discuss step (G) of Section 2.7: we show that the assumptions (7.2) and (7.3) contradict Theorem 3.2, which asserts that the singular set of N^b_∞ has $\mathcal{H}^{m-2+\alpha}$ measure zero.

Set

$$\Upsilon := \left\{x \in \bar{B}_1 : N^b_\infty(x) = Q\,[\![0]\!]\right\},$$

and note that, since $\eta\circ N^b_\infty \equiv 0$ and $\|N^b_\infty\|_{L^2(B_{\frac{3}{2}})} = 1$, from Theorem 3.2 it follows that $\mathcal{H}^{m-2+\alpha}_\infty(\Upsilon) = 0$.

The main line of the contradiction argument can be summarized in three steps.

1. By (7.2) and (7.3) there exists a set $\Lambda_k \subset \mathrm{Dir}_Q(\bar{N}_k)$ such that

$$\mathrm{dist}(\Lambda_k, \Upsilon) > c_1 > 0 \quad \text{and} \quad \mathcal{H}^{m-2+\alpha}_\infty(\Lambda_k) > c_2 > 0,$$

   for suitable constants $c_1, c_2 > 0$.
   The key aspect of the set Λ_k is the following: by the Hölder regularity of Dir-minimizing functions in Theorem 3.2, the normal approximation $\bar{N}_k$ must be big in modulus around any point in Λ_k.
2. Moreover, it follows from the Lipschitz approximation Theorem 3.9 (see Theorem 7.2 below that around any multiplicity Q point of the current the energy of the Lipschitz approximation is large enough with respect to the L^2 norm (cp. [17, Theorem 1.7]). This is what we call *persistence of Q-point* phenomenon, and is in fact the analytic core of this part of the proof.

We moreover stress that this part of the proof (even if it is not apparent from our exposition) also uses the *splitting-before-tilting* estimates.

3. Putting together the previous two steps, we then conclude that there is a big part of the current where the energy of the Lipschitz approximation is large enough: matching the constant in the previous estimates, one realizes that this cannot happen on a set of positive $\mathcal{H}^{m-2+\alpha}$ measure.

As usual, the actual proof is much more involved of the heuristic scheme above. In the following we try to give some more explanations, referring to [14,15,17] for the detailed proof.

Step (1). We cover Υ by balls $\{\mathbf{B}_{\sigma_i}(x_i)\}$ in such a way that

$$\sum_i \omega_{m-2+\alpha}(4\sigma_i)^{m-2+\alpha} \leq \frac{\eta'}{2},$$

where $\eta' > 0$ is the constant in (7.8). By the compactness of Υ, such a covering can be chosen finite. Let $\sigma > 0$ be a radius whose specific choice will be given only at the very end, and such that $0 < 40\,\sigma \leq \min \sigma_i$. Denote by Λ_k the set of Q points of $\bar{T}_k$ far away from the singular set Υ:

$$\Lambda_k := \big\{p \in D_Q(\bar{T}_k) \cap \mathbf{B}_1 : \operatorname{dist}(p, \Upsilon) > 4 \min \sigma_i\big\}.$$

Clearly, $\mathcal{H}^{m-2+\alpha}_\infty(\Lambda_k) \geq \frac{\eta'}{2}$. Let $\mathbf{V}$ denote the neighborhood of Υ of size $2 \min \sigma_i$. By the Hölder continuity of Dir-minimizing functions in Theorem 3.2 (ii), there is a positive constant $\vartheta > 0$ such that $|N^b_\infty(x)|^2 \geq 2\vartheta$ for every $x \notin \mathbf{V}$. It then follows that

$$2\,\vartheta \leq \fint_{B_{2\sigma}(x)} |N^b_\infty|^2 \qquad \forall\, x \in B_{\frac{5}{4}} \text{ with } \operatorname{dist}(x, \Upsilon) \geq 3 \min \sigma_i,$$

and therefore, for sufficiently large k's,

$$\vartheta\, \mathbf{h}_k^2 \leq \fint_{\mathcal{B}_{2\sigma}(x)} \mathcal{G}(\bar{N}_k, Q\,[\![\eta \circ \bar{N}_k]\!])^2, \tag{7.19}$$

for all $x \in \Gamma_k := \mathbf{p}_{\bar{\mathcal{M}}_k}(\Lambda_k)$. This is the claimed lower bound on the modulus of $\bar{N}_k$.

Step (2). This is the most important step of the proof. We start introducing the following notation. For every $p \in \Lambda_k$, consider $\bar{z}_k(p) = \mathbf{p}_{\pi_0}(p)$ and $\bar{x}_k(p) := \bar{\mathbf{\Phi}}_k \in \bar{\mathcal{M}}_k$, where $\bar{\mathbf{\Phi}}_k$ is the induced parametrization.

The key claim is the following: there exists a geometric constant $c_0 > 0$ (in particular, independent of σ) such that, when k is large enough, for each $p \in \Lambda_k$ there is a radius $\varrho_p \leq 2\sigma$ with the following properties:

$$\frac{c_0\,\vartheta}{\sigma^\alpha}\mathbf{h}_k^2 \leq \frac{1}{\varrho_p^{m-2+\alpha}}\int_{\mathcal{B}_{\varrho_p}(\bar{x}_k(p))} |D\bar{N}_k|^2, \tag{7.20}$$

$$\mathcal{B}_{\varrho_p}(\bar{x}_k(p)) \subset \mathbf{B}_{4\varrho_p}(p)\,. \tag{7.21}$$

We show here the main heuristics leading to (7.20) (and we warn the reader that these are not the complete arguments), referring to [15] for (7.21). The key estimate in this regard is the following: there exists a constant $\bar{s} < 1$ such that

$$\fint_{\mathcal{B}_{\bar{s}\ell(L_k)}(x_k)} \mathcal{G}(N_{j(k)}, Q\,[\![\eta \circ N_{j(k)}]\!])^2 \leq \frac{\vartheta}{4\omega_m \ell(L_k)^{m-2}}\int_{\mathcal{B}_{\ell(L_k)}(x_k)} |DN_{j(k)}|^2\,,$$

that is, rescaling to $\bar{\mathcal{M}}_k$, there exists $t(p) \leq \bar{\ell}_k$ such that

$$\fint_{\mathcal{B}_{\bar{s}t(p)}(\bar{x}_k)(p)} \mathcal{G}(\bar{N}_k, Q\,[\![\eta \circ \bar{N}_k]\!])^2 \leq \frac{\vartheta}{4\omega_m t(p)^{m-2}}\int_{\mathcal{B}_{t(p)}(\bar{x}_k(p))} |D\bar{N}_k|^2\,. \tag{7.22}$$

We show that we can choose $\varrho_p \in]\bar{s}\,t(p), 2\sigma[$ such that (7.20) follows from (7.22). To this aim we can distinguish two cases. Either

$$\frac{1}{\omega_m t(p)^{m-2}}\int_{\mathcal{B}_{t(p)}(\bar{x}_k(p))} |DN_k|^2 \geq \mathbf{h}_k^2\,, \tag{7.23}$$

and (7.20) follows with $\varrho_p = t(p)$. Or (7.23) does not hold, and we argue as follows. We use first (7.22) to get

$$\fint_{\mathcal{B}_{\bar{s}t(p)}(\bar{x}_k(p))} \mathcal{G}(\bar{N}_k, Q\,[\![\eta \circ \bar{N}_k]\!])^2 \leq \frac{\vartheta}{4}\mathbf{h}_k^2\,. \tag{7.24}$$

Then, we show by contradiction that there exists a radius $\varrho_y \in [\bar{s}t(p), 2\sigma]$ such that (7.20) holds. Indeed, if this were not the case, setting for simplicity $f := \mathcal{G}(\bar{N}_k, Q\,[\![\eta \circ \bar{N}_k]\!])$ and letting j be the smallest integer such that $2^{-j}\sigma \leq \bar{s}t(p)$, we can estimate as follows

$$\begin{aligned}
\fint_{\mathcal{B}_{2\sigma}(\bar{x}_k(p))} f^2 &\leq 2\fint_{\mathcal{B}_{\bar{s}t(p)}(\bar{x}_k(p))} f^2 + \sum_{i=0}^{j}\left(\fint_{\mathcal{B}_{2^{1-i}\sigma}(\bar{x}_k(p))} f^2 - \fint_{\mathcal{B}_{2^{-i}\sigma}(\bar{x}_k(p))} f^2\right) \\
&\overset{(7.24)}{\leq} \frac{\vartheta}{2}\mathbf{h}_k^2 + C\sum_{i=1}^{j}\frac{1}{(2^{-j}\sigma)^{m-2}}\int_{\mathcal{B}_{2^{1-i}\sigma}(\bar{x}_k(p))} |D\bar{N}_k|^2 \\
&\leq \frac{\vartheta}{2}\mathbf{h}_k^2 + C c_0 \frac{\vartheta}{\sigma^\alpha}\mathbf{h}_k^2 \sum_{i=1}^{j}(2^{-j}\sigma)^\alpha \leq \mathbf{h}_k^2\left(\frac{\vartheta}{2} + C(\alpha)c_0\vartheta\right).
\end{aligned}$$

In the second line we have used the simple Morrey inequality

$$\left| \fint_{\mathcal{B}_{2t}(\bar{x}_k(p))} f^2 - \fint_{\mathcal{B}_t(\bar{x}_k(p))} f^2 \right| \le \frac{C}{t^{m-2}} \int_{\mathcal{B}_{2t}(\bar{x}_k(p))} |Df|^2$$
$$\le \frac{C}{t^{m-2}} \int_{\mathcal{B}_{2t}(\bar{x}_k(p))} |D\bar{N}_k|^2 .$$

The constant C depends only upon the regularity of the underlying manifold $\bar{\mathcal{M}}_k$, and, hence, can assumed independent of k.

Since $C(\alpha)$ depends only on α, m and Q, for c_0 chosen sufficiently small the latter inequality would contradict (7.19).

Step (3). We collect the estimates (7.20) and (7.21) to infer the desired contradiction. We cover Λ_k with balls $\mathbb{B}^i := \mathbf{B}_{20\varrho_{p_i}}(p_i)$ such that $\mathbf{B}_{4\varrho_{p_i}}(p_i)$ are disjoint, and deduce

$$\frac{\eta'}{2} \le C(m) \sum_i \varrho_{p_i}^{m-2+\alpha} \overset{(7.20)}{\le} \frac{C(m)}{c_0} \frac{\sigma^\alpha}{\vartheta \mathbf{h}_k^2} \sum_i \int_{\mathcal{B}_{\varrho_{p_i}}(\bar{x}_k(p_i))} |D\bar{N}_k|^2$$
$$\le \frac{C(m)}{c_0} \frac{\sigma^\alpha}{\vartheta \mathbf{h}_k^2} \int_{\mathcal{B}_{\frac{3}{2}}} |D\bar{N}_k|^2 \overset{(7.10)}{\le} C \frac{\sigma^\alpha}{\vartheta},$$

where $C(m) > 0$ is a dimensional constant. We have used that the balls $\mathcal{B}_{\varrho_{p_i}}(\mathbf{p}_{\bar{\mathcal{M}}_k}(p_i))$ are pairwise disjoint by (7.21). Now note that ϑ and c_0 are independent of σ, and therefore we can finally choose σ small enough to lead to a contradiction.

7.2.1 Persistence of Q-points Here we explain a simple instance of estimate (7.22), reporting the following theorem from [17].

Theorem 7.2 (Persistence of Q-points). *For every $\hat{\delta} > 0$, there is $\bar{s} \in]0, \frac{1}{2}[$ such that, for every $s < \bar{s}$, there exists $\hat{\varepsilon}(s, \hat{\delta}) > 0$ with the following property. If T is as in Theorem 3.9, $E := \mathbf{E}(T, \bar{C}_{4r}(x)) < \hat{\varepsilon}$ and $\Theta(T, (p, q)) = Q$ at some $(p, q) \in \bar{C}_{r/2}(x)$, then the approximation f of Theorem 3.9 satisfies*

$$\int_{B_{sr}(p)} \mathcal{G}(f, Q[\![\eta \circ f]\!])^2 \le \hat{\delta} s^m r^{2+m} E . \tag{7.25}$$

This theorem states that, in the presence of multiplicity Q points of the current, the Lipschitz (and therefore also the normal) approximations must have a relatively small L^2 norm, compared to the excess; or, as explained above, if in the normal approximation the excess is linked to the Dirichlet energy (for example this is the case of (EX)-cubes in the Whitney decomposition), the energy needs to be relatively large with respect to the L^2 norm, thus vaguely explaining the link to (7.22).

Proof. By scaling and translating we assume $x = 0$ and $r = 1$; the choice of $\bar{s}$ will be specified at the very end, but for the moment we impose $\bar{s} < \frac{1}{4}$. Assume by contradiction that, for arbitrarily small $\hat{\varepsilon} > 0$, there are currents T and points $(p, q) \in \bar{C}_{1/2}$ satisfying: $E := \mathbf{E}(T, \bar{C}_4) < \hat{\varepsilon}$, $\Theta(T, (p, q)) = Q$ and, for f as in Theorem 3.9,

$$\int_{B_s(p)} \mathcal{G}(f, Q [\![\eta \circ f]\!])^2 > \hat{\delta} s^m E . \tag{7.26}$$

Set $\bar{\delta} = \frac{1}{4}$ and fix $\bar{\eta} > 0$ (whose choice will be specified later). For a suitably small $\hat{\varepsilon}$ we can apply Theorem 3.10, obtaining a Dir-minimizing approximation w. If $\bar{\eta}$ and $\hat{\varepsilon}$ are suitably small, we have

$$\int_{B_s(p)} \mathcal{G}(w, Q [\![\eta \circ w]\!])^2 \geq \tfrac{3\hat{\delta}}{4} s^m E ,$$

and $\sup\{\mathrm{Dir(f)}, \mathrm{Dir(w)}\} \leq \mathrm{CE}$. Then there exists $\bar{p} \in B_s(p)$ with

$$\mathcal{G}(w(\bar{p}), Q [\![\eta \circ w(\bar{p})]\!])^2 \geq \frac{3\hat{\delta}}{4\omega_m} E,$$

and, by the Hölder continuity in Theorem 3.2 (ii), we conclude

$$\begin{aligned} g(x) &:= \mathcal{G}(w(x), Q [\![\eta \circ w(x)]\!]) \\ &\geq \left(\tfrac{3\hat{\delta}}{4\omega_m} E\right)^{\frac{1}{2}} - 2(CE)^{\frac{1}{2}} \bar{C} \bar{s}^{\kappa} \geq \left(\tfrac{\hat{\delta}}{2} E\right)^{\frac{1}{2}} , \end{aligned} \tag{7.27}$$

where we assume that $\bar{s}$ is chosen small enough in order to satisfy the last inequality. Setting $h(x) := \mathcal{G}(f(x), Q [\![\eta \circ f(x)]\!])$, we recall that we have

$$\int_{B_s(p)} |h - g|^2 \leq C \bar{\eta} E .$$

Consider therefore the set $A := \left\{h > \left(\frac{\hat{\delta}}{4} E\right)^{\frac{1}{2}}\right\}$. If $\bar{\eta}$ is sufficiently small, we can assume that

$$|B_s(p) \setminus A| < \frac{1}{8} |B_s|.$$

Further, define $\bar{A} := A \cap K$, where K is the set of Theorem 3.9. Assuming $\hat{\varepsilon}$ is sufficiently small we ensure $|B_s(p) \setminus \bar{A}| < \frac{1}{4} |B_s|$. Let N be the smallest integer such that $N \frac{\hat{\delta} E}{64 Q s} \geq \frac{s}{2}$. Set

$$\sigma_i := s - i \frac{\hat{\delta} E}{64 Q s} \quad \text{for } i \in \{0, 1 \ldots, N\},$$

and consider, for $i \leq N-1$, the annuli $\mathcal{C}_i := B_{\sigma_i}(p) \setminus B_{\sigma_{i+1}}(p)$. If $\hat{\varepsilon}$ is sufficiently small, we can assume that $N \geq 2$ and $\sigma_N \geq \frac{s}{4}$. For at least one of these annuli we must have $|\bar{A} \cap \mathcal{C}_i| \geq \frac{1}{2}|\mathcal{C}_i|$. We then let $\sigma := \sigma_i$ be the corresponding outer radius and we denote by $\mathcal{C}$ the corresponding annulus.

Consider now a point $x \in \mathcal{C} \cap \bar{A}$ and let T_x be the slice $\langle T, \mathbf{p}, x\rangle$. Since $\bar{A} \subset K$, for a.e. $x \in \bar{A}$ we have $T_x = \sum_{i=1}^{Q} [\![(x, f_i(x))]\!]$. Moreover, there exist i and j such that $|f_i(x) - f_j(x)|^2 \geq \frac{1}{Q}\mathcal{G}(f(x), [\![\eta \circ f(x)]\!])^2 \geq \frac{\hat{\delta}}{4Q}E$ (recall that $x \in \bar{A} \subset A$). When $x \in \mathcal{C}$ and the points (x, y) and (x, z) belong both to $\mathbf{B}_\sigma((p, q))$, we must have

$$|y - z|^2 \leq 4\left(\sigma^2 - \left(\sigma - \tfrac{\hat{\delta}E}{64Qs}\right)^2\right) \leq \tfrac{\sigma\hat{\delta}E}{8Qs} \leq \tfrac{\hat{\delta}E}{8Q}\,.$$

Thus, for $x \in \bar{A} \cap \mathcal{C}$ at least one of the points $(x, f_i(x))$ is not contained in $\mathbf{B}_\sigma((p, q))$. We conclude therefore

$$\begin{aligned}\|T\|(\bar{\mathcal{C}}_\sigma(p) \setminus \mathbf{B}_\sigma((p, q))) &\geq |\mathcal{C} \cap \bar{A}| \geq \frac{1}{2}|\mathcal{C}| \\ &= \frac{\omega_m}{2}\left(\sigma^m - \left(\sigma - \tfrac{\hat{\delta}E}{64Qs}\right)^m\right) \\ &\geq \frac{\omega_m}{2}\sigma^m\left(1 - \left(1 - \tfrac{\hat{\delta}E}{64Qs\sigma}\right)^m\right). \qquad (7.28)\end{aligned}$$

Recall that, for τ sufficiently small, $(1-\tau)^m \leq 1 - \frac{m\tau}{2}$. Since $\sigma \geq \frac{s}{4}$, if $\hat{\varepsilon}$ is chosen sufficiently small we can therefore conclude

$$\|T\|(\bar{\mathcal{C}}_\sigma(p) \setminus \mathbf{B}_\sigma(p)) \geq \frac{\omega_m \sigma^m \hat{\delta} E}{256 Q s \sigma} \geq \frac{\omega_m}{1024 Q}\hat{\delta} E \sigma^{m-2} = c_0 \hat{\delta} E \sigma^{m-2}\,. \qquad (7.29)$$

Next, by Theorem 3.9 and Theorem 3.10,

$$\|T\|(\bar{\mathcal{C}}_\sigma(p)) \leq Q\omega_m \sigma^m + CE^{1+\gamma_1} + \bar{\eta}E + \int_{B_\sigma(p)} \frac{|Dw|^2}{2}. \qquad (7.30)$$

Moreover, as shown in [13, Proposition 3.10], we have

$$\int_{B_\sigma(p)} |Dw|^2 \leq C\mathrm{Dir}(w)\sigma^{m-2+2\kappa}, \qquad (7.31)$$

(for some constants κ and C depending only on m, n and Q; in fact the exponent κ is the one of Theorem 3.2 (ii)). Combining (7.29), (7.30) and (7.31), we conclude

$$\begin{aligned}\|T\|(\mathbf{B}_\sigma((p, q))) \leq\; & Q\omega_m \sigma^m + \bar{\eta}\,E + CE^{1+\gamma_1} \\ & + CE\sigma^{m-2+2\kappa} - c_0 \sigma^{m-2}\hat{\delta}E\,. \qquad (7.32)\end{aligned}$$

Next, by the monotonicity formula, $\rho \mapsto \rho^{-m}\|T\|(\mathbf{B}_\rho((p,q)))$ is a monotone function. Using $\Theta(T,(p,q)) = Q$, we conclude

$$\|T\|(\mathbf{B}_\sigma((p,q))) \geq Q\omega_m\sigma^m. \qquad (7.33)$$

Combining (7.32) and (7.33) we conclude

$$C\sigma^2 + (\bar\eta + C E_1^\gamma)\sigma^{2-m} + C\sigma^{2\kappa} \geq c_0\hat\delta. \qquad (7.34)$$

Recalling that $\sigma \leq s < \bar s$, we can, finally, specify $\bar s$: it is chosen so that $C\bar s^2 + C\bar s^{2\kappa}$ is smaller than $\frac{c_0}{2}\hat\delta$. Combined with (7.27) this choice of $\bar s$ depends, therefore, only upon $\hat\delta$. (7.34) becomes then

$$(\bar\eta + C E^{\gamma_1})\sigma^{2-m} \geq \tfrac{c_0}{2}\hat\delta. \qquad (7.35)$$

Next, recall that $\sigma \geq \frac{s}{4}$. We then choose $\hat\varepsilon$ and $\bar\eta$ so that $(\bar\eta + C\hat\varepsilon^{\gamma_1})(\frac{s}{4})^{2-m} \leq \frac{c_0}{4}\hat\delta$. This choice is incompatible with (7.35), thereby reaching a contradiction: for this choice of the parameter $\hat\varepsilon$ (which in fact depends only upon $\hat\delta$ and s) the conclusion of the theorem, *i.e.* (7.25), must then be valid. □

8 Open questions

We close this survey recalling some open problems concerning the regularity of area minimizing integer rectifiable currents. Some of them have been only slightly touched and would actually explain some of the complications that we met along the proof of the partial regularity result.

For more open problems and comments, we suggest the reading of [1,11].

(A) One of the main, perhaps the most well-known, open problems is the uniqueness of the tangent cones to an area minimizing current, *i.e.* the uniqueness of the limit $(\iota_{x,r})_\sharp T$ as $r \to 0$ for every $x \in \operatorname{spt}(T)$. The uniqueness is known for two dimensional currents (cp. [35]), and there are only partial results in the general case (see [4,30]).

We have run into this issue in dealing with the step (C) of Section 2.7, because it is one of the possible reasons why a center manifold may be sufficient in our proof.

(B) A related question is that of the uniqueness of the inhomogeneous blowup for Dir-minimizing Q-valued functions. Also in this case the uniqueness is known for two dimensional domains (cp. [13], following ideas of [7]).

Even if it does not play a role in the contradiction argument for the partial regularity, a positive answer to this question could indeed contribute to the solution of next two other major open problems.

(C) It is unknown whether the singular set of an area minimizing current has always locally finite $\mathcal{H}^{m-2}$ measure. This is the case for two dimensional currents (as proven by Chang [7]); note that in this result the uniqueness of the blowup Dir-minimizing map plays a fundamental role.

(D) It is unknown whether the singular set of an area minimizing current has some geometric structure, *e.g.*. if it is rectifiable (*i.e.*, roughly speaking, if it is contained in lower dimensional $(m-2)$-dimensional submanifolds). Once again it is known the positive answer for two dimensional currents, where the singularities are known to be locally isolated, and the uniqueness of the tangent map is one of the fundamental steps in the proof.

(E) We mention also the problem of finding more example of area minimizing currents, other than those coming from complex varieties or similar calibrations. Indeed, our understanding of the possible pathological behaviors of such currents is pretty much limited by the few examples we have at disposal. In particular, it would be extremely interesting to understand if there could be minimizing currents with weird singular set (*e.g*.., of Cantor type).

(F) Finally, we mention the problem of boundary regularity for higher codimension area minimizing currents, which to our knowledge is mostly open.

ACKNOWLEDGEMENTS. I am very grateful to A. Marchese, for reading a first draft of these lecture notes and suggesting many precious improvements.

References

[1] *Some open problems in geometric measure theory and its applications suggested by participants of the 1984 AMS summer institute*, In: "Geometric Measure Theory and the Calculus of Variations" (Arcata, Calif., 1984), J. E. Brothers (ed.), Proc. Sympos. Pure Math., Vol.44, Amer. Math. Soc., Providence, RI, 1986, 441–464.

[2] W. K. ALLARD, *On the first variation of a varifold*, Ann. of Math. (2) **95** (1972), 417–491.

[3] W. K. ALLARD, *On the first variation of a varifold: boundary behavior*, Ann. of Math. (2) **101** (1975), 418–446.

[4] W. K. ALLARD and F. J. ALMGREN, JR., *On the radial behavior of minimal surfaces and the uniqueness of their tangent cones*, Ann. of Math. (2) **113** (1981), 215–265.

[5] F. J. ALMGREN, JR., "Almgren's big Regularity Paper", World Scientific Monograph Series in Mathematics, Vol. 1, World Scientific Publishing Co. Inc., River Edge, NJ, 2000.

[6] L. AMBROSIO, *Metric space valued functions of bounded variation*, Ann. Scuola Norm. Sup. Pisa Cl. Sci. (4) **17** (1990), 439–478.

[7] S. XU-DONG CHANG, *Two-dimensional area minimizing integral currents are classical minimal surfaces*, J. Amer. Math. Soc. (4) **1** (1988), 699–778.

[8] E. DE GIORGI, *Su una teoria generale della misura* $(r-1)$-*dimensionale in uno spazio ad r dimensioni*, Ann. Mat. Pura Appl. (4) **36** (1954), 191–213.

[9] E. DE GIORGI, *Nuovi teoremi relativi alle misure* $(r-1)$-*dimensionali in uno spazio ad r dimensioni*, Ricerche Mat. **4** (1955), 95–113.

[10] E. DE GIORGI, *Frontiere orientate di misura minima*, Seminario di Matematica della Scuola Normale Superiore di Pisa, 1960-61, Editrice Tecnico Scientifica, Pisa 1961.

[11] C. DE LELLIS, *Almgren's Q-valued functions revisited*, In: "Proceedings of the International Congress of Mathematicians", Volume III, Hindustan Book Agency, New Delhi, 2010, 1910–1933.

[12] C. DE LELLIS and E. SPADARO, *Center manifold: a case study*, Discrete Contin. Dyn. Syst. **31** (2011), 1249–1272.

[13] C. DE LELLIS and E. SPADARO, *Q-valued functions revisited*, Mem. Amer. Math. Soc. **211** (2011), vi+79.

[14] C. DE LELLIS and E. SPADARO, *Regularity of area-minimizing currents II: center manifold*, preprint, 2013.

[15] C. DE LELLIS and E. SPADARO, *Regularity of area-minimizing currents III: blow-up*, preprint, 2013.

[16] C. DE LELLIS and E. SPADARO, *Multiple valued functions and integral currents*, Ann. Sc. Norm. Super. Pisa Cl. Sci. (5) (2014), to appear.

[17] C. DE LELLIS and E. SPADARO, *Regularity of area-minimizing currents I: gradient* L^p *estimates*, GAFA (2014), to appear.

[18] H. FEDERER, "Geometric Measure Theory", Die Grundlehren der mathematischen Wissenschaften, Band 153, Springer-Verlag New York Inc., New York, 1969, xiv+676.

[19] H. FEDERER and W. H. FLEMING, *Normal and integral currents*, Ann. of Math. (2) **72** (1960), 458–520.

[20] W. H. FLEMING, *On the oriented Plateau problem*, Rend. Circ. Mat. Palermo (2) **11** (1962), 69–90.

[21] M. FOCARDI, A. MARCHESE and E. SPADARO, *Improved estimate of the singular set of Dir-minimizing Q-valued functions*, preprint 2014.

[22] M. GROMOV and R. SCHOEN, *Harmonic maps into singular spaces and p-adic superrigidity for lattices in groups of rank one*, Inst. Hautes Études Sci. Publ. Math. **76** (1992), 165–246.

[23] R. HARDT and L. SIMON, *Boundary regularity and embedded solutions for the oriented Plateau problem*, Ann. of Math. (2) **110** (1979), 439–486.

[24] J. HIRSCH, *Boundary regularity of Dirichlet minimizing Q-valued fucntions*, preprint 2014.

[25] J. JOST, *Generalized Dirichlet forms and harmonic maps*, Calc. Var. Partial Differential Equations **5** (1997), 1–19.

[26] N. J. KOREVAAR and R. M. SCHOEN, *Sobolev spaces and harmonic maps for metric space targets*, Comm. Anal. Geom. **1** (1993), 561–659.

[27] P. LOGARITSCH and E. SPADARO, *A representation formula for the p-energy of metric space-valued Sobolev maps*, Commun. Contemp. Math. **14** (2012), 10 pp.

[28] E. R. REIFENBERG, *On the analyticity of minimal surfaces*, Ann. of Math. (2) **80** (1964), 15–21.

[29] T. RIVIÈRE, *A lower-epiperimetric inequality for area-minimizing surfaces*, Comm. Pure Appl. Math. **57** (2004), 1673–1685.

[30] L. SIMON, *Asymptotics for a class of nonlinear evolution equations, with applications to geometric problems*, Ann. of Math. (2) **118** (1983), 525–571.

[31] L. SIMON, "Lectures on Geometric Measure Theory", Proceedings of the Centre for Mathematical Analysis, Australian National University, Vol. 3, Australian National University Centre for Mathematical Analysis, Canberra, 1983, vii+272.

[32] L. SIMON, *Rectifiability of the singular sets of multiplicity* 1 *minimal surfaces and energy minimizing maps*, In: "Surveys in Differential Geometry", Vol. II (Cambridge, MA, 1993), Int. Press, Cambridge, MA, 1995, 246–305.

[33] J. SIMONS, *Minimal varieties in riemannian manifolds*, Ann. of Math. (2) **88** (1968), 62–105.

[34] E. SPADARO, *Complex varieties and higher integrability of Dir-minimizing Q-valued functions*, Manuscripta Math. **132** (2010), 415–429.

[35] B. WHITE, *Tangent cones to two-dimensional area-minimizing integral currents are unique*, Duke Math. J. **50** (1983), 143–160.

The regularity problem for sub-Riemannian geodesics

Davide Vittone

Abstract. We study the regularity problem for sub-Riemannian geodesics, *i.e.*, for those curves that minimize length among all curves joining two fixed endpoints and whose derivatives are tangent to a given, smooth distribution of planes with constant rank. We review necessary conditions for optimality and we introduce extremals and the Goh condition. The regularity problem is nontrivial due to the presence of the so-called abnormal extremals, *i.e.*, of certain curves that satisfy the necessary conditions and that may develop singularities. We focus, in particular, on the case of Carnot groups and we present a characterization of abnormal extremals, that was recently obtained in collaboration with E. Le Donne, G. P. Leonardi and R. Monti, in terms of horizontal curves contained in certain algebraic varieties. Applications to the problem of geodesics' regularity are provided.

1 Introduction

A *sub-Riemannian manifold* is a smooth, connected n-dimensional manifold M endowed with a smooth, bracket-generating sub-bundle $\Delta \subset TM$ (called *horizontal*), having constant rank r, and with a smooth metric g on Δ. In these notes, we give a brief overview on the problem of the regularity of length minimizers, *i.e.*, of the shortest (with respect to g) curves among all curves that join two fixed endpoints and are *horizontal*, *i.e.*, tangent to Δ. We also present some results on the problem recently obtained, in the framework of *Carnot groups*, in collaboration with E. Le Donne, G. P. Leonardi and R. Monti [17,18]. These notes are based on a course given by the author on the occasion of the ERC School *Geometric Measure Theory and Real Analysis* held at the Centro De Giorgi, Pisa, in October 2013.

It is well-known (see *e.g.* the basic references [3,4,27]) that length minimizers are *extremals*, *i.e.*, satisfy certain necessary conditions given

The author is supported by PRIN 2010-11 Project "Calculus of Variations" of MIUR (Italy), GNAMPA of INdAM (Italy), University of Padova, and Fondazione CaRiPaRo Project "Nonlinear Partial Differential Equations: models, analysis, and control-theoretic problems".

by the Pontryagin Maximum Principle of Optimal Control Theory. Extremals may be either *normal* or *abnormal*: while normal extremals are always smooth, abnormal ones may develop singularities. Hence, the regularity problem for length minimizers is reduced to the regularity of abnormal minimizers.

Let us spend a few words about the literature and the state of the art on the problem. We do not claim to be exhaustive and we refer to the beautiful introductions in [23,30] for a more comprehensive account.

It was originally claimed in [35] that length minimizing curves are smooth, all of them being normal extremals. The wrong argument relied upon an incorrect application of Pontryagin Maximum Principle, ignoring the possibility of abnormal extremals; see also [13]. A correction to [35] appeared in [36], where it was proved that minimizers in *strong bracket-generating* distributions are always normal and, hence, smooth.

The first example of a strictly abnormal length minimizer was provided by R. Montgomery in [26]. Other examples in the same vein are studied in [22,37]. Distributions of rank 2 are rich of abnormal geodesics: in [23], W. Liu and H. J. Sussmann introduced a class of abnormal extremals, called *regular abnormal* extremals, that are always locally length minimizing. Strictly abnormal length minimizers appear also in the setting of Carnot groups, see [11]. Notice, however, that all known examples of abnormal minimizers are smooth, so that the regularity problem is widely open.

As we said, abnormal extremals may have singularities. In the paper [21], G. P. Leonardi and R. Monti developed an elaborate *cutting-the-corner* technique (see also [2]) to show that, when the horizontal bundle satisfies a certain technical condition, length minimizers do not have corner-type singularities. In several interesting structures (among them, Carnot groups of rank 2 and nilpotency *step* at most 4), this is enough to conclude that length minimizers are smooth. More recently, R. Monti [29] was able to exclude certain singularities of higher order for length minimizers in structures satisfying the same condition introduced in [21].

Finally, a complete characterization of extremals in Carnot groups was recently obtained in [17, 18]. In particular, abnormal extremals in this setting are characterized as horizontal curves contained in certain algebraic varieties; the key tool here is represented by *extremal polynomials*. This allows for several applications; let us only mention the results discussed in these notes. First, one can give a very short proof of the regularity of length minimizers in Carnot groups of step not greater than 3 (a result first proved in [38]). Second, we describe a new technique for proving the negligibility of the endpoints of abnormal extremals; for the motivations behind this problem, which are only sketched in Remark 3.23, see [27, Section 10.2] and [2]. This technique cannot be applied to general Carnot groups; however, it is likely to work in many specific examples.

A few words about the organization of these notes. In Section 2, we introduce the sub-Riemannian (or *Carnot-Carathéodory*) distance. In Section 3, we derive the necessary conditions of extremality for length minimizers; the properties of normal and abnormal extremals are briefly described in Sections 3.3 and 3.4. In Section 4, we introduce Carnot groups and present the characterization of extremals obtained in [17,18]. In Section 5, we apply our results to prove the smoothness of minimizers in Carnot groups of step at most 3. Finally, in Section 6 we describe the technique connected with the negligibility problem for the endpoints of abnormal extremals.

2 The Carnot-Carathéodory distance

2.1 Definition of Carnot-Carathéodory distance

A sub-Riemannian manifold is a smooth, connected n-dimensional manifold M endowed with a smooth, bracket-generating sub-bundle $\Delta \subset TM$ (called *horizontal sub-bundle*) of constant rank r and with a smooth metric g on Δ. Without loss of generality, the regularity problem for length minimizers can be localized. Namely, we can assume that $M = \mathbb{R}^n$ and that the horizontal bundle Δ is generated by smooth, linearly independent vector fields $X_1, \dots, X_r$ which form an orthonormal system with respect to g.

A Lipschitz continuous curve $\gamma : [0, 1] \to \mathbb{R}^n$ is said to be *horizontal* if $\dot\gamma(t) \in \Delta_{\gamma(t)}$ for a.e. $t \in [0, 1]$, *i.e.*, if

$$\dot\gamma(t) = \sum_{j=1}^{r} h_j(t) X_j(\gamma(t)) \quad \text{for a.e. } t \in [0, 1] \tag{2.1}$$

for suitable functions $h = (h_1, \dots, h_r) \in L^\infty([0, 1], \mathbb{R}^r)$. We will refer to the functions h_j as to the *controls* associated with γ. The length of γ is

$$L(\gamma) := \int_0^1 |h(t)|\, dt\,.$$

The fact that the length L is defined by integrating $|h(t)| := (h_1(t)^2 + \cdots + h_r(t)^2)^{1/2}$ corresponds to the fact that $X_1, \dots, X_r$ are orthonormal.

Definition 2.1. The *Carnot-Carathéodory (CC) distance* between $x, y \in \mathbb{R}^n$ is defined as

$$d(x, y) := \inf \{L(\gamma) : \gamma \text{ is horizontal}, \gamma(0) = x \text{ and } \gamma(1) = y\}. \quad (2.2)$$

The structure induced by the Carnot-Carathéodory distance is often called *sub-Riemannian* because, intuitively speaking, the "allowed" directions form only a subspace of the whole tangent bundle.

Exercise 2.2. Given γ and h as above, define $L_2(\gamma) := \left(\int_0^1 |h(t)|^2\, dt\right)^{1/2}$; prove that, for any $x, y \in \mathbb{R}^n$, the CC distance $d(x, y)$ is equal to

$$d_2(x, y) := \inf \{L_2(\gamma) : \gamma \text{ is horizontal}, \gamma(0) = x \text{ and } \gamma(1) = y\}.$$

2.2 The Chow-Rashevski theorem

The family of curves in the right hand side of (2.2) might be empty (*i.e.*, no horizontal curve joins x and y), hence d is not necessarily a distance. Consider, for instance, $\mathbb{R}^3$ with horizontal distribution generated by the vector fields $X_1 := (1, 0, 0)$ and $X_2 := (0, 1, 0)$: clearly, in this case there is no horizontal curve joining the origin and the point $(0, 0, 1)$.

On the contrary, it is immediate to check that d is a distance whenever we can guarantee that any couple of points can be connected by horizontal curves. Sufficient conditions for connectivity are well-known; they are usually based on the following observation, which is a consequence of the Baker-Campbell-Hausdorff formula (see *e.g.* [40]). Here and in the sequel, we adopt the standard identification between vector fields and first-order derivations.

Fact. Given a point $p \in \mathbb{R}^n$, two vector fields X, Y and a positive real number $t \ll 1$, one has

$$e^{-tY} e^{-tX} e^{tY} e^{tX}(p) = e^{t^2[X,Y]}(p) + o(t^2), \quad (2.3)$$

where we define $e^{tX}(p) := c(t)$ as the curve c solving the problem $\dot{c} = X(c)$, $c(0) = p$, and where the *commutator* (or *bracket*) $[X, Y]$ is the vector field $XY - YX$.

Roughly speaking, if we are allowed to move along both X and Y, then we are also allowed to move in the direction of their commutator. This holds also for iterated brackets and suggests the following result, which we state without proof. Here and in the sequel, we denote by $\mathcal{L}(X_1, \ldots, X_r)$ the Lie algebra of vector fields (with Lie product $[\cdot, \cdot]$) generated by $X_1, \ldots, X_r$.

Theorem 2.3. *Assume that the* bracket-generating condition

$$\text{rank } \mathcal{L}(X_1, \ldots, X_r)(x) = n \quad \forall x \in \mathbb{R}^n \tag{2.4}$$

holds. Then, for any $x, y \in \mathbb{R}^n$ there exists a horizontal curve joining x and y; in particular, the Carnot-Carathéodory distance d is an actual distance.

Theorem 2.3 was proved independently by W. L. Chow [10] and P. K. Rashevski [33]; see also [8].

Condition (2.4) is also known as *Hörmander condition*, as it was used by L. Hörmander in the seminal paper [14] on hypoelliptic equations. In what follows, we will always assume that (2.4) is satisfied.

2.3 The Ball-Box Theorem

In this section, we state the classical Ball-Box Theorem by A. Nagel, E. M. Stein and S. Wainger [32], that allows to compare (small) CC balls $B(x, r)$ with suitable anisotropic boxes. See also [31].

If $\Omega \subset \mathbb{R}^n$ is an open bounded set, then there exists an integer κ such that condition (2.4) is verified at every $x \in \Omega$ by commutators of $X_1, \ldots, X_r$ with length at most κ (the length of a commutator $[\cdots[X_{j_1}, X_{j_2}], X_{j_3}], \ldots, X_{j_m}]$ is by definition m). Let $Y_1, \ldots, Y_q$ be a fixed enumeration of all the commutators of length at most κ and let $d(Y_k) \in \{1, \ldots, \kappa\}$ denote the length of Y_k.

Given $x \in \mathbb{R}^n$ and a multi-index $\mathcal{I} = (i_1, \ldots, i_n) \in \{1, \ldots, q\}^n$, define

$$\begin{aligned} &d(\mathcal{I}) := d(Y_{i_1}) + \cdots + d(Y_{i_n}) \\ &\lambda_{\mathcal{I}}(x) := \det \operatorname{col}\left[Y_{i_1}(x) \mid Y_{i_2}(x) \mid \cdots \mid Y_{i_n}(x)\right] \end{aligned}$$

and the map

$$E_{\mathcal{I}}(x, h) := e^{h_1 Y_{i_1} + h_2 Y_{i_2} + \cdots + h_n Y_{i_n}}(x), \quad h \in \mathbb{R}^n.$$

Let us define the *box*

$$\mathscr{B}_{\mathcal{I}}(x, r) := \{E_{\mathcal{I}}(x, h) : h \in \mathbb{R}^n \text{ and } \max_{k=1,\ldots,n} |h_k|^{1/d(Y_{i_k})} < r\}.$$

We can then state the following result.

Theorem 2.4. *Let $K \subset \Omega$ be a compact set; then, there exist positive numbers $\hat{r}, \alpha, \beta$, with $\beta < \alpha < 1$, such that the following holds. If $x \in K, r \in (0, \hat{r})$ and $\mathcal{I}$ are such that*

$$|\lambda_{\mathcal{I}}(x)| r^{d(\mathcal{I})} > \tfrac{1}{2} \max_{\mathcal{J}} |\lambda_{\mathcal{J}}(x)| r^{d(\mathcal{J})}, \tag{2.5}$$

then

$$B(x, \beta r) \subset \mathscr{B}_{\mathcal{I}}(x, \alpha r) \subset B(x, r).$$

In particular, there exists $C = C(K) > 0$ *such that* $d(x, y) \leqslant C|x-y|^{1/\kappa}$ *for any* $x, y \in K$.

Remark 2.5. As an important consequence, one can deduce from Theorem 2.4 that the topology induced by d is the standard one on $\mathbb{R}^n$.

3 Length minimizers and extremals

This section is devoted to the derivation of necessary conditions for length minimizing curves. Usually, such conditions are obtained by making use of the Pontryagin Maximum Principle of Optimal Control Theory; however, we will not directly refer to it. Our presentation is not meant to be exhaustive; the basic references [3,4,27] can be consulted for a more detailed account on these and related topics.

3.1 Length minimizers, existence and non-uniqueness

Definition 3.1. A horizontal curve $\gamma : [0, 1] \to \mathbb{R}^n$ is a *length minimizer* if it realizes the distance between its endpoints, *i.e.*, if $L(\gamma) = d(\gamma(0), \gamma(1))$.

As a preliminary result, we prove the local existence of minimizers.

Theorem 3.2. *For any* $x \in \mathbb{R}^n$, *there exists* $\rho > 0$ *with the following property: if* $d(x, y) < \rho$, *then there exists a length minimizer connecting* x *and* y.

Proof. Let $x \in \mathbb{R}^n$ be fixed and let $\rho > 0$ be such that the CC ball $B(x, \rho)$ is a bounded open subset of $\mathbb{R}^n$. The existence of such a ρ is guaranteed by Remark 2.5. We are going to prove that, for any point $y \in B(x, \rho)$, there exists a length minimizer from x to y.

Let then x, ρ, y be as above and consider a sequence of horizontal curves $\gamma^k : [0, 1] \to \mathbb{R}^n$, $k \in \mathbb{N}$, such that

$$\gamma^k(0) = x, \quad \gamma^k(1) = y \quad \text{and} \quad L(\gamma^k) \to d(x, y) \text{ as } k \to \infty .$$

In particular, for large k we have Im $\gamma^k \subset B(x, \rho) \Subset \mathbb{R}^n$. Let $h^k : [0, 1] \to \mathbb{R}^r$ be the controls associated with γ^k; we can assume that, for any k, $|h^k| \equiv L(\gamma^k)$ is constant on $[0, 1]$. Thus, for large k, the Euclidean norm $\|\dot\gamma^k\|_{L^\infty}$ is bounded uniformly in k; by Ascoli-Arzelà's Theorem we deduce that, up to a subsequence, there exists a Lipschitz curve $\gamma : [0, 1] \to \overline{B(x, \rho)}$ such that $\gamma^k \to \gamma$ uniformly on $[0, 1]$. Now, by the Dunford-Pettis theorem, up to a further subsequence we have that $h^k \rightharpoonup h$ in $L^1([0,1],\mathbb{R}^r)$. For any $t \in [0,1]$ there holds

$$\gamma^k(t) = \int_0^t \sum_{j=1}^r h_j^k(s)\, X_j(\gamma^k(s))\, ds\,.$$

Taking into account the uniform convergence of γ^k and the weak convergence of h^k, on passing to the limit as $k \to \infty$ we get

$$\gamma(t) = \int_0^t \sum_{j=1}^r h_j(s)\, X_j(\gamma(s))\, ds\,,$$

i.e., the curve γ is horizontal with associated controls h. In particular we have

$$\gamma(0)=x, \quad \gamma(1)=y \quad \text{and} \quad L(\gamma)=\|h\|_{L^1} \leqslant |\liminf_{k\to\infty} \|h^k\|_{L^1} = d(x,y),$$

i.e., γ is a length minimizer connecting x and y. This concludes the proof. □

Unlike Riemannian geodesics, sub-Riemannian length minimizers are not unique, even locally. To illustrate this situation, we consider the sub-Riemannian *Heisenberg group*, *i.e.*, the space $\mathbb{R}^3$ with horizontal distribution generated by the linearly independent vector fields

$$X_1 := \partial_1 - \frac{x_2}{2}\partial_3, \quad X_2 := \partial_2 + \frac{x_1}{2}\partial_3\,.$$

Notice that the bracket-generating condition is trivially satisfied because $[X_1, X_2] = \partial_3$. Our aim it to describe length minimizers starting from the origin; a more detailed study can be found in [5].

It can be easily checked that a Lipschitz curve $\gamma = (\gamma_1, \gamma_2, \gamma_3) : [0,1] \to \mathbb{R}^3$ is horizontal if and only if

$$\dot\gamma_3 = -\frac{\gamma_2}{2}\dot\gamma_1 + \frac{\gamma_1}{2}\dot\gamma_2 \quad \text{a.e. on } [0,1].$$

In particular, if c denotes the planar curve $c(t) := (\gamma_1(t), \gamma_2(t))$ and Σ is the planar region bounded by c and by the (oriented) segment σ joining $c(1)$ to the origin, one can recover $\gamma_3(1)$ as

$$\gamma_3(1) = \int_c \Big(-\frac{x_2}{2}dx_1 + \frac{x_1}{2}dx_2\Big) = \int_{c\cup\sigma} \Big(-\frac{x_2}{2}dx_1 + \frac{x_1}{2}dx_2\Big) = \int_\Sigma dx_1 \wedge dx_2,$$

where we have used Stokes' theorem. Hence, the problem of connecting the origin $(0,0,0)$ to (x,y,t) with a length minimizing horizontal curve amounts to the problem of connecting $(0,0)$ to (x_1,x_2) with the shortest planar curve enclosing (algebraic) area x_3. This is (a version of) Dido's problem and it is well-known that, if $x_3 \neq 0$, its solutions are arcs of circles. The corresponding horizontal curves are spirals which can be parametrized by arclength by the formulae

$$\begin{cases} x_1(t) = \dfrac{A(1-\cos\varphi t) + B\sin\varphi t}{\varphi} \\ x_2(t) = \dfrac{-B(1-\cos\varphi t) + A\sin\varphi t}{\varphi} \\ x_3(t) = -\dfrac{\varphi t - \sin\varphi t}{2\varphi^2}, \end{cases} \tag{3.1}$$

for suitable $(A,B) \in S^1 \subset \mathbb{R}^2$ and $\varphi \neq 0$. If $x_3 = 0$ we have instead the straight lines $\gamma(t) = (Bt, At, 0)$.

It can be proved that the spirals in (3.1) are length minimizing up to time $t = 2\pi/\varphi$, when they reach the point $(0,0,\pi/\varphi^2)$. In particular, for any $\varepsilon > 0$ there exists a family of length minimizers joining the origin and $(0,0,\varepsilon)$: this family is parametrized by $(A,B) \in S^1$ with the choice $\varphi = \sqrt{\pi/\varepsilon}$.

3.2 First-order necessary conditions

We want to derive necessary conditions for a horizontal curve to be length minimizing. To this end, we fix a length minimizer $\gamma : [0,1] \to \mathbb{R}^n$ with associated optimal controls h. Without loss of generality, we may assume that $\gamma(0) = 0$ and that γ is parametrized by constant speed, *i.e.*, that $|h| = c$ a.e. on $[0,1]$. In particular, by Exercise 2.2, γ is also a minimizer for the problem

$$\inf\{L_2(\tilde\gamma) : \tilde\gamma \text{ is horizontal}, \tilde\gamma(0) = \gamma(0) \text{ and } \tilde\gamma(1) = \gamma(1)\}.$$

For any fixed $x \in \mathbb{R}^n$, let $\gamma_x : [0,1] \to \mathbb{R}^n$ be the solution of

$$\begin{cases} \dot\gamma_x = h \cdot X(\gamma_x) \\ \gamma_x(0) = x, \end{cases}$$

where we write $h \cdot X(\gamma_x)$ to denote the function $\sum_{j=1}^r h_j X_j(\gamma_x)$ defined on $[0,1]$.

For any fixed $t \in [0,1]$, let us define $F_t : \mathbb{R}^n \to \mathbb{R}^n$ by

$$F_t(x) := \gamma_x(t). \tag{3.2}$$

Notice that F_t is well defined in a neighbourhood of the origin and that it is a diffeomorphism from such neighbourhood to its image.

Given another control $k \in L^\infty([0,1],\mathbb{R}^r)$ we denote by q_k the horizontal curve solving

$$\begin{cases} \dot q_k = k \cdot X(q_k) \\ q_k(0) = 0\,, \end{cases} \tag{3.3}$$

Finally, given $v \in L^\infty([0,1],\mathbb{R}^r)$, we define the *(extended) endpoint map* $\varphi_v : \mathbb{R} \to \mathbb{R}^{n+1}$

$$\varphi_v(s) := \left(F_1^{-1}(q_{h+sv}(1)), \textstyle\int_0^1 (h+sv)^2\right). \tag{3.4}$$

The first component of $\varphi_v(s)$ is (up to a diffeomorphism) the endpoint of the horizontal curve q_{h+sv}, while the last component is (the square of) its 2-length $L_2(q_{h+sv})$.

Lemma 3.3. *If γ is length minimizing and parametrized by constant speed, then there exists $\overline{\xi} \in \mathbb{R}^{n+1} \setminus \{0\}$ such that*

$$\langle \overline{\xi}, \varphi_v'(0)\rangle = 0 \quad \forall v \in L^\infty([0,1],\mathbb{R}^r)\,. \tag{3.5}$$

Proof. Assume not: then, there exist $v_1,\dots,v_{n+1} \in L^\infty([0,1],\mathbb{R}^r)$ such that the vectors $\varphi_{v_1}'(0),\dots,\varphi_{v_{n+1}}'(0) \in \mathbb{R}^{n+1}$ are linearly independent. Writing $s \cdot v := s_1 v_1 + \dots + s_{n+1} v_{n+1}$, it follows that the map

$$\begin{aligned} \Phi &: \mathbb{R}^{n+1} \to \mathbb{R}^{n+1} \\ \Phi(s_1,\dots,s_{n+1}) &:= \left(F_1^{-1}(q_{h+s\cdot v}(1)), L_2(q_{h+s\cdot v})^2\right) \end{aligned}$$

is such that $\nabla\Phi(0)$ is invertible because $\frac{\partial\Phi}{\partial s_i}(0) = \varphi_{v_i}'(0)$. In particular, Φ is an open map (in a neighbourhood of 0), hence one can find $\bar s \in \mathbb{R}^{n+1}$ such that the control $\bar h := h + \bar s_1 v_1 + \dots + \bar s_{n+1} v_{n+1}$ satisfies

$$\begin{aligned} F_1^{-1}(q_{\bar h}(1)) &= F_1^{-1}(q_h(1)) \\ L_2(q_{\bar h})^2 &< L_2(q_h)^2\,. \end{aligned}$$

Since F_1 is a diffeomorphism, if $\bar s$ is close enough to 0, the first equality above implies that $q_{\bar h}(1) = q_h(1)$. This contradicts the minimality of $\gamma = q_h$. □

Remark 3.4. An important role in the derivation of the necessary conditions in Theorem 3.6 will be played by the previous lemma. A key point in its proof is the fact that the extended endpoint map cannot be an open map in any neighbourhood of length minimizers. This suggests a sort of recipe to produce necessary conditions for optimality: in principle, any open mapping theorem might be used to derive necessary conditions. Also the Goh condition in the subsequent Theorem 3.20 is obtained by exploiting a suitable open mapping theorem.

Lemma 3.5. *If $v \in L^\infty([0,1],\mathbb{R}^r)$ is fixed and φ_v is as in (3.4), then*

$$\varphi_v'(0) = \left(\int_0^1 JF_t(0)^{-1}(v\cdot X(\gamma(t)))\,dt,\, 2\int_0^1 \langle h(t), v(t)\rangle\, dt\right) \in \mathbb{R}^n \times \mathbb{R}, \tag{3.6}$$

where JF_t is the Jacobian matrix of F_t and, again, $v\cdot X = v_1X_1+\cdots+v_rX_r$.

Proof. Let $s\in\mathbb{R}$ be fixed and, for any $t\in[0,1]$, define $x_{h+sv}(t) := F_t^{-1}(q_{h+sv}(t))$; equivalently,

$$q_{h+sv}(t) = F_t(x_{h+sv}(t))\,. \tag{3.7}$$

In particular, the first n components in the definition of $\varphi_v(s)$ are equal to $x_{h+sv}(1)$.

We can differentiate (3.7) with respect to t to obtain

$$(h+sv)\cdot X(q_{h+sv}) = h\cdot X(q_{h+sv}) + JF_t(x_{h+sv})\dot{x}_{h+sv}\,,$$

hence

$$\dot{x}_{h+sv} = s\, JF_t(x_{h+sv})^{-1}[v\cdot X(q_{h+sv})]\,.$$

It follows that

$$x_{h+sv}(t) = s\int_0^t JF_\tau(x_{h+sv}(\tau))^{-1}[v\cdot X(F_\tau(x_{h+sv}(\tau)))]\,d\tau\,,$$

i.e.,

$$\begin{aligned}\varphi_v'(0) &= \left(\left.\frac{\partial x_{h+sv}(1)}{\partial s}\right|_{s=0},\, 2\int_0^1 \langle h(\tau), v(\tau)\rangle\, d\tau\right)\\ &= \left(\int_0^1 JF_\tau(x_h(\tau))^{-1}[v\cdot X(F_\tau(x_h(\tau)))]\,d\tau,\, 2\int_0^1 \langle h(\tau), v(\tau)\rangle d\tau\right).\end{aligned}$$

The desired equality (3.6) easily follows on noticing that $x_h(\tau) = F_\tau^{-1}(q_h(\tau)) = F_\tau^{-1}(\gamma(\tau)) = 0$. □

We can now pass to the main result of this section.

Theorem 3.6 (First-order necessary conditions). *Let $\gamma : [0,1]\to\mathbb{R}^n$ be a length minimizer with $\gamma(0) = 0$ and with associated controls h; assume that γ is parametrized by constant speed,* i.e., *$|h| \equiv c$. Then, there exist $\xi_0 \in \{0,1\}$ and $\xi \in \mathrm{Lip}([0,1],\mathbb{R}^n)$ such that*

(i) $(\xi(t), \xi_0) \neq 0$ *for any* $t \in [0, 1]$;
(ii) *for any* $j = 1, \ldots, r$, *the equality* $\xi_0 h_j + \langle \xi, X_j(\gamma) \rangle = 0$ *holds a.e. on* $[0, 1]$;
(iii) $\dot{\xi} = -(\sum_{j=1}^r h_j \, JX_j(\gamma))^T \xi$ *a.e. on* $[0, 1]$,

where JX_j *denotes the* $n \times n$ *Jacobian matrix of* $X_j : \mathbb{R}^n \to \mathbb{R}^n$ *and the superscript* ^{T} *denotes matrix transposition.*

Proof. Let $\overline{\xi} \in \mathbb{R}^{n+1} \setminus \{0\}$ be as in Lemma 3.3; write $\overline{\xi} =: (\xi(0), \xi_0/2) \in \mathbb{R}^n \times \mathbb{R}$. Using Lemma 3.5, we deduce from (3.5) the following necessary condition:

$$0 = \int_0^1 \sum_{j=1}^r v_j(t) \left\{ \langle \xi(0), JF_t(0)^{-1}(X_j(\gamma(t))) \rangle + \xi_0 h_j \right\} dt$$
$$= \int_0^1 \sum_{j=1}^r v_j(t) \left\{ \langle [JF_t(0)^{-1}]^T \xi(0), X_j(\gamma(t)) \rangle + \xi_0 h_j \right\} dt$$
$$\forall v \in L^\infty([0, 1], \mathbb{R}^r).$$

Upon defining $\xi(t) := [JF_t(0)^{-1}]^T \xi(0)$, the Fundamental lemma of the Calculus of Variations immediately implies statement (ii).

Statement (i) is clearly true if $\xi_0 \neq 0$ (notice that, in this case, one can also normalize $\overline{\xi}$ to have $\xi_0 = 1$); on the contrary, if $\xi_0 = 0$ we have $\xi(0) \neq 0$, hence $\xi(t) \neq 0$ for all $t \in [0, 1]$ because $JF_t(0)^{-1}$ is invertible. Hence, also (i) is proved.

We are left with statement (iii). By definition of $\xi(t)$, we have $\xi(0) = JF_t(0)^T \xi(t)$ and, on differentiating with respect to t,

$$0 = \left(\frac{d}{dt} JF_t(0)^T \right) \xi(t) + JF_t(0)^T \dot{\xi}(t) \quad \text{a.e. on } [0, 1]. \qquad (3.8)$$

Let us compute

$$\frac{d}{dt} JF_t(0) = J \left. \frac{d}{dt} F_t(x) \right|_{x=0} = J \left. \left(\sum_{j=1}^r h_j(t) X_j(F_t(x)) \right) \right|_{x=0}$$
$$= \sum_{j=1}^r h_j(t) \, JX_j(F_t(0)) \, JF_t(0)$$
$$= \left(\sum_{j=1}^r h_j(t) \, JX_j(\gamma(t)) \right) JF_t(0) \quad \text{a.e. on } [0, 1].$$

Recalling (3.8) and the fact that $JF_t(0)^T$ is invertible, we obtain

$$\dot{\xi}(t) = -\left(\sum_{j=1}^{r} h_j(t)\, JX_j(\gamma(t))\right)^T \xi(t) \quad \text{for a.e. } t \in [0,1],$$

as desired. □

Definition 3.7. A horizontal curve $\gamma : [0,1] \to \mathbb{R}^n$ with $\gamma(0) = 0$ and with associated controls h is said to be an *extremal* if there exist $\xi_0 \in \{0,1\}$ and $\xi \in \mathrm{Lip}([0,1], \mathbb{R}^n)$ such that statements (i), (ii) and (iii) in Theorem 3.6 hold. The function ξ is called *dual curve* (or *dual variable*).
If $\xi_0 = 1$, we say that γ is a *normal extremal*.
If $\xi_0 = 0$, we say that γ is an *abnormal extremal*.

Theorem 3.6 states that length minimizers parametrized by constant speed are also extremals; on the contrary, there exist extremals that are not minimizers, see Section 3.5. We do not require extremals to be parametrized by constant speed because this is automatically satisfied for normal extremals (see Exercise 3.11), while for abnormal extremals the parametrization plays essentially no role (see Exercise 3.17).

We will review the main properties of normal and abnormal extremals in Sections 3.3 and 3.4; now, a few observations are in order.

Remark 3.8. An extremal γ might be normal and abnormal at the same time, in the sense that it could possess two different dual curves that make γ normal and abnormal. An example of this phenomenon is given in Exercise 4.5. An extremal which is normal but not abnormal is called *strictly normal*; on the contrary, we call *strictly abnormal* an extremal which is abnormal but not also normal.

Exercise 3.9. Prove that, if γ is strictly normal, then it possesses a unique dual curve $\xi(t)$.

Hint: assume that $\xi_1(t), \xi_2(t)$ are dual curves making γ normal; prove that γ is abnormal with associated dual curve $\xi_1 - \xi_2$.

Theorem 3.6 possesses also an Hamiltonian formulation. Define the Hamiltonian

$$H(x,\xi) := \sum_{j=1}^{r} \langle X_j(x), \xi \rangle^2 \quad x, \xi \in \mathbb{R}^n .$$

Then, the following result holds.

Exercise 3.10. If γ is a normal extremal with dual variable ξ, then the couple (γ, ξ) solves the system of Hamiltonian equations

$$\begin{cases} \dot{\gamma} = -\dfrac{1}{2}\dfrac{\partial H}{\partial \xi}(\gamma, \xi) \\ \dot{\xi} = \dfrac{1}{2}\dfrac{\partial H}{\partial x}(\gamma, \xi). \end{cases}$$

If γ is an abnormal extremal with dual variable ξ, then $H(\gamma, \xi) \equiv 0$.

3.3 Normal extremals

In this section we deal with basic properties and facts about normal extremals. We begin with the following exercise.

Exercise 3.11. Let γ be a normal extremal; then, γ is parametrized by constant speed.

Hint: use Exercise 3.10 and the fact that, if h denotes the controls associated with γ, then $|h(t)|^2 = H(\gamma(t), \xi(t))$.

The most important result in this subsection is the following one.

Proposition 3.12. *Normal extremals are C^∞ smooth.*

Proof. Let $\gamma : [0, 1] \to \mathbb{R}^n$ be a normal extremal with associated controls h and dual curve ξ. Using (2.1) and (ii), (iii) in Theorem 3.6 we easily obtain the following chain of implications

$$\begin{aligned} & \gamma, \xi \in C^0([0,1]) \overset{\text{(ii)}}{\Longrightarrow} h_j \in C^0([0,1])\ \forall j = 1, \dots, r \\ \overset{(2.1),\text{(iii)}}{\Longrightarrow}\ & \gamma, \xi \in C^1([0,1]) \overset{\text{(ii)}}{\Longrightarrow} h_j \in C^1([0,1])\ \forall j = 1, \dots, r \\ \overset{(2.1),\text{(iii)}}{\Longrightarrow}\ & \gamma, \xi \in C^2([0,1]) \overset{\text{(ii)}}{\Longrightarrow} h_j \in C^2([0,1])\ \forall j = 1, \dots, r \\ \Longrightarrow\ & \dots \end{aligned}$$

□

Exercise 3.13. Prove that, if γ is a normal extremal with dual curve ξ, then condition (ii) in Theorem 3.6 holds on the whole interval $[0, 1]$ (and not only almost everywhere).

The following results, as well as the Proposition 3.12, show that normal minimizers/extremals share several common features with Riemannian geodesics.

Remark 3.14. When $r = n$ (*i.e.*, the CC structure is indeed Riemannian), any length minimizer/extremal γ is strictly normal. Otherwise, there would exist a dual curve ξ such that $\langle \xi, X_j(\gamma) \rangle = 0$ for any $j = 1, \dots, n$. Since $X_1, \dots, X_n$ now form a basis of $\mathbb{R}^n$, we obtain that $\xi \equiv 0$, which contradicts (i) in Theorem 3.6.

Exercise 3.15. Assume again that we are in the Riemannian case $r = n$. Then, by (2.1) and (ii) in Theorem 3.6, there is a natural way of identifying $\dot\gamma$, h and ξ, in the sense that any of the three uniquely determines the others. Prove that equation (iii) in Theorem 3.6 corresponds to the ODE of Riemannian geodesics.

The following important result is a special case of more general results in Optimal Control Theory, see for instance [7,13,20] and [23, Appendix C].

Theorem 3.16. *Every normal extremal is locally length minimizing.*

On the contrary, strictly abnormal extremals might not be length minimizers, see Section 3.5.

3.4 Abnormal extremals

By Theorem 3.6 (ii), an abnormal extremal γ and its dual variable ξ satisfy

$$\langle \xi, X_j(\gamma) \rangle = 0 \text{ on } [0,1] \quad \forall j = 1, \dots, r\,. \tag{3.9}$$

The compact notation $\xi \perp \Delta_\gamma$ will often be used to abbreviate the previous formula.

When dealing with abnormal extremals, it is not necessary to require that they are parametrized by constant speed; this is justified by the following fact.

Exercise 3.17. Assume that $\tilde\gamma : [0,1] \to \mathbb{R}^n$ is an abnormal extremal parametrized by constant speed and with dual curve $\tilde\xi$. Let γ be a different parametrization of the same curve; namely, let $\gamma := \tilde\gamma \circ f$ for an increasing, Lipschitz continuous homeomorphism $f : [0,1] \to [0,1]$. Then, γ satisfies (i), (ii) and (iii) in Theorem 3.6 with $\xi := \tilde\xi \circ f$.

Exercise 3.18. Prove that, if γ is an abnormal extremal with dual curve ξ, then condition (ii) in Theorem 3.6 holds on the whole interval $[0,1]$ (and not only almost everywhere).

Abnormal extremals are often introduced in the literature as singular points of the endpoint map; a few comments on this viewpoint are in order.

Going back to Section 3.2, let $\gamma : [0,1] \to \mathbb{R}^n$ be an extremal with $\gamma(0) = 0$ and associated optimal controls $h \in L^\infty([0,1], \mathbb{R}^r)$. Define the *endpoint map* End : $L^\infty([0,1], \mathbb{R}^r) \to \mathbb{R}^n$ by

$$\mathrm{End}(k) := q_k(1), \quad k \in L^\infty([0,1], \mathbb{R}^r)\,,$$

the curve q_k being defined as in (3.3). For any $v \in L^\infty([0,1],\mathbb{R}^r)$, the map $\varphi_v(s)$ in (3.4) can then be rewritten as

$$\varphi_v(s) = \left(F_1^{-1} \circ \mathrm{End}(h+sv), L_2(q_{h+sv})^2\right),$$

where the diffeomorphism F_1 is defined as in (3.2).

Now, if γ is an abnormal extremal, then the vector $\overline{\xi} = (\xi(0), \xi_0/2) \in \mathbb{R}^n \times \mathbb{R}$ provided by Lemma 3.3 is such that $\xi_0 = 0$. Hence (again by Lemma 3.3), the vector $\xi(0) \neq 0$ is such that

$$\xi(0) \perp \frac{d}{ds}(F_1^{-1} \circ \mathrm{End}(h+sv))\Big|_{s=0} \qquad \forall v \in L^\infty([0,1],\mathbb{R}^r).$$

Since F_1^{-1} is a diffeomorphism, there exists also a vector $\eta \neq 0$ such that

$$\eta \perp \frac{d}{ds}(\mathrm{End}(h+sv))\Big|_{s=0} = d\mathrm{End}(h)[v] \quad \forall v \in L^\infty([0,1],\mathbb{R}^r),$$

where $d\mathrm{End}(h)[v]$ denotes the differential of End at h in direction v. In particular, the image of $d\mathrm{End}(h)$ does not contain the vector η; equivalently, h is a point where the differential of the endpoint map is not surjective.

We have proved that (the controls associated with) abnormal extremals are singular points of End; the following exercise shows that the converse is also true.

Exercise 3.19. Prove that, if the differential of the endpoint map End is not surjective at some controls h associated with an horizontal curve γ, then γ is an abnormal extremal.

As already pointed out in Remark 3.8, an extremal might be normal and abnormal at the same time. By Proposition 3.12 any minimizer/extremal is C^∞ smooth unless it is strictly abnormal; hence, the relevant curves in the regularity problem for length minimizers are precisely the strictly abnormal ones. For such minimizers, a further necessary condition, the so-called *Goh condition*, can be proved.

Theorem 3.20 (Goh condition). *Let $\gamma : [0,1] \to \mathbb{R}^n$ be a strictly abnormal length minimizer. Then, there exists an associated dual curve ξ that satisfies*

$$\langle \xi, [X_i, X_j](\gamma)\rangle = 0 \textit{ on } [0,1] \quad \textit{for any } i,j = 1,\dots,r. \tag{3.10}$$

We refer to [4, Chapter 20] for the proof of Theorem 3.20. The proof is in the spirit of Remark 3.4: if (3.10) does not hold for any dual curve ξ, then a suitable open mapping theorem allows to conclude that a certain mapping of endpoint-type is open at γ, contradicting its minimality.

Corollary 3.21. *If the horizontal distribution $X_1, \dots, X_r$ is of step 2,* i.e., *if*

$$\textit{dim span}\,\{X_i, [X_i, X_j] : i, j \in \{1, \dots, r\}\}(x) = n \quad \forall x \in \mathbb{R}^n,$$

then any length minimizer is C^∞ smooth.

Proof. Assume by contradiction that there exists a length minimizer γ that is not of class C^∞; then, by Proposition 3.12, γ is strictly abnormal. By (3.9) and Theorem 3.20, there exists a dual variable ξ that is orthogonal (at points of γ) to X_i and $[X_i, X_j]$ for any $i, j \in \{1, \dots, r\}$. Since, by assumption, these elements generates all the tangent space at any point, then we have necessarily $\xi \equiv 0$, which contradicts Theorem 3.6 (i). □

We stress the fact that the minimality assumption is crucial in Theorem 3.20. In general, (3.10) might not hold for strictly abnormal extremals, with the following remarkable exception concerning general abnormal extremals in structures with rank 2.

Remark 3.22. If the horizontal distribution has rank $r = 2$, then any abnormal extremal $\gamma : [0, 1] \to \mathbb{R}^n$ and any associated dual curve ξ satisfy

$$\langle \xi(t), [X_1, X_2](\gamma(t)) \rangle = 0 \quad \forall t \in [0, 1] \tag{3.11}$$

Let us prove (3.11); we claim that it is enough to show that

$$\begin{aligned} &\langle \xi(t), [X_1, X_2](\gamma(t)) \rangle = 0 \\ &\text{for a.e. } t \in [0, 1] \text{ such that } \dot\gamma(t) \text{ exists and } \dot\gamma(t) \neq 0\,. \end{aligned} \tag{3.12}$$

Indeed, (3.12) and the continuity of ξ imply (3.11) for any γ such that $\dot\gamma \neq 0$ a.e. on $[0, 1]$; for instance, whenever γ is parametrized by constant speed. For different parametrizations of γ, it is enough to reason as in Exercise 3.17.

Let us prove (3.12). By equation (iii) in Theorem 3.6 and the abnormality of γ, we get

$$\begin{aligned} 0 &= \frac{d}{dt} \langle \xi, X_1(\gamma) \rangle \\ &= -\langle (h_1\, JX_1(\gamma) + h_2\, JX_2(\gamma))^T \xi, X_1(\gamma) \rangle + \langle \xi, JX_1(\gamma)[\dot\gamma] \rangle \\ &= -\langle \xi, (h_1\, JX_1(\gamma) + h_2\, JX_2(\gamma))[X_1(\gamma)] \rangle \\ &\quad + \langle \xi, JX_1(\gamma)[h_1 X_1(\gamma) + h_2 X_2(\gamma)] \rangle \\ &= h_2 \langle \xi, -JX_2(\gamma)[X_1(\gamma)] + JX_1(\gamma)[X_2(\gamma)] \rangle \\ &= -h_2 \langle \xi, [X_1, X_2](\gamma) \rangle \qquad \text{a.e. on } [0, 1]. \end{aligned}$$

With similar computations one gets

$$0 = \frac{d}{dt}\langle \xi, X_2(\gamma)\rangle = h_1\langle \xi, [X_1, X_2](\gamma)\rangle \quad \text{a.e. on } [0, 1].$$

In particular, if $t \in [0, 1]$ is such that $\dot{\gamma}(t) \neq 0$, then either $h_1(t) \neq 0$ or $h_2(t) \neq 0$, and this is enough to obtain (3.12).

As done before, for notational convenience we write $\xi \perp (\Delta \cup [\Delta, \Delta])_\gamma$ whenever the Goh condition holds for the couple (γ, ξ), to mean that the dual variable ξ is orthogonal to both horizontal vectors and brackets of horizontal vector fields. With a slight change of notation, we could also introduce the time-dependent 1-form

$$\xi^*(t) := \xi_1(t)dx_1 + \cdots + \xi_n(t)dx_n \tag{3.13}$$

and write $\xi^* \in \Delta_\gamma^\perp$ (for abnormal extremals) or $\xi^* \in \Delta_\gamma^\perp \cap [\Delta, \Delta]_\gamma^\perp$ (when the Goh condition is in force). The 1-form ξ^* is going to appear again later in these notes.

Remark 3.23. Another important fact about abnormal minimizers has been proved in [1] (see also [34]) in connection with the smoothness problem for the CC distance d: if the horizontal vectors $X_1, \dots, X_r$ are analytic, then the set Σ of point in $\mathbb{R}^n$ that can be connected to the origin (or to any other base point) with abnormal length minimizers is a closed set with empty interior. An important open question is the following Morse-Sard problem for the endpoint map: does Σ have measure zero? We refer to [27, Section 10.2] and to the recent preprint [2] for more detailed discussions on this and other topics.

3.5 An interesting family of extremals

An interesting sub-Riemannian structure was proposed by A. Agrachev and J. P. Gauthier during the meeting "Geometric control and sub-Riemannian geometry" held in Cortona in May 2012. Consider the CC structure of rank 2 induced on $\mathbb{R}^4$ by the vector fields

$$X_1(x) := \partial_1 + 2x_2\partial_3 + x_3^2\partial_4, \quad X_2(x) := \partial_2 - 2x_1\partial_3\,.$$

It can be easily checked that the bracket-generating condition holds and that, for any $\alpha \in \mathbb{R}$, the curve $\gamma^\alpha(t) := (t, \alpha|t|, 0, 0)$, $t \in \mathbb{R}$, is a strictly abnormal extremal with dual curve $\xi(t) = (0, 0, 0, 1)$. It is fairly easy to show that γ^0 is a minimizer; R. Monti has proved with a cutting-the-corner technique that γ^α is not a minimizer when $\alpha \notin \{0, 1, -1\}$. Using a different and much simpler argument, the remaining case $\alpha = \pm 1$ was recently settled in [19], where it is also proved that all length minimizers in the CC structure under consideration are smooth.

Exercise 3.24. Prove that γ^0 is uniquely length minimizing.

4 Carnot groups

4.1 Stratified groups

In this section we are going to describe a few basic facts on stratified groups. Recall that the Lie algebra $\mathfrak{g}$ associated with a Lie group $\mathbb{G}$ is defined as the Lie algebra of left-invariant vector fields on $\mathbb{G}$. A vector field X on $\mathbb{G}$ is said to be *left-invariant* if

$$X(p) = d\ell_p(X(0)) \quad \forall p \in \mathbb{G},$$

where $d\ell_p$ denotes the differential of the left-translation $\ell_p(z) = p \cdot z$ by p, $\cdot$ denotes the group product and 0 denotes the identity of $\mathbb{G}$. Equivalently, X is left-invariant if $(Xf)(\ell_p(x)) = X(f \circ \ell_p)(x)$ for any $p, x \in \mathbb{G}$ and any $f \in C^\infty(\mathbb{G})$.

Definition 4.1. A *stratified group* $\mathbb{G}$ is a connected, simply connected and nilpotent Lie group whose Lie algebra $\mathfrak{g}$ admits a *stratification*, *i.e.*, a decomposition

$$\mathfrak{g} = V_1 \oplus V_2 \oplus \cdots \oplus V_s$$

with the properties that $V_i = [V_1, V_{i-1}]$ for any $i = 2, \ldots, s$ and $[V_1, V_s] = \{0\}$.

A few comments are in order:

- the Lie algebra $\mathfrak{g}$ is nilpotent of *step s*;
- one can easily see that $[V_i, V_j] \subset V_{i+j}$ for any $i, j \geqslant 1$ such that $i + j \leqslant s$;
- if $i + j \geqslant s + 1$, then $[V_i, V_j] = \{0\}$.

Moreover, the exponential map $\exp : \mathfrak{g} \to \mathbb{G}$ induces a diffeomorphism between $\mathbb{G}$ and $\mathbb{R}^n \equiv \mathfrak{g}$, n being the dimension of $\mathfrak{g}$. However, in the sequel we will identify $\mathbb{G}$ with $\mathbb{R}^n$ by means of a different set of coordinates, the so-called exponential coordinates of the second-type (see (4.1) and (4.2) below).

Let us fix an *adapted basis* of $\mathfrak{g}$, *i.e.*, a basis $X_1, \ldots, X_n$ whose order is coherent with the stratification:

$$\underbrace{X_1, \ldots, X_r}_{\text{basis of } V_1}, \underbrace{X_{r+1}, \ldots, X_{r_2}}_{\text{basis of } V_2}, \underbrace{X_{r_2+1}, \ldots}_{\text{basis of } V_3} \ldots\ldots \underbrace{\ldots, X_n}_{\text{basis of } V_s}.$$

The integer $r_2 := \dim V_1 + \dim V_2$ will be used also in the sequel. We can then identify $\mathbb{G}$ with $\mathbb{R}^n$ by introducing *exponential coordinates of the second type*

$$\mathbb{R}^n \ni (x_1, \ldots, x_n) \longleftrightarrow \exp(x_n X_n) \cdot \exp(x_{n-1} X_{n-1}) \cdots \exp(x_1 X_1) \in \mathbb{G} \tag{4.1}$$

or, equivalently, by using flows of vector fields

$$\mathbb{R}^n \ni (x_1, \ldots, x_n) \longleftrightarrow e^{x_1 X_1} \circ \cdots \circ e^{x_{n-1}X_{n-1}} \circ e^{x_n X_n}(0) \in \mathbb{G}. \quad (4.2)$$

As a matter of fact (see *e.g.* [18]), one can prove that in these coordinates

$$\begin{aligned} X_1 &= \partial_1 \\ X_i(x) &= \partial_i + \textstyle\sum_{j=r+1}^n f_{ij}(x)\partial_j \quad \forall i = 2, \ldots, r \end{aligned} \quad (4.3)$$

for suitable analytic functions $f_{ij} : \mathbb{R}^n \to \mathbb{R}$.

The stratification of $\mathfrak{g}$ allows to define a family of intrinsic dilations on $\mathbb{G}$. For any $i = 1, \ldots, r$, let us define its *degree* $d(i) \in \{1, \ldots, s\}$ by

$$d(i) = k \iff X_i \in V_k .$$

One can define a one-parameter family of dilations on $\mathfrak{g}$ in the following way. For any $r > 0$, let $\delta_r : \mathfrak{g} \to \mathfrak{g}$ be the unique linear map such that

$$\delta_r(X_i) = r^{d(i)} X_i .$$

Then, by the stratification assumption, δ_r is a Lie algebra isomorphism. One can also define dilations on the group (in coordinates) by

$$\delta_r(x_1, \ldots, x_n) := (r x_1, \ldots, r^{d(i)} x_i, \ldots, r^s x_n) .$$

Clearly, $\delta_r : \mathbb{G} \to \mathbb{G}$ defines a one-parameter family of group isomorphisms.

Example 4.2. The Heisenberg group (see also Section 3.1, where it is presented in a different set of coordinates) is the stratified group associated with the Lie algebra of step 2 $\mathfrak{g} := V_1 \oplus V_2$, where $V_1 = \text{span}\{X_1, X_2\}$, $V_2 = \text{span}\{X_3\}$ and

$$[X_2, X_1] = X_3, \quad [X_3, X_1] = [X_3, X_2] = 0 .$$

The Heisenberg group can be represented in exponential coordinates of the second type as $\mathbb{R}^3$ with

$$X_1 = \partial_1, \quad X_2 = \partial_2 - x_1\partial_3, \quad X_3 = \partial_3 .$$

Group dilations read as $\delta_r(x_1, x_2, x_3) = (r x_1, r x_2, r^2 x_3)$.

Example 4.3. The *Engel group* is the stratified group associated with the Lie algebra of step 3 $\mathfrak{g} := V_1 \oplus V_2 \oplus V_3$, where $V_1 = \text{span}\{X_1, X_2\}$, $V_2 = \text{span}\{X_3\}$, $V_3 = \text{span}\{X_4\}$ and

$$\begin{aligned} &[X_2, X_1] = X_3, \\ &[X_3, X_1] = X_4, \\ &[X_3, X_2] = [X_4, X_1] = [X_4, X_2] = [X_4, X_3] = 0 . \end{aligned}$$

The Engel group can be represented in exponential coordinates of the second type as $\mathbb{R}^4$ with

$$X_1 = \partial_1, \quad X_2 = \partial_2 - x_1\partial_3 + \frac{x_1^2}{2}\partial_4, \quad X_3 = \partial_3 - x_1\partial_4, \quad X_4 = \partial_4 .$$

Group dilations read as $\delta_r(x_1, x_2, x_3, x_4) = (rx_1, rx_2, r^2x_3, r^3x_4)$.

4.2 Carnot groups

Stratified groups can be endowed with a canonical CC structure induced by a basis $X_1, \ldots, X_r$ of the first layer V_1. Notice that the horizontal sub-bundle $\Delta := V_1$ is left-invariant and bracket-generating (by the stratification assumption), hence the CC distance d is well defined. We refer to [15] for a metric characterization of Carnot groups and to [16] for an introduction to sub-Riemannian geometry on groups.

Exercise 4.4. Prove that, for any $p, x, y \in \mathbb{G}$ and any $r > 0$, there holds

$$d(p \cdot x, p \cdot y) = d(x, y) \quad \text{and} \quad d(\delta_r x, \delta_r y) = r d(x, y) .$$

Exercise 4.5. Prove that the horizontal curve $\gamma(t) = (0, t, 0, 0)$ in the Engel group (represented in the coordinates of Example 4.3) is an extremal that is normal and abnormal at the same time.

Our interest in Carnot groups is motivated by the well-known fact that the tangent metric space (in the Gromov-Hausdorff sense) to a CC space at a "generic" point is a Carnot group: roughly speaking, Carnot groups are the infinitesimal models of CC spaces. See *e.g.* [6,24,25].

4.3 The dual curve and extremal polynomials

Let $\gamma : [0, 1] \to \mathbb{G}$ be an extremal with associated controls $h \in L^\infty([0, 1], \mathbb{R}^r)$ and dual curve $\xi \in \mathrm{Lip}([0, 1], \mathbb{R}^n)$; assume that $\gamma(0) = 0$. Recall that ξ induces a time-dependent 1-form ξ^* as in (3.13); we are going to write ξ^* in a different system of coordinates for 1-forms.

The group structure allows to define a frame $\theta_1, \ldots, \theta_n$ of left-invariant 1-forms, dual to the adapted basis $X_1, \ldots, X_n$, by imposing that

$$\theta_i(X_j) = \delta_{ij} \quad \text{on } \mathbb{G}, \tag{4.4}$$

δ_{ij} denoting the Kronecker delta. We can therefore define $\lambda \in \mathrm{Lip}([0,1], \mathbb{R}^n)$ by imposing that

$$\begin{aligned} \xi^*(t) &= \xi_1(t)dx_1 + \cdots + \xi_n(t)dx_n \\ &= \big(\lambda_1(t)\theta_1 + \cdots + \lambda_n(t)\theta_n\big)(\gamma(t)) \quad \forall t \in [0, 1] . \end{aligned} \tag{4.5}$$

We use the term *dual curve* also for the function λ. One can immediately notice that, by (4.4), statement (iii) in Theorem 3.6 is equivalent to

$$\xi_0 h_i + \lambda_i = 0 \quad \text{a.e. on } [0, 1], \qquad \forall\, i = 1, \dots, r\,. \tag{4.6}$$

Moreover, the differential equation (iii) of Theorem 3.6 is equivalent to the following ODE for λ (we refer to [17, Theorem 2.6] for details). For any $i = 1, \dots, n$, there holds

$$\dot\lambda_i = -\sum_{j=1}^{r}\sum_{k=1}^{n} c_{ij}^k h_j \lambda_k \quad \text{a.e. on } [0, 1]\,, \tag{4.7}$$

where the constants c_{ij}^k are the *structure constants* of the Lie algebra $\mathfrak{g}$ defined by

$$[X_i, X_j] = \sum_{k=1}^{n} c_{ij}^k X_k \quad \forall\, i, j = 1, \dots, n\,.$$

Exercise 4.6. Prove the implication

$$d(k) \neq d(i) + d(j) \Rightarrow c_{ij}^k = 0 \quad \forall i, j, k = 1, \dots, n\,. \tag{4.8}$$

Hint: recall that $[X_i, X_j] \in V_{d(i)+d(j)}$.

Deduce, as a consequence, that (4.7) is equivalent to

$$\dot\lambda_i = -\sum_{j=1}^{r} \sum_{\substack{k=1,\dots,n \\ d(k)=d(i)+1}} c_{ij}^k h_j \lambda_k \quad \text{a.e. on } [0, 1]\,. \tag{4.9}$$

From the technical viewpoint, the main achievement of [17, 18] is an explicit formula for the dual curve λ as a function of γ, see Theorem 4.11 below. This is obtained through the integration of the ODE (4.7), which is in turn based on the following result.

Lemma 4.7. *Let $\gamma : [0, 1] \to \mathbb{G}$ be an extremal with $\gamma(0) = 0$; let $h \in L^\infty([0, 1], \mathbb{R}^r)$ be the associated controls and $\lambda \in \mathrm{Lip}([0, 1], \mathbb{R}^n)$ be its dual curve. Suppose that there exist functions $P_i \in C^1(\mathbb{G})$, $i = 1, \dots, n$, such that*

$$P_i(0) = \lambda_i(0) \quad \textit{and} \quad X_j P_i = \sum_{k=1}^{n} c_{ji}^k P_k \textit{ on } \mathbb{G} \tag{4.10}$$

for any $i, j = 1, \dots, n$. Then, for any $i = 1, \dots, n$, there holds

$$\lambda_i(t) = P_i(\gamma(t)) \quad \forall\, t \in [0, 1]. \tag{4.11}$$

Proof. The proof is based on a reverse-order inductive argument on i; we start by proving (4.11) for $i = n$. Since $X_n \in V_s$ is in the kernel of $\mathfrak{g}$, we have $[X_j, X_n] = 0$ for any $j = 1, \dots, n$, *i.e.*, $c^k_{jn} = -c^k_{nj} = 0$. In particular, by (4.7) and (4.10)

- $\dot{\lambda}_n = -\sum_{j=1}^r \sum_{k=1}^n c^k_{nj} h_j \lambda_k = 0$, hence λ_n is constant on $[0, 1]$;
- for any $j = 1, \dots, n$, $X_j P_n = \sum_{k=1}^n c^k_{jn} P_k = 0$, hence P_n is constant on $\mathbb{G}$.

Since, by assumption, $P_n(0) = \lambda_n(0)$, we obtain that $\lambda_n(t) = P_n(\gamma(t))$ for any $t \in [0, 1]$.

Assume now that $\lambda_k = P_k(\gamma)$ for any $k \geqslant i + 1$; recalling that $\dot{\gamma} = \sum_{j=1}^r h_j X_j(\gamma)$, we have

$$\frac{d}{dt}(P_i \circ \gamma) = \sum_{j=1}^r h_j\, X_j P_i(\gamma) = \sum_{j=1}^r \sum_{k=1}^n h_j c^k_{ji} P_k(\gamma)$$

$$\overset{(4.8)}{=} \sum_{j=1}^r \sum_{\substack{k=1,\dots,n \\ d(k)=d(i)+1}} h_j c^k_{ji} P_k(\gamma)\,.$$

We can now use the inductive assumption together with the equality $c^k_{ji} = -c^k_{ij}$ to get

$$\frac{d}{dt}(P_i \circ \gamma) = -\sum_{j=1}^r \sum_{\substack{k=1,\dots,n \\ d(k)=d(i)+1}} h_j c^k_{ij} \lambda_k(\gamma) \overset{(4.9)}{=} \dot{\lambda}_i\,.$$

In particular, the Lipschitz functions λ_i and $P_i \circ \gamma$ have the same derivative and, by assumption, they coincide at time $t = 0$. This is sufficient to conclude the validity of (4.11). □

The integration of the dual variable λ is thus reduced to the search for functions P_i satisfying (4.10); these functions are provided by the *extremal polynomials* introduced below in Definition 4.8. Let us introduce some preliminary notation. Given a multi-index $\alpha = (\alpha_1, \dots, \alpha_n) \in \mathbb{N}^n$ and $x \in \mathbb{R}^n \equiv \mathbb{G}$, we write

$$x^\alpha = x_1^{\alpha_1} x_2^{\alpha_2} \cdots x_n^{\alpha_n}$$
$$|\alpha| = \alpha_1 + \cdots + \alpha_n$$
$$\alpha! = \alpha_1!\, \alpha_2! \cdots \alpha_n!\,.$$

For the sake of precision: we agree that $0 \in \mathbb{N}$, hence the null multi-index $\alpha = 0$ is admissible. If $x = 0$ and $\alpha = 0$, we agree that $x^\alpha = 1$.

Definition 4.8. For any $v \in \mathbb{R}^n$ and $i = 1, \dots, n$, we define the *extremal polynomial* $P_i^v : \mathbb{G} \to \mathbb{R}$ by

$$P_i^v(x) = \sum_{\alpha \in \mathbb{N}^n} \sum_{k=1}^{n} \frac{(-1)^{|\alpha|}}{\alpha!} c_{i\alpha}^k \, v_k \, x^\alpha , \tag{4.12}$$

where the symbols $c_{i\alpha}^k$ denote the *generalized structure constants* of $\mathfrak{g}$ defined by

$$[\cdots[X_i, \underbrace{X_1], X_1], \dots X_1]}_{\alpha_1 \text{ times}}, \underbrace{X_2], \dots X_2]}_{\alpha_2 \text{ times}}, X_3], \dots] \dots] = \sum_{k=1}^{n} c_{i\alpha}^k X_k .$$

Exercise 4.9. Prove that the summation in (4.12) is finite and, more precisely, that P_i^v is a polynomial of both degree and *homogeneous degree* (see *e.g.* [17, Remark 4.2]) at most $s - d(i)$.
Hint: define $d(\alpha) := \sum_{j=1}^{n} \alpha_i d(i)$ and prove the implication

$$d(k) \neq d(i) + d(\alpha) \Rightarrow c_{i\alpha}^k = 0 .$$

As already mentioned, extremal polynomials satisfy (4.10) in Lemma 4.7.

Theorem 4.10. *For any $v \in \mathbb{R}^n$ and $i = 1, \dots, n$, the extremal polynomials satisfy*

$$P_i^v(0) = v_i \quad \text{and} \quad X_j P_i^v = \sum_{k=1}^{n} c_{ji}^k P_k^v \text{ on } \mathbb{G}. \tag{4.13}$$

While the first equality in (4.13) can be easily checked, the formulae for the derivatives of the P_i^v's are not trivial at all. Their proof is however beyond the scopes of these notes. In the framework of *free* Carnot groups, the second equality in (4.13) was first proved in [17] as a consequence of certain algebraic identities obtained along the proof of [17, Theorem 4.6]. The latter is nothing but Theorem 4.11 below (in the special case of free groups), but its proof follows a completely different line from the one presented here, being based on explicit formulae for the horizontal vector fields (see [12]) rather than on Lemma 4.7. For general Carnot groups, the proof of (4.13) was achieved in [18] with an argument of a differential-geometric flavour involving also non-trivial algebraic identities.

Lemma 4.7 and Theorem 4.10 have the following, immediate consequence.

Theorem 4.11. *Let $\gamma : [0, 1] \to \mathbb{G}$ be an extremal with $\gamma(0) = 0$; let $\lambda \in \mathrm{Lip}([0, 1], \mathbb{R}^n)$ be an associated dual curve and set $v := \lambda(0) \in \mathbb{R}^n$. Then,*

$$\lambda_i(t) = P_i^v(\gamma(t)) \quad \text{for any } t \in [0, 1] .$$

4.4 Extremals in Carnot groups

Theorem 4.11 is our main result from a technical viewpoint. Its consequences, however, are probably even more interesting; let us start by discussing its implications in the case of normal extremals.

Theorem 4.12. (Characterization of normal extremals in Carnot groups). *Let $\gamma : [0,1] \to \mathbb{G}$ be an horizontal curve with $\gamma(0) = 0$. Then, the following conditions are equivalent:*

(a) *γ is a normal extremal;*
(b) *there exists $v \in \mathbb{R}^n$ such that $\dot\gamma = -\sum_{i=1}^r P_i^v(\gamma)X_i(\gamma)$.*

In particular, the sum $P_1^v(\gamma)^2 + \cdots + P_r^v(\gamma)^2$ is constant on $[0,1]$.

Proof. Let us begin with the implication (a)⇒(b). Let h be the controls associated with γ and λ be the dual variable; set $v := \lambda(0) \in \mathbb{R}^n$. By (4.6) and Theorem 4.11, we have

$$h_i = -\lambda_i = -P_i^v(\gamma) \quad \text{on } [0,1], \qquad \forall i = 1,\dots,n,$$

and (b) immediately follows. The fact that $P_1^v(\gamma)^2 + \cdots + P_r^v(\gamma)^2$ is constant is equivalent to $|h|$ being constant.

Concerning the implication (b)⇒(a), notice that the controls $h_i = -P_i^v(\gamma)$, together with the functions $\lambda_i := P_i^v(\gamma)$, satisfy (4.6) (with $\xi_0 = 1$) and (4.7), because

$$\begin{aligned}\dot\lambda_i &= \frac{d}{dt}\left(P_i^v(\gamma)\right) \overset{(b)}{=} -\sum_{j=1}^r P_j^v(\gamma)\, X_j P_i^v(\gamma)\\ &\overset{(4.13)}{=} \sum_{j=1}^r \sum_{k=1}^n (-P_j^v(\gamma)) c_{ji}^k P_k^v(\gamma) = -\sum_{j=1}^r \sum_{k=1}^n c_{ij}^k\, h_j\, \lambda_k\,.\end{aligned}$$

This proves that γ is a normal extremal with dual curve λ, as desired. □

Theorem 4.12 characterizes normal extremals as solutions to a certain ODE: notice that we have reduced the $2n$-variables Hamiltonian system of Exercise 3.10 to a system of ODEs in n variables.

Recalling that left-invariant vector fields are analytic, by Theorem 4.12 one can improve Proposition 3.12 on the regularity of normal extremals.

Corollary 4.13. *Let $\gamma : [0,1] \to \mathbb{G}$ be a normal extremal; then, γ is analytic regular.*

Let us now examine the case of abnormal extremals. If λ is the dual curve associated with an abnormal extremal γ, then (4.6) and Theorem 4.11 imply that

$$\lambda_i = P_i^v(\gamma) = 0 \quad \text{on } [0,1] \qquad \forall i = 1, \ldots, r,$$

provided $v := \lambda(0) \in \mathbb{R}^n$. Moreover, by (4.4), (4.5) and the fact that the basis $X_1, \ldots, X_n$ is adapted to the stratification, the Goh condition (3.10) is equivalent to

$$\lambda_i = P_i^v(\gamma) = 0 \text{ on } [0,1] \quad \text{for any } i = r+1, \ldots, r_2 .$$

Recall that the integer r_2 has been defined as $\dim V_1 + \dim V_2$. We have therefore

Theorem 4.14. (Characterization of abnormal extremals in Carnot groups). *Let $\gamma : [0,1] \to \mathbb{G}$ be an horizontal curve with $\gamma(0) = 0$. Then, the following conditions are equivalent:*

(a) *γ is an abnormal extremal;*
(b) *there exists $v \in \mathbb{R}^n \setminus \{0\}$ such that $P_1^v(\gamma) = \cdots = P_r^v(\gamma) = 0$.*

Moreover, the Goh condition (3.10) *holds if and only if $P_{r+1}^v(\gamma) = \cdots = P_{r_2}^v(\gamma) = 0$.*

The proof is left as an exercise to the reader, who will also notice that the parameter $v \in \mathbb{R}^n$ is equal to $\lambda(0)$, which is not zero due to Theorem 3.6 (i).

Remark 4.15. The fact that $v \neq 0$ implies that there exist at least one index $i \in \{1, \ldots, r\}$ and another index $j \in \{r+1, \ldots, r_2\}$ such that neither P_i^v nor P_j^v are the null polynomial; see [18, Proposition 2.6] for more details. In particular, any abnormal extremal γ belongs to an algebraic variety (the one defined by the equalities in Theorem 4.14 (b)) that is not trivial.

The characterization of abnormal extremals in Carnot groups allows for several applications; here, we are going to recall a few of those presented in [17] and [18].

It is possible to construct very irregular abnormal extremals satisfying also the Goh condition. For instance, there exists a 32-dimensional Carnot group $\mathbb{G}$ such that, for *any* Lipschitz function $\varphi : [0,1] \to \mathbb{G}$, there exists a Goh abnormal extremal of the form

$$\gamma(t) = (t^2, t, \varphi(t), *, \ldots, *).$$

See [17, Section 6.4] for more details. In the same spirit, a "spiral-like" abnormal Goh extremal has been provided in [18, Section 5]. These examples somehow suggest that a finer analysis of necessary conditions is needed if one aims at proving smoothness of minimizers, since even second-order necessary conditions (the Goh one) are not enough to ensure regularity.

W. Liu and H. J. Sussman have proved in [23] that, if γ is an abnormal extremal in a CC structure of rank $r = 2$ with dual curve ξ satisfying

$$\xi(t) \not\perp [[\Delta, \Delta], \Delta]_{\gamma(t)} \quad \text{for any } t \in [0, 1],$$

then γ is smooth. Abnormal extremals satisfying the previous condition are called *regular abnormal* and are somehow "generic"; let us recall that the Goh condition $\xi \perp (\Delta \cup [\Delta, \Delta])_\gamma$ holds for abnormal extremals because of Remark 3.22. When working in Carnot groups of rank 2, the regularity of such extremals can be proved in a plain way by using extremal polynomials, see [17, Section 6.2]. The results in [23] are anyway much finer, as they show (in a more general framework) that regular abnormal extremals are also locally minimizing.

In the following sections we analyze with more details two further applications of our machinery.

5 Minimizers in step 3 Carnot group

In this section, we review the proof given in [17, Section 6.1] of the following result, that was first proved by K. Tan and X. Yang in [38].

Theorem 5.1. *Any minimizer in a Carnot group of step* 3 *is* C^∞ *smooth.*

Proof. By contradiction, assume that there exists a length minimizing curve $\gamma : [0, 1] \to \mathbb{G}$ that is not of class C^∞; then, γ is a strictly abnormal minimizer and satisfies the Goh condition. By left invariance, we can assume that $\gamma(0) = 0$. By Theorem 4.14 and Remark 4.15, there exist $v \in \mathbb{R}^n \setminus \{0\}$ and $j \in \{r + 1, \dots, r_2\}$ such that P_j^v is not the null polynomial and

$$P_j^v(\gamma) = 0 \quad \text{on } [0, 1]. \tag{5.1}$$

By Exercise 4.9, P_j^v has homogeneous degree at most 1, hence there exists $(a_1, \dots, a_r) \in \mathbb{R}^r \setminus \{0\}$ such that

$$P_j^v(x) = a_1 x_1 + \dots + a_r x_r , \tag{5.2}$$

where we have also used the fact that $P_j^v(0) = 0$. Define the left-invariant horizontal vector field $Y_1 := a_1 X_1 + \dots + a_r X_r$ and complete it to a basis

$Y_1, \dots, Y_r$ of V_1. Using (4.3), (5.1) and (5.2), we obtain that $\dot\gamma$ is of the form

$$\dot\gamma = h_2 Y_2(\gamma) + \cdots + h_r Y_r(\gamma).$$

Hence, γ is contained in the subgroup of $\mathbb{G}$ associated with the Lie subalgebra of $\mathfrak{g}$ generated by $Y_2, \dots, Y_r$ and, in particular, it is contained in a Carnot group of rank $r-1$ and step (at most) 3. An easy argument by induction on the rank of the group allows to conclude. □

6 On the negligibility of the abnormal set

In this Section we review the results contained in [18, Section 4]; to this end, we have to introduce some preliminary notions.

The *Tanaka prolongation* Prol $\mathfrak{g}$ of a stratified Lie algebra $\mathfrak{g} = V_1 \oplus \cdots \oplus V_s$ is the largest stratified Lie algebra which can be written in the form

$$\text{Prol}\,\mathfrak{g} = \cdots \oplus V_{-1} \oplus V_0 \oplus V_1 \oplus \cdots \oplus V_s$$

and with the property that $[V_i, V_j] \subset V_{i+j}$ for any $i \leqslant s$, $j \leqslant s$. Here, "largest" means that any other extension of $\mathfrak{g}$ with these properties is (isomorphic to) a sub-algebra of Prol $\mathfrak{g}$. The explicit construction of Prol $\mathfrak{g}$ was provided by N. Tanaka in [39]. The prolongation is never trivial, in the sense that Prol $\mathfrak{g} \neq \mathfrak{g}$; indeed, it can be proved that $\dim V_0 \geqslant 1$. Notice that the number of layers in Prol $\mathfrak{g}$ in not necessarily finite; when Prol $\mathfrak{g}$ is infinite dimensional we say that $\mathbb{G}$ is *nonrigid*.

Let $X_1, \dots, X_n$ be an adapted basis of $\mathfrak{g}$; let us extend it to an adapted basis of Prol $\mathfrak{g}$

$$\underbrace{\dots\dots, X_{-j}, \dots,}_{\text{basis of } V_{-1}} \underbrace{\dots, X_{-1}, X_0,}_{\text{basis of } V_0} \underbrace{X_1, \dots, X_r,}_{\text{basis of } V_1} \dots\dots, \underbrace{\dots, X_n}_{\text{basis of } V_s}.$$

With a slight abuse of notation, we denote this basis by $(X_i)_{i\leqslant n}$, where the notation "$i \leqslant n$" means

- either $-\infty < i \leqslant n$, if $\dim \text{Prol}\,\mathfrak{g} = \infty$;
- or $-m \leqslant i \leqslant n$, if $m \in \mathbb{N}$ is such that $\dim \text{Prol}\,\mathfrak{g} = m + n + 1$.

We will adopt a similar convention for notations like "$i \leqslant r$" and "$i \leqslant 0$".

The Lie algebra Prol $\mathfrak{g}$ possesses its own structure constants and generalized structure constants. We still denote these constants (that are defined for $i, j, k \leqslant n$ and $\alpha \in \mathbb{N}^n$) by c_{ij}^k and $c_{i\alpha}^k$ because they clearly coincide with those of $\mathfrak{g}$ when $1 \leqslant i, j, k \leqslant n$. Hence, as in Definition

4.8, for any fixed $v \in \mathbb{R}^n$ and any $i \leqslant n$ (*i.e.*, also for $i \leqslant 0$) one can define the extremal polynomial

$$P_i^v(x) := \sum_{\alpha \in \mathbb{N}^n} \sum_{k=1}^{n} \frac{(-1)^{|\alpha|}}{\alpha!} c_{i\alpha}^k \, v_k \, x^\alpha, \qquad x \in \mathbb{G}, \tag{6.1}$$

where we agree that $v_k = 0$ whenever $k \leqslant 0$. As in Theorem 4.10, one can prove that

$$\begin{aligned} &P_i^v(0) = v_i \text{ for any } i \leqslant n \\ &X_j P_i^v = \sum_{k \leqslant n} c_{ji}^k P_k^v \quad \text{for any } i \leqslant n, 1 \leqslant j \leqslant n. \end{aligned} \tag{6.2}$$

These formulae are key tools in the proof of the following result; see [18] for more details.

Theorem 6.1. *Let $\gamma : [0,1] \to \mathbb{G}$ be an abnormal extremal with $\gamma(0) = 0$. Then, there exists $v \in \mathbb{R}^n$ such that*

$$P_i^v(\gamma) = 0 \text{ on } [0,1] \qquad \text{for any } i \leqslant r\,. \tag{6.3}$$

If the Goh condition holds, then the previous formula holds for any $i \leqslant r_2$.

Proof. Given the formulae (6.2), the proof is quite elementary and similar to that of Lemma 4.7. We prove (6.3) by reverse induction on the homogeneous degree[1] $d(i)$ of i. We set again $v := \lambda(0)$, λ being the dual curve associated with γ.

The base of the induction is the case $d(i) = 1$, where (6.3) holds by Theorem 4.14.

Assume then that $P_k^v(\gamma) \equiv 0$ for any k such that $d(i) < d(k) \leqslant 1$. Let $h \in L^\infty([0,1], \mathbb{R}^r)$ be the controls associated with γ, so that $\dot{\gamma} = \sum_{j=1}^r h_j X_j(\gamma)$. Then

$$\begin{aligned} \frac{d}{dt} P_i^v \circ \gamma &= \sum_{j=1}^{r} h_j \, X_j P_i^v(\gamma) \overset{(6.2)}{=} \sum_{j=1}^{r} \sum_{k \leqslant n} h_j c_{ji}^k P_k^v(\gamma) \\ &= \sum_{j=1}^{r} \sum_{\substack{k \leqslant n \\ d(k) = d(i)+1}} h_j c_{ji}^k P_k^v(\gamma) = 0\,, \end{aligned}$$

i.e., $P_i^v(\gamma)$ is constant and equal to $P_i^v(\gamma(0)) = 0$. □

[1] Clearly, the homogeneous degree is defined by $d(i) = k \Leftrightarrow X_i \in V_k$ also for $i \leqslant 0$.

Theorem 4.14 states that abnormal extremals in Carnot groups are contained in certain algebraic varieties (of a very specific type). Theorem 6.1 improves it because it states that these algebraic varieties can be made smaller, as there are more polynomials (than in Theorem 4.14) that vanish along γ.

We show an application of our techniques to the Morse-Sard problem for abnormal extremals. In our opinion, the strategy we follow has chances to be adapted to many Carnot groups; however, we present it only in a specific group.

Let us consider the free[2] Carnot group $\mathbb{G}$ of rank 2 and step 4, *i.e.*, the group associated with the stratified Lie algebra

$$\mathfrak{g} = V_1 \oplus V_2 \oplus V_3 \oplus V_4$$

with

$$V_1 = \text{span}\{X_1, X_2\}, \quad V_2 = \text{span}\{X_3\},$$
$$V_3 = \text{span}\{X_4, X_5\}, \quad V_4 = \text{span}\{X_6, X_7, X_8\}$$

and commutation relations

$$[X_2, X_1] = X_3$$
$$[X_3, X_1] = X_4, \ [X_3, X_2] = X_5$$
$$[X_4, X_1] = X_6, \ [X_4, X_2] = [X_5, X_1] = X_7, \ [X_5, X_2] = X_8 .$$

Using exponential coordinates of the second type (see [12]), $\mathbb{G}$ can be identified with $\mathbb{R}^8$ in such a way that

$$X_1 = \partial_1$$

$$X_2 = \partial_2 - x_1\partial_3 + \frac{x_1^2}{2}\partial_4 + x_1x_2\partial_5 - \frac{x_1^3}{6}\partial_6 - \frac{x_1^2x_2}{2}\partial_7 - \frac{x_1x_2^2}{2}\partial_8 .$$

We are going to prove the following result.

Theorem 6.2. *Let $\mathbb{G} \equiv \mathbb{R}^8$ be the free Carnot group of rank 2 and step 4. Then, there exists a non-zero polynomial in 8 variables $Q : \mathbb{R}^8 \to \mathbb{R}$ such that the following holds: if $p \in \mathbb{G}$ is the endpoint of an abnormal extremal starting from 0, then $Q(p) = 0$.*

In particular, the set of points in $\mathbb{G}$ that can be connected to the origin with abnormal extremals is contained in the algebraic variety $\{x \in \mathbb{R}^8 : Q(x) = 0\}$ and has measure zero.

---

[2] *Free* means, roughly speaking, that it is the Carnot group with largest dimension among those with rank 4 and step 2; equivalently, that any other such group is (isomorphic to) a quotient of the free one.

Remark 6.3. Theorems 4.14 and 6.1 show that any abnormal extremal is contained in an algebraic variety whose definition depends on a parameter v, *i.e.*, on the extremal itself. On the contrary, by Theorem 6.2 there exists a *universal* algebraic variety containing *all* abnormal extremals.

Proof. As proved in [41], the Tanaka prolongation of $\mathbb{G}$ is of the form $\mathrm{Prol}\,\mathfrak{g} = V_0 \oplus \mathfrak{g}$ with $\dim V_0 = 4$. Let us extend $X_1, \dots, X_8$ to an adapted basis $\{X_i\}_{-3 \leqslant i \leqslant 8}$ of $\mathrm{Prol}\,\mathfrak{g}$. By Theorem 6.1 we know that, for any abnormal extremal $\gamma : [0,1] \to \mathbb{G}$ with $\gamma(0) = 0$, there exists $v \in \mathbb{R}^8$ such that

$$P_i^v(\gamma) = 0 \text{ on } [0,1] \quad \text{for any } i = -3, \dots, 3\,. \tag{6.4}$$

We have also used Remark 3.22, *i.e.*, the fact that the Goh condition holds. In particular,

$$v_i = P_i^v(0) = 0 \quad \text{for } i = 1, 2, 3.$$

Therefore, recalling (6.1), any P_i^v can be written in the form

$$P_i^v(x) = \sum_{k=4}^{8} v_k Q_{ik}(x), \quad i = -3, \dots, 8$$

for suitable polynomials $Q_{ik}(x)$ that are independent from v. For any $i = -3, \dots, 3$, let us define the map $Q_i : \mathbb{G} \to \mathbb{R}^5$ by

$$Q_i(x) = \big(Q_{i4}(x), Q_{i5}(x), Q_{i6}(x), Q_{i7}(x), Q_{i8}(x)\big),$$

so that $P_i^v(x) = \langle (v_4, \dots, v_8), Q_i(x) \rangle$. Hence, (6.4) can be rewritten as

$$Q_i(\gamma(t)) \perp (v_4, \dots, v_8) \quad \forall t \in [0,1],\ \forall\, i = -3, \dots, 3\,.$$

In particular, for any $t \in [0,1]$, the seven 5-dimensional vectors $Q_i(\gamma(t))$, $-3 \leqslant i \leqslant 3$, belong to the vector space $(v_4, \dots, v_8)^\perp \subset \mathbb{R}^5$; this vector space has dimension 4 because $(v_4, \dots, v_8) \neq 0$ due to Theorem 3.6 (i). Hence, any 5 of these 7 vectors are linearly dependent, *i.e.*, any 5×5 minor of the 5×7 matrix

$$(Q_{ik}(x))_{\substack{-3 \leqslant i \leqslant 3 \\ 4 \leqslant k \leqslant 8}} = \mathrm{col}\,\big[Q_{-3}|Q_{-2}|\cdots|Q_3\big](x) \tag{6.5}$$

has determinant 0 at any point x on γ. In particular, the determinant of the minor

$$\mathrm{col}\,\big[Q_{-1}|Q_0|Q_1|Q_2|Q_3\big](x)$$

is a polynomial $Q(x)$ (independent from v) which vanish along γ.

It is now a boring task to prove that Q is not the null polynomial; we refer to the proof of [18, Theorem 4.1] for details. This concludes the proof. □

Remark 6.4. The determinant of *any* 5×5 minor of the matrix in (6.5) has to vanish along abnormal extremals; hence, in principle, one could produce $\binom{7}{5} = 21$ polynomials as in the statement of Theorem 6.2. See [18, Remark 4.2.] for a more detailed discussion on these and other considerations.

ACKNOWLEDGEMENTS. It is a pleasure to thank R. Monti for many invaluable comments, remarks and suggestions. We have to thank G. P. Leonardi for suggesting the characterization of normal extremals in Carnot groups contained in Theorem 4.12. We are indebted with E. Le Donne and E. Pasqualetto for their careful reading of a preliminary version of these notes. Finally, we want to thank the organizers G. Alberti, L. Ambrosio and C. De Lellis, as well as all the participants, for the nice time and the pleasant atmosphere during the ERC School on Geometric Measure Theory and Real Analysis.

References

[1] A. A. AGRACHEV, *Any sub-Riemannian metric has points of smoothness* (Russian). Dokl. Akad. Nauk **424** (2009), no. 3, 295–298; translation in Dokl. Math. **79** (2009), no. 1, 45–47.

[2] A. A. AGRACHEV, *Some open problems*, In: "Geometric Control Theory and Sub-Riemannian Geometry", Springer INdAM Series **5** (2014), 1–14.

[3] A. A. AGRACHEV, D. BARILARI and U. BOSCAIN, *Introduction to Riemannian and Sub-Riemannian geometry*. Preprint available at www.math.jussieu.fr/ barilari/

[4] A. A. AGRACHEV and Y. L. SACHKOV, *Control theory from the geometric viewpoint*, In: "Encyclopaedia of Mathematical Sciences", Vol. 87, Control Theory and Optimization, II. Springer-Verlag, Berlin , 2004, xiv+412.

[5] L. AMBROSIO and S. RIGOT, *Optimal mass transportation in the Heisenberg group*, J. Funct. Anal. **208** (2004), no. 2, 261–301.

[6] A. BELLAÏCHE, *The tangent space in subriemannian geometry*. In: "Subriemannian Geometry", Progress in Mathematics 144, A. Bellaiche and J. Risler (eds.), Birkhauser Verlag, Basel, 1996.

[7] V. G. BOLTYANSKII, *Sufficient conditions for optimality and the justification of the dynamic programming method*. SIAM J. Control **4** (1966), 326–361.

[8] C. CARATHÉODORY, *Untersuchungen über die Grundlagen der Thermodynamik*, Math. Ann. **67** (1909), 355–386.

[9] Y. CHITOUR, F. JEAN & E. TRÉLAT, *Genericity results for singular curves*, J. Differential Geom. **73** (2006), no. 1, 45–73.

[10] W. L. CHOW, *Über Systeme von linearen partiellen Differentialgleichungen erster Ordnung*, Math. Ann. **117** (1939), 98–105.

[11] C. GOLÉ and R. KARIDI, *A note on Carnot geodesics in nilpotent Lie groups*, J. Dynam. Control Systems **1** (1995), no. 4, 535–549.

[12] M. GRAYSON and R. GROSSMAN, *Models for free nilpotent Lie algebras*, J. Algebra **135** (1990), no. 1, 177–191.

[13] U. HAMENSTÄDT, *Some regularity theorems for Carnot-Carathéodory metrics*, J. Differential Geom. **32** (1990), no. 3, 819–850.

[14] L. HÖRMANDER, *Hypoelliptic second-order differential equations*, Acta Math. **121** (1968), 147–171.

[15] E. LE DONNE, *A metric characterization of Carnot groups*, Proc. Amer. Math. Soc., to appear. Preprint available at arxiv.org/abs/1304.7493.

[16] E. LE DONNE, *Lecture notes on sub-Riemannian geometry*. Preprint available at sites.google.com/site/enricoledonne/.

[17] E. LE DONNE, G. P. LEONARDI, R. MONTI and D. VITTONE, *Extremal curves in nilpotent Lie groups*, Geom. Funct. Anal. **23** (2013), 1371–1401.

[18] E. LE DONNE, G. P. LEONARDI, R. MONTI and D. VITTONE, *Extremal polynomials in stratified groups*. Preprint available at cvgmt.sns.it.

[19] E. LE DONNE, G. P. LEONARDI, R. MONTI and D. VITTONE, *Corners in non-equiregular sub-Riemannian manifolds*, ESAIM Control Optim. Calc. Var., to appear. Preprint available at cvgmt.sns.it.

[20] E. B. LEE and L. MARKUS, "Foundations of Optimal Control Theory", second edition, Robert E. Krieger Publishing Co., Inc., Melbourne, FL, 1986. xii+576 pp.

[21] G. P. LEONARDI and R. MONTI, *End-point equations and regularity of sub-Riemannian geodesics*, Geom. Funct. Anal. **18** (2008), no. 2, 552–582.

[22] W. LIU and H. J. SUSSMAN, *Abnormal sub-Riemannian minimizers*, Differential equations, dynamical systems, and control science, Lecture Notes in Pure and Appl. Math., Vol. 152, Dekker, New York, 1994, 705–716.

[23] W. LIU and H. J. SUSSMAN, *Shortest paths for sub-Riemannian metrics on rank-two distributions*, Mem. Amer. Math. Soc. 118 (1995), no. 564, x+104 pp.

[24] G. A. MARGULIS and G. D. MOSTOW, *Some remarks on the definition of tangent cones in a Carnot-Carathéodory space*, J. Anal. Math. **80** (2000), 299–317.
[25] J. MITCHELL, *On Carnot-Carathèodory metrics*. J. Differential Geom. **21** (1985), 35–45.
[26] R. MONTGOMERY, *Abnormal minimizers*, SIAM J. Control Optim. **32** (1994), no. 6, 1605–1620.
[27] R. MONTGOMERY, "A Tour of Subriemannian Geometries, their Geodesics and Applications", Mathematical Surveys and Monographs, Vol. 91, American Mathematical Society, Providence, RI, 2002. xx+259.
[28] R. MONTI, *A family of nonminimizing abnormal curves*, Ann. Mat. Pura Appl. (online). Preprint available at cvgmt.sns.it.
[29] R. MONTI, *Regularity results for sub-Riemannian geodesics*, Calc. Var. Partial Differential Equations **49** (2014), no. 1-2, 549–582.
[30] R. MONTI, *The regularity problem for sub-Riemannian geodesics*, In: "Geometric Control Theory and Sub-Riemannian Geometry", Springer INdAM Series **5** (2014), 313–332.
[31] D. MORBIDELLI, *Fractional Sobolev norms and structure of Carnot-Carathèodory balls for Hörmander vector fields*, Studia Math. **139** (2000), no. 3, 213–244.
[32] A. NAGEL, E. M. STEIN and S. WAINGER, *Balls and metrics defined by vector fields I: basic properties*, Acta Math. **155** (1985), 103–147.
[33] P. K. RASHEVSKY, *Any two points of a totally nonholonomic space may be connected by an admissible line*, Uch. Zap. Ped. Inst. im. Liebknechta, Ser. Phys.Math. **2** (1938) 83–94.
[34] L. RIFFORD and E. TRÉLAT, *Morse-Sard type results in sub-Riemannian geometry*, Math. Ann. **332** (2005), no. 1, 145–159.
[35] R. STRICHARTZ, *Sub-Riemannian geometry*, J. Differential Geom. **24** (1986), no. 2, 221–263.
[36] R. STRICHARTZ, *Corrections to: Sub-Riemannian geometry*, J. Differential Geom. **30** (1989), no. 2, 595–596.
[37] H. J. SUSSMANN, *A cornucopia of four-dimensional abnormal sub-Riemannian minimizers*. In: "Subriemannian Geometry", Progress in Mathematics 144, A. Bellaiche and J. Risler (eds.), Birkhäuser Verlag, Basel, 1996.
[38] K. TAN and X. YANG, *Subriemannian geodesics of Carnot groups of step 3*, ESAIM Control Optim. Calc. Var. **19** (2013), no. 1, 274–287.
[39] N. TANAKA, *On differential systems, graded Lie algebras and pseudogroups*, J. Math. Kyoto Univ. **10** (1970), 1–82.

[40] V. S. VARADARAJAN, "Lie Groups, Lie Algebras, and their Representations", Graduate Texts in Mathematics, Springer-Verlag, New York, 1974.

[41] B. WARHURST, *Tanaka prolongation of free Lie algebras*, Geom. Dedicata **130** (2007), 59–69.

CRM Series

Publications by the Ennio De Giorgi Mathematical Research Center Pisa

The Ennio De Giorgi Mathematical Research Center in Pisa, Italy, was established in 2001 and organizes research periods focusing on specific fields of current interest, including pure mathematics as well as applications in the natural and social sciences like physics, biology, finance and economics. The CRM series publishes volumes originating from these research periods, thus advancing particular areas of mathematics and their application to problems in the industrial and technological arena.

Published volumes

1. Matematica, cultura e società 2004 (2005). ISBN 88-7642-158-0
2. Matematica, cultura e società 2005 (2006). ISBN 88-7642-188-2
3. M. Giaquinta, D. Mucci, *Maps into Manifolds and Currents: Area and* $W^{1,2}$-, $W^{1/2}$-, BV*-Energies*, 2006. ISBN 88-7642-200-5
4. U. Zannier (editor), *Diophantine Geometry*. Proceedings, 2005 (2007). ISBN 978-88-7642-206-5
5. G. Métivier, *Para-Differential Calculus and Applications to the Cauchy Problem for Nonlinear Systems*, 2008. ISBN 978-88-7642-329-1
6. F. Guerra, N. Robotti, *Ettore Majorana. Aspects of his Scientific and Academic Activity*, 2008. ISBN 978-88-7642-331-4
7. Y. Censor, M. Jiang, A.K. Louisr (editors), *Mathematical Methods in Biomedical Imaging and Intensity-Modulated Radiation Therapy (IMRT)*, 2008. ISBN 978-88-7642-314-7
8. M. Ericsson, S. Montangero (editors), *Quantum Information and Many Body Quantum systems*. Proceedings, 2007 (2008). ISBN 978-88-7642-307-9
9. M. Novaga, G. Orlandi (editors), *Singularities in Nonlinear Evolution Phenomena and Applications*. Proceedings, 2008 (2009). ISBN 978-88-7642-343-7

– Matematica, cultura e società 2006 (2009). ISBN 88-7642-315-4

10. H. HOSNI, F. MONTAGNA (editors), *Probability, Uncertainty and Rationality*, 2010. ISBN 978-88-7642-347-5
11. L. AMBROSIO (editor), *Optimal Transportation, Geometry and Functional Inequalities*, 2010. ISBN 978-88-7642-373-4
12*. O. COSTIN, F. FAUVET, F. MENOUS, D. SAUZIN (editors), *Asymptotics in Dynamics, Geometry and PDEs; Generalized Borel Summation*, vol. I, 2011. ISBN 978-88-7642-374-1, e-ISBN 978-88-7642-379-6
12**. O. COSTIN, F. FAUVET, F. MENOUS, D. SAUZIN (editors), *Asymptotics in Dynamics, Geometry and PDEs; Generalized Borel Summation*, vol. II, 2011. ISBN 978-88-7642-376-5, e-ISBN 978-88-7642-377-2
13. G. MINGIONE (editor), *Topics in Modern Regularity Theory*, 2011. ISBN 978-88-7642-426-7, e-ISBN 978-88-7642-427-4
– Matematica, cultura e società 2007-2008 (2012). ISBN 978-88-7642-382-6
14. A. BJORNER, F. COHEN, C. DE CONCINI, C. PROCESI, M. SALVETTI (editors), *Configuration Spaces*, Geometry, Combinatorics and Topology, 2012. ISBN 978-88-7642-430-4, e-ISBN 978-88-7642-431-1
15 A. CHAMBOLLE, M. NOVAGA E. VALDINOCI (editors), *Geometric Partial Differential Equations*, 2013. ISBN 978-88-7642-343-7, e-ISBN 978-88-7642-473-1
16 J. NEŠETŘIL AND M. PELLEGRINI (editors), *The Seventh European Conference on Combinatorics, Graph Theory and Applications*, EuroComb 2013. ISBN 978-88-7642-474-8, e-ISBN 978-88-7642-475-5
17 L. AMBROSIO (editor), *Geometric Measure Theory and Real Analysis*, 2014. ISBN 978-88-7642-522-6 , e-ISBN 978-88-7642-523-3

Volumes published earlier

Dynamical Systems. Proceedings, 2002 (2003)
Part I: *Hamiltonian Systems and Celestial Mechanics*. ISBN 978-88-7642-259-1
Part II: *Topological, Geometrical and Ergodic Properties of Dynamics*. ISBN 978-88-7642-260-1

Matematica, cultura e società 2003 (2004). ISBN 88-7642-129-7

Ricordando Franco Conti, 2004. ISBN 88-7642-137-8

N.V. KRYLOV, *Probabilistic Methods of Investigating Interior Smoothness of Harmonic Functions Associated with Degenerate Elliptic Operators*, 2004. ISBN 978-88-7642-261-1

Phase Space Analysis of Partial Differential Equations. Proceedings, vol. I, 2004 (2005). ISBN 978-88-7642-263-1

Phase Space Analysis of Partial Differential Equations. Proceedings, vol. II, 2004 (2005). ISBN 978-88-7642-263-1

Printed in the United States
By Bookmasters